II International Scientific and Practical Conference "Actual Problems of the Energy Complex: Mining, Production, Transmission, Processing and Environmental Protection"

IOP Conference Series: Materials Science and Engineering
Volume 976

Moscow, Russia
16 - 17 April 2020

ISBN: 978-1-7138-3297-3
ISSN: 1757-8981

Printed from e-media with permission by:

Curran Associates, Inc.
57 Morehouse Lane
Red Hook, NY 12571

CURRAN ASSOCIATES INC.
proceedings
.com

Some format issues inherent in the e-media version may also appear in this print version.

Institute of Physics
Dirac House, Temple Back
Bristol BS1 6BE UK

Phone: 44 1 17 929 7481
Fax: 44 1 17 920 0979

techtracking@iop.org

Additional copies of this publication are available from:

Curran Associates, Inc.
57 Morehouse Lane
Red Hook, NY 12571 USA
Phone: 845-758-0400
Fax: 845-758-2633
Email: curran@proceedings.com
Web: www.proceedings.com

TABLE OF CONTENTS

IOP Conf. Series: Materials Science and Engineering **976** (2020) 011001 doi:10.1088/1757-899X/976/1/011001

II International scientific and practical conference "Actual problems of the energy complex: mining, production, transmission, processing and environmental protection"

R V Lutoshkin[1], D E Morkovkin[2] and M I Sadriddinov[3]

[1] OOO "RID", apt. 74, 102A, Yamashev street, Kazan, 420133, Russian Federation
[2] Financial University under the Government of the Russian Federation, 49, Leningradsky avenue, Moscow, 125993, Russian Federation
[3] Tajik State Finance and Economics University, 64/14, Nakhimova Street, Dushanbe, 734067, Republic of Tajikistan

E-mail: lrv@ridlaw.ru

Abstract. Within the framework of the presented material, the results of the conducted II International scientific and practical conference "Actual problems of the energy complex: mining, production, transmission, processing and environmental protection".

We are pleased to present a collection of materials to IOP Publishing II International scientific and practical conference "Actual problems of the energy complex: mining, production, transmission, processing and environmental protection", which took place in Moscow (Russian Federation) and was devoted to the functioning and development of the fuel and energy complex.

The conference was held to summarize international experience in the field of the fuel and energy complex, namely:

- Activation of the scientific and practical potential of the energy complex in order to form new proposals for the technological, technical, innovative, organizational and digital development of the fuel and energy complex;
- Popularization of fundamental and applied research in the fuel and energy complex;
- Accelerated growth of technical and technological recommendations aimed at improving production, production, transmission, processing and ensuring sustainable development of the energy sector in the long term [1].

The topics of the conference were as follows:

- Oil and gas, environmental, chemical and energy engineering;
- Technical and technological development of the energy sector;
- Intelligent, digital and automated technologies in the energy sector;
- Materials science and technology of materials in power engineering;
- Coal, gas and oil in the fuel and energy complex;
- Renewable and alternative energy sources;

- Power systems and complexes;
- Environmental protection and ecological safety of energy facilities;
- Reliability and safety of the energy system;
- Organization of functioning of energy facilities and energy economics;
- Management of energy production in the context of the transition to sustainable development;
- World energy infrastructure and power complexes;
- Transportation of fuel and energy resources and transmission of electrical energy;
- System analysis, physical and mathematical modeling, computer and information technologies in the energy sector [1].

The international program committee and organizing committee included famous scientists who worked on the collection of the conference and contributed to the holding of this conference.

The collection includes materials presented by scientists from various research and educational institutions, industrial enterprises, energy organizations, as well as representatives of government agencies.

- National Research University "MPEI";
- Kazan State Power Engineering University;
- Platov South-Russian State Polytechnic University;
- Bratsk State University;
- Irkutsk State Transport University;
- Irkutsk National Research Technical University;
- North Caucasian Mining and Metallurgical Institute (State Technological University);
- Tver State Technical University;
- Astrakhan State University;
- Volgograd State Technical University;
- Transbaikal State University;
- Kemerovo State University;
- Kazan State Energetics University;
- Socio-economic Institute of Yuri Gagarin State Technical University of Saratov;
- Siberian state industrial University;
- State University of Management;
- Ufa State Petroleum Technological University;
- Peter the Great St. Petersburg Polytechnic University;
- Petrozavodsk State University;
- People’s Friendship University of Russia;
- Mendeleev University of Chemical Technology;
- Russian state Academy of intellectual property;
- Far East Investment and Export Agency;
- Tula Regional Dispatching Office, a branch of System Operator of the Unified Power System;
- Vladimir Regional Dispatching Office, a branch of System Operator of the Unified Power System;
- Oil and Gas Research Institute RAS;
- Orenburg federal research center Ural Branch of the Russian Academy of Sciences;
- LLC EPC "Inzhekhim";
- Joint Stock Company Tomsk Research and Design Institute of Oil and Gas;
- Northern Energetics Research Centre Kola Science Centre of the Russian Academy of Sciences;
- Tajik state university of finance and economics;
- Omsk State Transport University;
- Russian State University of Oil and Gas (NRU) named after I.M. Gubkina;
- National Research Tomsk Polytechnic University;
- Orenburg State University;
- Ufa State Aviation Technical University;
- Chuvash State University;
- Chuvash State Pedagogical University;
- Obninsk Institute for Nuclear Power Engineering of NRNU «MEPhI»;
- Saratov State Law Academy;
- Financial University under the Government of the Russian Federation.

Thus, the conference allowed formulating scientific recommendations for fuel and energy companies, state and regional authorities, international and supranational organizations, the scientific and professional community [2].

Acknowledgments
The organizing committee of the conference expresses its gratitude to the staff of IOP Publishing, who helps to place the proceedings of scientific conferences in international journals and provide support for the publication of materials.

In addition, the organizing committee would like to thank the conference participants; reviewers and everyone who helped organize this conference and form the presented volume for publication in the IOP Conference Series: Materials Science and Engineering.

References

[1] *Official site OOO "RID"* Retrieved from: http://ridlaw.ru
[2] 2020 IOP Conf. Ser.: Mater. Sci. Eng. **837** 011001

Peer review declaration

All papers published in this volume of *IOP Conference Series: Materials Science and Engineering* has been peer reviewed through processes administered by the Editors. Reviews were conducted by expert referees to the professional and scientific standards expected of a proceedings journal published by IOP Publishing.

- **Type of peer review:** Double-blind

- **Conference submission management system:**
  1. Sending materials to the organizers of the conference;
  2. Verification of compliance with the subject of the conference;
  3. Checking materials for plagiarism using two systems;
  4. Submitting for double blind review;
  5. Checking the quality of translation of the article into English;
  6. Checking for the correct design of materials;
  7. Submission of materials to the Publisher.

- **Number of submissions received:** 53

- **Number of submissions sent for review:** 53

- **Number of submissions accepted:** 39

- **Acceptance Rate (Number of Submissions Accepted / Number of Submissions Received X 100):** 74%

- **Average number of reviews per paper:** 2

- **Total number of reviewers involved:** 10

- **Any additional info on review process:** To be accepted for publication, the material must have the following requirements: compliance with the conference theme; a high level of originality of the text (at least 80%); positive expertise of reviewers; good level of English and well-formed material.

- **Contact person for queries:** Ruslan V Lutoshkin (lrv@ridlaw.ru)

Expert system for predicting the state of transformers based on fuzzy logic

A P Khlebtsov[1], I A Rybakov[1], N D Svishchev[1] and A N Shilin[2]

[1] Astrakhan State University (ASU), 20a, st. Tatishcheva, Astrakhan, 414056, Russian Federation
[2] Volgograd State Technical University (VolgSTU), 28, Lenin Ave., Volgograd, 400005, Russian Federation

E-mail: gergines@mail.ru

Abstract. The article examines methods for monitoring the operation of oil-filled transformers and identifies methods for evaluating the state of the transformer using a fuzzy logic device. Solutions for remote analysis of transformer operation have been formed. The methods that were used for diagnostics are presented. Conclusions are drawn on which equipment is more effective for diagnostics of power transformers.

1. Introduction

Currently, substations use physically and morally outdated equipment, which can lead to failures and loss of electricity for consumers. One of the most important basic elements of TP electrical equipment is a power transformer. Its repair in case of damage is associated with a large expenditure of labor and money. Emergency foci occur primarily due to long-term operating time and significant physical wear and tear.

2. Materials and methods

An expert system for predicting the state of transformers based on fuzzy logic should solve a number of tasks aimed at improving the accuracy of the diagnostic result. The sequence of actions of the system is presented as follows:

- Processing of the survey results;
- Comparison procedure with the results of previous diagnostics in different operating modes;
- The calculation procedure of the fuzzy data in numerical format;
- The final output.

One of the main advantages of an expert system based on fuzzy logic is the ability to represent the knowledge base using a linguistic language. Thus, the database can be supplemented, expanded, and edited by a human expert without appropriate programming training.

The introduction of additional characteristics, of course, will require special knowledge in this area, but working with existing characteristics is possible at a simpler level [1-2].

Based on the results of the system, a report is generated that describes each analyzed situation from the fuzzy knowledge base, shows the results of mathematical processing of diagnostic data, and indicates the numerical values of the degree of confidence for each answer to the questions asked to the system.

Because the diagnostic features are random, the system uses modified functions based on the Gaussian membership function:

$$\mu(u) = \exp\left(-\frac{(u-b)^2}{2c^2}\right) \quad (1)$$

where: u is the abscissa, b is the maximum coordinate, and c is the concentration coefficient.

The degrees of significance of various parameters obtained in the table for evaluating the state of transformers already allow the expert to draw conclusions about the state of transformers, but it is possible to get a much more accurate expert assessment based on fuzzy logic methods. To develop such a system, we chose the SciLab software.

SciLab has a special Fuzzy Logic Toollbox add-on for working with fuzzy logic tools.

For the features highlighted in the table, we will set the linguistic values of variables corresponding to the degree of overheating according to thermal imaging data, the moisture content in the oil, the flash point, and the pH factor of the oil. Additionally, to improve the accuracy of evaluating the state of transformers, we will re-evaluate these characteristics on a 10-point scale.

When assessing the condition of equipment by excess temperature, the following areas are distinguished by the degree of malfunction:

Excessive temperature-(5...10) °C. With a mean initial degree of fault that should be kept under control and take steps to eliminate it during repairs, scheduled according to the schedule. Excessive temperature – (10...30) °C, indicates a developed defect. In this case, it is necessary to take measures to eliminate the fault when the electrical equipment is taken out of operation as soon as possible.

Excessive temperature - more than 30°C means an emergency defect. Requires an immediate solution.

Because of the analysis of table 1, the transformer membership functions were obtained and entered into SciLab. Similarly, we will identify the linguistic rules.

3. Results

For oil-filled transformers, these rules will be as follows:

- Degree of overheating according to thermal monitoring data: initial degree, developed defect, emergency defect;
- Oil moisture concentration – acceptable, risk area, unacceptable;
- Flash point – acceptable and unacceptable;
- pH factor of the oil – acceptable and unacceptable.

Figures 1-2 show the appearance of these functions for the various variables considered in the table. The type of selected function is trapezoidal.

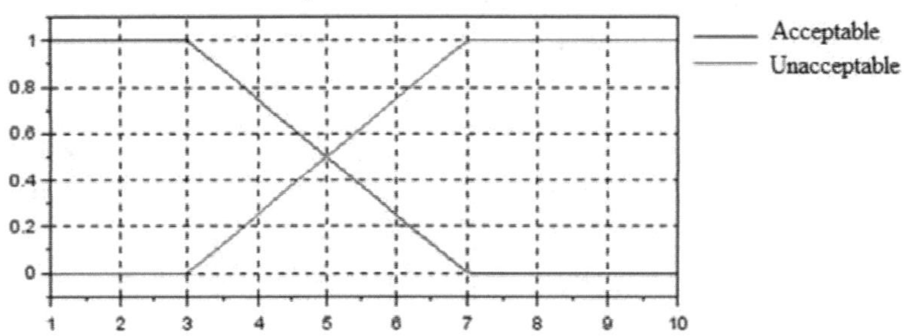

Figure 1. Fuzzy sets for the variable "Flash temperature".

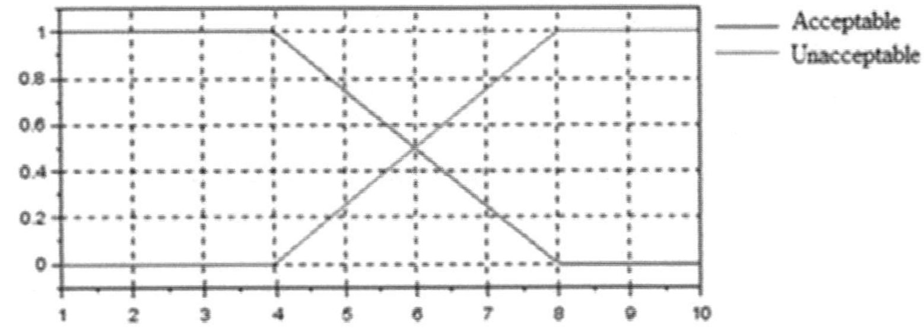

Figure 2. Fuzzy sets for the " pH factor" variable.

Based on the specified input functions, the membership functions for the output were formed. The type of selected function – Gaussian – emphasizes the probabilistic nature of the decision about the state of the transformer. In fact, this is the probability of its replacement, expressed as a percentage (figure 3).

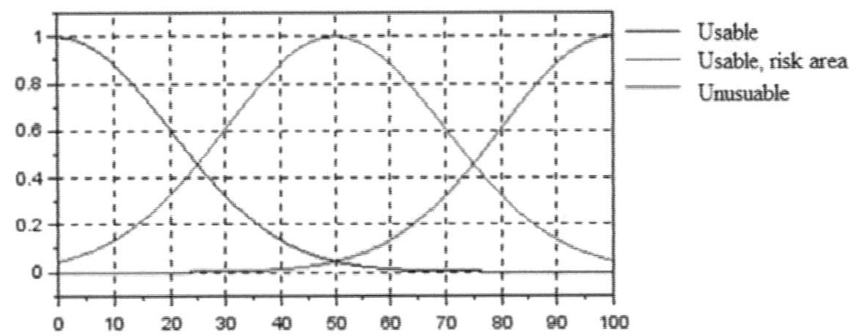

Figure 3. Output accessory functions that reflect the state of the transformer.

In the first case, we selected only the first two characteristics from the table in Appendix A, so it is possible to visualize the input state using the built-in SciLab graphical tools. Figure 4 shows a surface that reflects the state of the fuzzifier output depending on the input values.

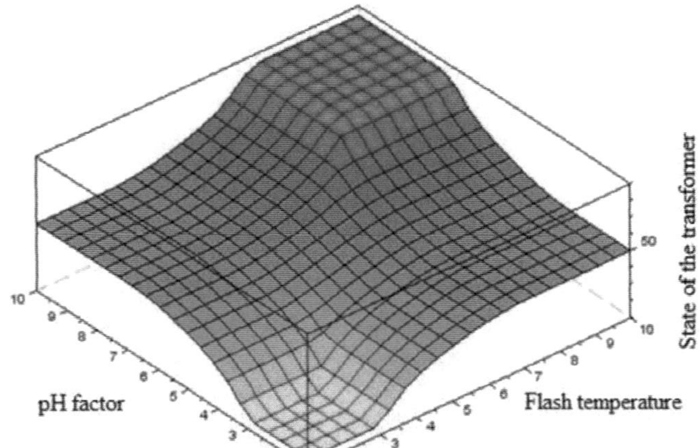

Figure 4. Function of dependence of the output state on the input values.

For the other two characteristics that reflect the features detected during transformer diagnostics, membership functions were also compiled. Their views are presented in figures 19-20. Figure 5-6 on the y-axis shows the weight of input status of fuzzifier. Comparing them with the previous ones, you can notice the appearance of the third state.

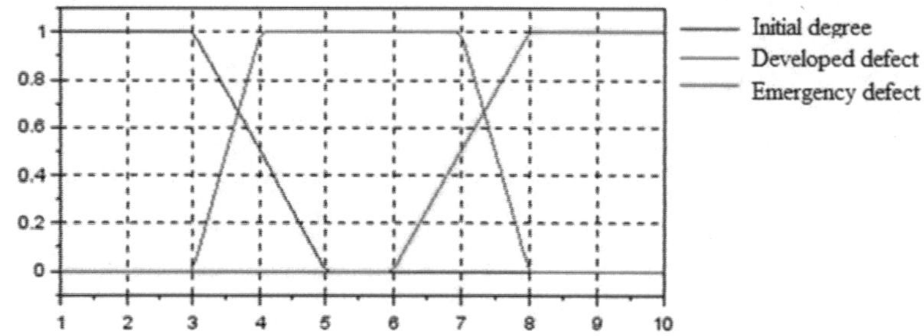

Figure 5. Fuzzy sets for the variable "Degree of overheating".

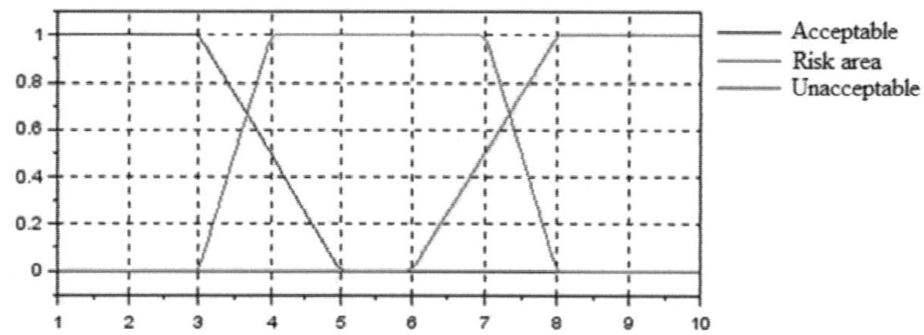

Figure 6. Fuzzy sets for the variable "Oil moisture concentration".

The membership functions for the output were set as in the previous case using Gaussian functions.

The surface of possible solutions is shown in figure 7. On the graph, you can see 4 characteristic plateaus that reflect the state of the transformers [3-6].

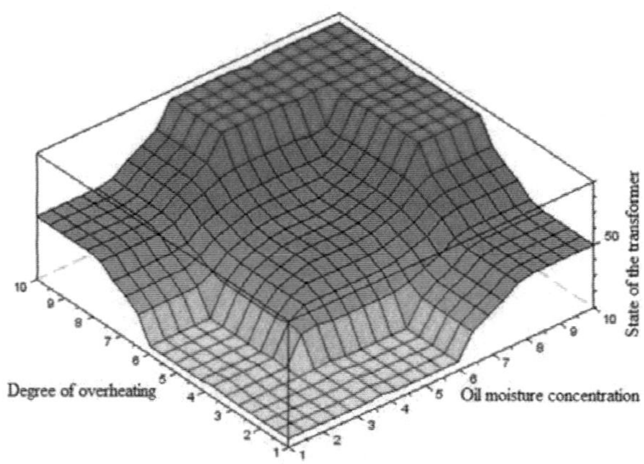

Figure 7. Function of dependence of the output state on the input values.

4. Discussion

After entering all the necessary data and setting the rules that connect the inputs and outputs, it is possible to perform aggregation and defuzzification to obtain an accurate value for the probability of replacing the transformer.

Aggregation is a procedure for determining the degree of truth of conditions for each of the rules of the fuzzy inference system. In this case, the values of the membership functions of the terms of linguistic variables that make up the above-mentioned conditions (antecedents) of the cores of fuzzy production rules obtained at the fuzzification stage are used [7-8].

Currently, the following algorithms are most used in fuzzy inference systems.

The Mamdani algorithm found application in the first fuzzy automatic control systems. It was proposed in 1975 by the English mathematician E. Mamdani to control a steam engine.

Defuzzification is performed using the center of gravity or center of area method. Defuzzification using the center of gravity method is described by the following formula:

$$\frac{\sum_{i=1}^{k} u_i \mu_A(u_i)}{\sum_{i=1}^{k} \mu_A(u_i)} \quad (2)$$

In our case, this method was used.

Figure 8 shows a view of the function of aggregated output values and a clear value of the probability of replacing the transformer in the case when the expert entered 2 and 4 ratings for mechanical damage and discharge density, respectively, in the system.

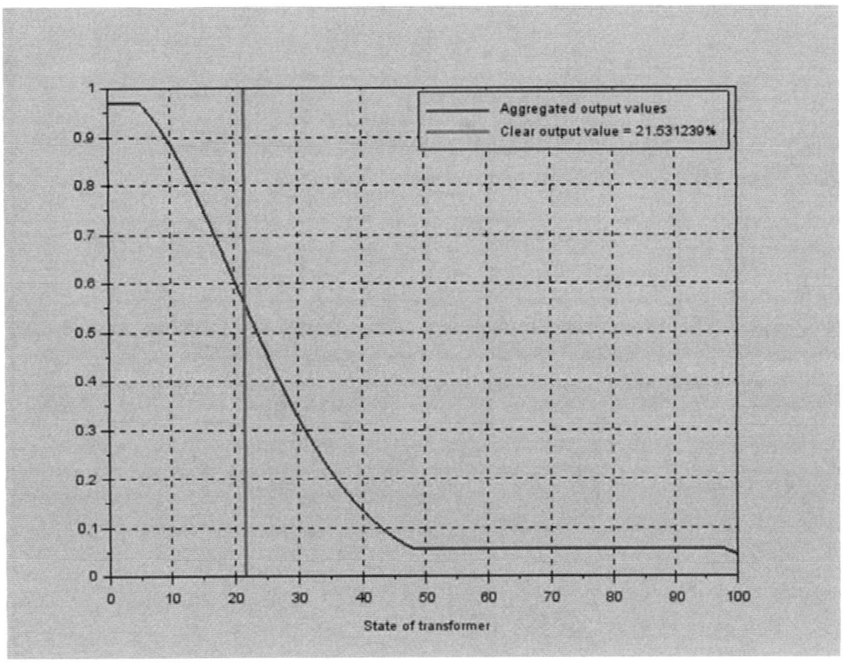

Figure 8. View of the function of aggregated output values and a clear value of the probability of replacing the transformer in the first case for the values of inputs 2 and 3.

Further, the expert evaluation method consists in applying the received values of the probability of transformer failure to the input of a new fuzzifier and obtaining one value for the state of the transformers based on four input data. The fuzzifier inputs are configured as Gaussian functions. In this case, the information is no longer entered as an estimate on a scale from 1 to 10, but as a percentage of the probability of finding the transformer in a particular state. The outputs of the final solution fuzzifier also have Gaussian membership functions, and the set of final solutions can be represented as a 3D graph [9-10].

5. Conclusion

The technical condition of the power transformer is characterized by a large number of parameters. With a large number of different data, the assessment of the technical condition of the power transformer is much more complicated. To solve this problem, you can use a fuzzy logic device. In this paper, based on a mathematical model, a method is developed for evaluating the technical condition of power transformers. This mathematical model is based on fuzzy logic, which can be trained using genetic algorithms. Using the device based on a fuzzy model of the technical condition of the power transformer will improve the quality of the assessment of the technical condition of the equipment by excluding a person when analyzing data. In addition, this will allow us to develop a number of similar programs for complex diagnostics, covering the entire range of electrical equipment of transformer substations, as well as to develop a program for predicting the remaining life, based on the assessment of a comprehensive analysis of the technical condition of electrical equipment of transformer substations.

References

[1] Zadeh L A 1968 Fuzzy algorithms. *Information and Control* **12(2)** 94-102
[2] Zadeh L A 1965 Fuzzy sets. *Information and Control* **8(3)** 338-353
[3] Jin L and Zhang D 2015 Contamination Grades Recognition of Ceramic Insulators Using Fused Features of Infrared and Ultraviolet Images. *Energies* **8** 837-58
[4] Da Costa E G, Ferreira T V, Neri M G G, Queiroz I B and Germano A D 2009 Characterization of polymeric insulators using thermal and UV imaging under laboratory conditions. *IEEE Trans. Dielectr. Electr. Insul.* **16** 985-92.
[5] Han S, Hao R and Lee J 2009 Inspection of insulators on high-voltage power transmission lines. *IEEE Trans. Power Deliv* **24** 2319-27
[6] Moreno V M and Gorur R S 2003 Impact of corona on the long-term performance of nonceramic insulators. *IEEE Trans. Dielectr. Electr. Insul* **10** 80-95
[7] Sawada J, Kusumoto K and Maikawa T 1991 A mobile robot for inspection of power transmission lines. *IEEE Trans. Power Deliv.* **6** 309-15
[8] Phillips A J, Childs D J and Schneider M H 1999 Water drop corona effects on full-scale 500 kV non-ceramic insulators. *IEEE Trans. Power Deliv* **14** 258-65
[9] Gubanski S M, Dernfalk A and Andersson J 2007 Diagnostic methods for outdoor polymeric insulators. *IEEE Trans. Dielectr. Electr. Insul.* **14** 1065-80
[10] Guo J, Wu G L and Zhang X Q 2005 The actuality and perspective of partial discharge detection techniques. *Trans. China Electrotech. Soc.* **20** 29-35

IOP Conf. Series: Materials Science and Engineering **976** (2020) 012002 doi:10.1088/1757-899X/976/1/012002

Ways to improve the energy efficiency of renewable energy sources

Yu S Petrov, A K Muzaev and S P Maskov

North Caucasian Mining and Metallurgical Institute (State Technological University), 44, st. Nikolaeva, Vladikavkaz, 362021, Russian Federation

E-mail: Alexandr.muzaev@mail.ru

Abstract. The ways of increasing the energy efficiency of the use of renewable energy sources are considered. The descriptions of universal energy complexes that simultaneously use converters of various types: solar rays, wind flows into electrical energy, patented at SKGMI (GTU) are given. The prospect of using renewable energy sources for autonomous power supply of objects in mountainous regions has been substantiated.

1. Introduction

Currently, there is a widespread introduction of renewable energy sources (RES) in various areas of energy supply [1-2], in connection with which the increase in the energy efficiency of the use of RES becomes important. The use of RES is characterized by different types of efficiency: economic, social, environmental, energy. These types are inextricably linked, and a separate consideration of each type is of a methodological nature, designed to accentuate certain parameters of the single process of using RES.

Next, three types of RES will be considered: solar panels, wind farms and low-power hydroelectric power plants.

2. Materials and methods

The energy efficiency of the use of RES is determined by the following main parameters: the energy payback period; the price and quality of the generated energy; stability of output parameters; reliability of power supply, etc.

The main ways to increase the energy efficiency of renewable energy sources are shown in Figure 1. The most significant ways are: increasing the efficiency of primary converters of natural energy flows (sun rays, wind, water flow) into electrical energy; reduction of the energy payback period; prevention of anomalous impact of natural environmental factors on RES facilities - manifestations of various kinds of environmental risks. These methods are integral, generalizing in nature and determine the general state of the facility that generates renewable energy. In figure 1, in addition to those already listed, other methods that are important for practice are also indicated.

One of the main ways to increase the energy efficiency of renewable energy sources is to reduce the energy payback period of power plants based on renewable energy sources (photovoltaic, wind and micro HPPs). In this case, the full life cycle of the object is considered: production of the installation, including all auxiliary operations; installation with the necessary change in natural conditions; operation and disposal of used parts of the installation and further - the installation itself. For the implementation of all components of the full life cycle of an object, a certain amount of energy is spent, which is sometimes called "connected" [3].

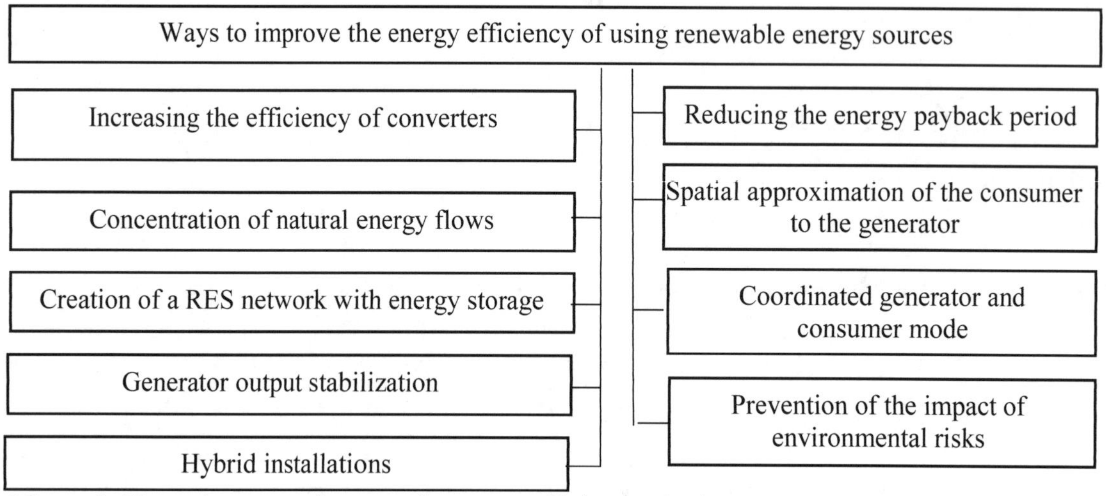

Figure 1. The main ways to improve the energy efficiency of using renewable energy sources.

The return period of the generation by the object spent on it during the entire life cycle of energy, during the return period of "bound" energy is called the term of the object's energy recoupment. The smaller the amount of "bound" energy and the greater the productivity (the volume of generated energy per unit time) of the facility, the shorter its energy payback period, therefore, the greater the efficiency of using the corresponding installation (facility).

Reducing the energy payback period (usually measured in years) can be achieved by implementing measures in two directions: rationalizing the auxiliary operations of the life cycle; increasing the efficiency of the object itself, i.e. increased energy generation.

If we consider the full life cycle of the use of solar cells, then it is necessary to highlight the following stages: extraction of the initial material (silicon), its purification, the technological process of manufacturing solar cells, installation of solar cells in a specific place, operation, dismantling, disposal. All stages are characterized by energy consumption. To the listed steps it is necessary to add transport operations and the corresponding energy consumption.

You can write the following equation to determine the energy payback period:

$$\sum_{k=1}^{k=N} W_k = W_{sr} \cdot t_{ok} \qquad (1)$$

where W_k - energy consumption at the "k" stage of the life cycle of the object under consideration, kW·h;

W_{av} - average annual energy production by the facility, kW·h/year;

t_{ok} - is the term of the object's energy recoupment, year.

In the formula (1), W_k and W_{av} are determined by calculation or experiment for the object under consideration; the energy payback period is defined as the ratio:

$$T_{ok} = \frac{\sum_{k=1}^{k=N} W_k}{W_{sr}} \qquad (2)$$

As already mentioned, to reduce the energy payback period, the energy intensity (energy consumption) should be reduced at all stages of the complete life cycle of RES use. In addition, it is necessary to increase the level of energy output by the facility for the development of renewable energy sources, increasing its operational and technical characteristics and implementing various methods of increasing energy efficiency.

One of the ways to increase the energy efficiency of using renewable energy sources, as shown in figure 1, is the concentration of natural energy flows, an increase in their intensity. The sun's rays and wind are characterized by a low concentration of energy (the amount of energy per unit area of the working surface of the converter installation). This leads to the need to use various kinds of concentrators.

3. Results

In the installations developed at the North Caucasian Mining and Metallurgical Institute (State Technological University), Vladikavkaz [4-8] concentrators, respectively, of solar rays and air flow are used. To increase the intensity of solar irradiation in the wind turbine, reflectors are used that direct the sun's rays to the working surfaces (solar panels) of the installation; to increase the effect of the wind, a guiding device is used that changes the direction of air movement towards the working body - the blades of the wind generator drive. The use of concentrators can increase the productivity of the installation by approximately 1.2÷1.5 times.

One of the problems associated with the use of renewable energy sources is the inconstancy of the parameters of natural energy flows used in energy generation. The wind can change speed and direction; solar irradiation can change with changes in weather conditions (clouds), with a change in the time of day, etc. All this leads to significant fluctuations in the output power of the installation. However, the power supply to consumers should not depend on fluctuations in the output power, therefore, wind and solar generators must have appropriate devices that stabilize their output parameters.

One of the effective ways to stabilize the output parameters of generating sets using RES is the use of intermediate energy storage devices (figure 2).

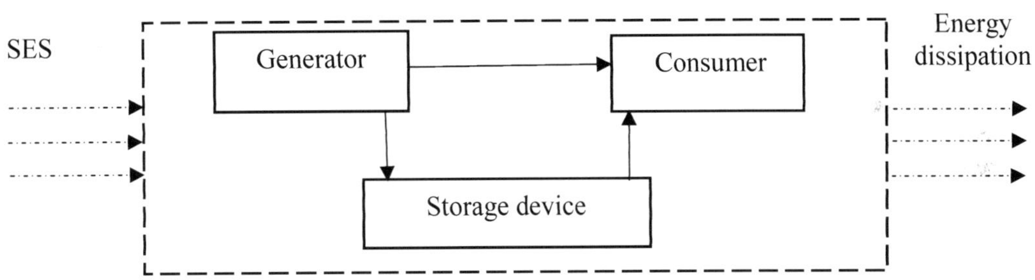

Figure 2. Scheme of using a storage device in a RES generation plant.

In figure 2 SES - natural energy flows of renewable energy sources, which in the primary converter (generator) are converted into electrical energy, which can be transferred partially to the consumer and partially to the storage device. The energy balance equation in general form will be:

$$S_G = S_P \pm S_N \tag{3}$$

S_G - output power of the converter (solar panel, wind generator, etc.);
S_P - the power consumed by the load;
S_N - the power transmitted to the storage device.

With an excess of generated power, part of it is transferred to the storage device, and it operates in the consumer mode ("+" sign). If the generated power is insufficient (in comparison with the one required for normal operation of the consumer), it comes from the storage device and the power SH has a "-" sign (goes with a plus to the first part of equation (3)).

In some cases, all the power from the generator goes to the storage device, the energy consumption of which must be significantly higher than the power consumption of the load in order to create a stability margin in the general energy system using RES.

More formally, a real electrical circuit with power supply from renewable energy sources can be represented as a circuit with an active four-pole, shown in figure 3.

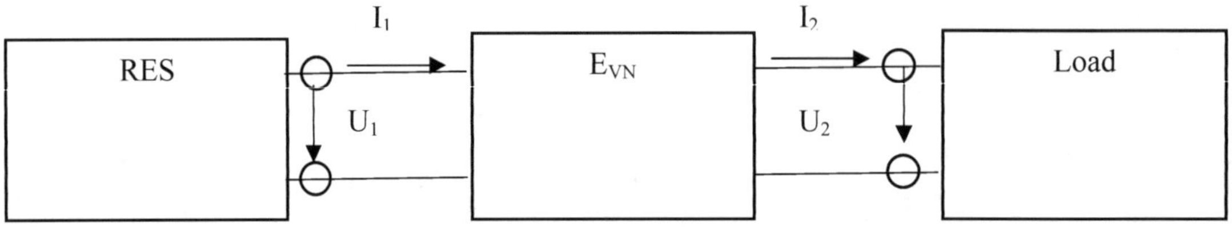

Figure 3. Equivalent electrical circuit of power supply from renewable energy sources.
E_{VN} - a source of EMF of an active four-terminal network.

An essential point when using an active four-port network is the condition that its power and energy capacity could compensate for both fluctuations in energy supply from RES and possible fluctuations in energy consumption by the load. To do this, its energy parameters in the calculation should be selected with a margin that takes into account the possibility of these fluctuations.

As you know, an active four-pole, i.e. containing the EMF E_{VN} inside itself, can be replaced by a passive one, if you short-circuit all the EMF contained inside the four-pole, replacing them with internal resistances, and add additional EMF into the equations of the four-pole: EMF E_{01} into the primary circuit, and E_{02} into the secondary EMF, which are equal to the voltages at the ends open clamps of this active four-port network.

Using this technique, we write the equations of an active four-port network in symbolic form, for example, for a fairly common Z-form:

$$\dot{U}_1 = \underline{Z}_{11} \cdot I_1' \cdot \underline{Z}_{21} \cdot I_2' + \dot{E}_{01} \tag{4}$$

$$\dot{U}_2 = \underline{Z}_{21} \cdot I_1' \cdot \underline{Z}_{22} \cdot I_2' + \dot{E}_{02} \tag{5}$$

When deriving equations (4), (5), it was assumed that the EMF of the sources does not depend on the currents in them. Based on the above reasoning, you can write down the equation of the four-port network in other forms and use them in accordance with the specifics of the task. The introduction of an intermediate four-port network with an energy storage device between the generator and the consumer is an effective means of stabilizing the output parameters of a RES power plant.

For individual objects using energy supply from renewable energy sources, an important circumstance is the possibility of placing a generator set in close proximity to the consumer (for example, placing solar panels on the roof of a house to supply electricity to consumers inside it). At the same time, there are practically no energy losses for its transportation, which significantly increases the energy efficiency of the use of RES.

In general, you can write:

$$P_{GENH} = P_{PR} + P_{TR} + P_N \tag{6}$$

R_{GEN} - full power of the generator set;
P_{PR} - power of converting devices;
P_{TR} - power of transmission (transmission) of energy to the consumer;
P_N - power consumed directly by the load.

To transport energy, appropriate converting devices and transmission lines or cables are required. Lack of transportation means direct consumption of the output power at the place of its production. At the same time, there are no losses both in converting devices and in power transmission lines. In addition, the reliability of the system increases, the production of energy becomes cheaper; the accompanying environmental effects decrease in general - the energy efficiency of the use of renewable energy sources increases.

4. Discussion

As an example of the spatial combination of a generator and a consumer, a road dividing block with an illumination system [8], which is supplied with electricity from renewable energy sources - wind and photovoltaic generators located directly on the block (Patent No. 196315 published on February 25, 2020, authors Yu.S. Petrov, M.K. Khadikov, A.K. Muzaev).

On the upper part of the block there is a bladed wind wheel, using the effect of oppositely moving air flows (in accordance with the oncoming traffic of traffic flows on opposite sides of the block, which is part of a separation road barrier); solar panels are applied to the sides of the block; inside the block there are switching, storage and conversion devices, from which electricity is transmitted directly to the luminaire, fixed on the side surfaces of the block.

Self-sufficiency in the lighting system of the road separation barrier allows for a reliable and efficient lighting system on motorways.

The features of energy generation from renewable energy sources allow the development of hybrid plants that implement various generation principles in one device. An example is developed at the North Caucasian Mining and Metallurgical Institute (State Technological University), Vladikavkaz, wind turbine plant [7] (patent No. 188444 published on 12.04.2019, Yu.S. Petrov, M.K. Khadikov, A. A. Sokolov, A.K. Muzaev).

The principle of operation of a wind turbine consists in the simultaneous conversion of wind energy and solar energy into electricity. The benefits of this transformation are clear. Firstly, the power of the power plant increases with practically unchanged dimensions in comparison with the sizes of installations using only one type of conversion. Secondly, the reliability of generation increases, since in the absence of one source of energy, another can operate; termination of the action of two energy sources of different types at once is unlikely. Thirdly, the total density of energy taken from a unit volume of total space increases, i.e. there is a concentration of the used energy flows by superimposing them on each other in different parts of the general converter installation. Fourthly, the environmental damage from the operation of a hybrid plant is reduced in comparison with plants using only one type of RES (single-profile type).

And, finally, it should be noted an obvious decrease in the energy payback period for hybrid installations - due to a significant increase in power.

Theoretically, hybrid installations that combine renewable energy sources of any type (wind, solar, hydro and other converters) have the right to exist. The development of hybrid plants of a specific type depends on the availability of this type of renewable energy in one place, on the characteristics of the consumer, the environmental situation, etc. According to some scientists [9-11], hybrid installations for the simultaneous use of RES of various types have a great future.

An increase in the energy efficiency of a RES installation can also be achieved by implementing a coordinated regime between the energy source and the load. As you know, in the case of a linear active two-terminal network with a known internal resistance R_{VN} - for direct current and for alternating current (in the general case $Z_{VN} = R_{VN} + jX_{VN}$, i.e. it consists of active R_{VN} and reactive X_{VN} components), the conditions for the release of maximum power in the load are as follows :

- For direct current:

$$R_N = R_{VN} \tag{7}$$

- For alternating current:

$$R_N = R_{VN},\ X_N = X_{VN},\ \underline{Z}_N = Z^*_{VN} \tag{8}$$

That is, the complex resistance of the load must be equal to the conjugate complex internal resistance of the energy source.

In the event that the internal resistance of the energy source (active two-terminal network) is nonlinear, as, for example, for a solar battery, the relationship between the resistances of the source and the load becomes more complicated and for the analytical determination of the parameters of the matched mode one should have an analytical expression for the output current-voltage characteristic of the source.

To increase the reliability of power supply to consumers, it is possible to create a network of autonomous power plants that use renewable energy sources, and form a single energy system that autonomously supplies a certain number of consumers. In addition to generating sets and consumers, such a network will include general energy storage and distribution system, the energy intensity of which will depend both on the number and power of consumers and on fluctuations in the generated power in the system.

The use of various types of converters (wind, hydro, solar batteries) in the general generation network will significantly reduce the risks of critical fluctuations in the generated power in the general system (simultaneous failure of various installations is unlikely), and will increase its resistance to changes in external factors.

Figure 4 shows a diagram of the application of a network of installations containing wind, hydro generators, and solar batteries.

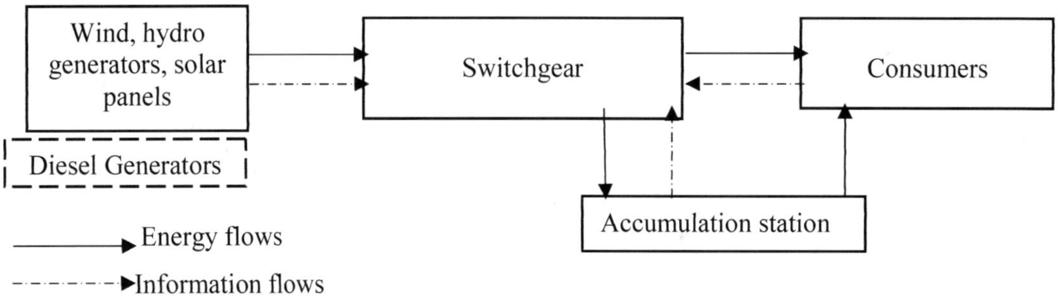

Figure 4. Scheme of the use of a network of generating sets on RES.

The network works as follows. The energy generated by the different installations is transmitted to the general switchgear together with the necessary information about the operation mode of the installations. From consumers to the switchgear, information is transmitted about the power required for the normal operation of consumers. The generated power is distributed among consumers; in case of its lack, the storage station is activated. With an excess of generated power, it replenishes the energy reserves of the storage station. The switchgear operation is organized in accordance with the information received from consumers, storage station and generator sets.

In some cases, to increase the overall reliability of the network of complex RES facilities, the system can be supplemented with a diesel generator. It will be turned on in abnormal situations of a sharp decrease in the capacity of generating renewable energy.

The operation of various generating sets using renewable energy sources on a common (autonomous) power grid with the corresponding consumers and a storage station will significantly increase the operational parameters of the system; improve its energy performance, and, consequently, the energy efficiency of using renewable energy sources.

Mountain areas have wide opportunities for using renewable energy sources. However, in mountainous conditions there is a real danger of manifestation of environmental risks of various natures. The most typical of them are: river floods, mudflows in summer, snowfalls and avalanches in winter; showers, strong winds, rockfalls, etc. up to environmental disasters (for example, the gas-dynamic discharge of the Kolka glacier into the Genaldon river gorge in 2002 (North Ossetia - Alania)).

In a number of cases, and not only in mountainous areas, natural factors can cause accidents both in the centralized electric power system and in autonomous power supply systems using renewable energy sources. Energy efficiency, the stability of the operation of electrical installations in this case will significantly depend on the possible hazardous impact on them of environmental risks, in particular, mountainous areas, especially since the bulk of installations on renewable energy sources are located, usually, in mountain conditions due to the presence in them a large number of renewable energy sources of various types.

For renewable energy installations, abnormal natural impacts will have the most destructive consequences, because these installations directly use natural energy flows (wind, water flow in a mountain river) in their own designs. Therefore, the prevention of the impact of natural hazards on generating sets using RES is of utmost importance.

Environmental safety, and, consequently, the energy efficiency of installations using renewable energy sources, can be ensured by implementing appropriate measures both at the design stage and at the stage of plant operation. When designing, it is necessary to take into account the degree of environmental risks in the place of the future location of the installation, the dynamics of changes in the environmental situation in the relevant area. During operation, it is necessary to observe measures to prevent the impact of environmental risks on the generating sets.

The algorithm of actions to prevent the dangerous impact of environmental risks on generating sets using RES is shown in figure 5. To ensure environmental safety in the area under consideration, first of all, information about the ecological state of the natural environment (OPS) is needed, which can be obtained, in particular, using a GIS (geographic information system) for the relevant area, as well as information on the parameters of natural energy flows (CES) - wind, river flows, sun rays, which are involved in the corresponding generator sets.

The assessment of the degree of hazard of environmental risks is made by comparing information on the environmental parameters P_i of the environment with their normalized P_{iN} values:

$$\frac{P_i - P_{iN}}{P_{iN}} = \delta_i \leq \delta_{iD} \tag{9}$$

where δ_i and δ_{iD} - real and permissible, respectively, relative deviations of the "i" -th ecological parameter P_i of the environment from the normalized P_{iN} value.

Cases of excess of P_i values over P_{iN} are considered, that is $P_i \geq P_{iN}$. Cases $P_i < P_{iN}$ certainly satisfy the safety conditions and therefore are not involved in further analysis.

If all real deviations δ_i satisfy condition (9), then the ecological situation is recognized as normal, not dangerous for the operation of generating sets with renewable energy sources. If any one parameter δ_k (or several parameters) does not satisfy condition (9), then the situation is recognized as dangerous (according to one or several parameters) and the implementation of measures is required to reduce the influence of relevant natural factors on autonomous generating sets.

After the implementation of the required measures, the monitoring of the ecological state of the environment and the parameters of natural energy flows is carried out again. If necessary, the described cycle is repeated. Prevention of the hazardous effects of abnormal manifestations of natural factors is a guarantee of stable operation of generating sets based on renewable energy sources, a significant way to increase the energy efficiency of renewable energy sources.

5. Conclusion

The use of renewable energy sources takes an increasing part in the general energy of almost any industrialized country and is increasing every year. To increase the energy efficiency of the use of renewable energy sources, it is necessary to: increase the efficiency of conversion plants for energy generation based on renewable energy sources; take measures to reduce the energy payback period of installations, focusing on their entire life cycle (production, operation, disposal); using concentrators of various kinds to increase the concentration of natural energy flows, converted into electrical energy; stabilize the output parameters of generating sets, in particular through intermediate energy storage. A promising direction in the development of renewable energy generators is the creation of hybrid plants that combine converters of various types. The practical advantage of autonomous conversion plants based on renewable energy sources is the possibility of their spatial location in the immediate vicinity of the consumer and the possibility of implementing a coordinated mode of the generator and load. However, it should also be noted that the reliability of the operation of autonomous generating sets powered by renewable energy sources depends on external conditions - on the ecological situation in the natural environment, in connection with which it is necessary to take timely measures to prevent the dangerous impact of environmental risks on the power supply system of autonomous generating sets using renewable energy sources.

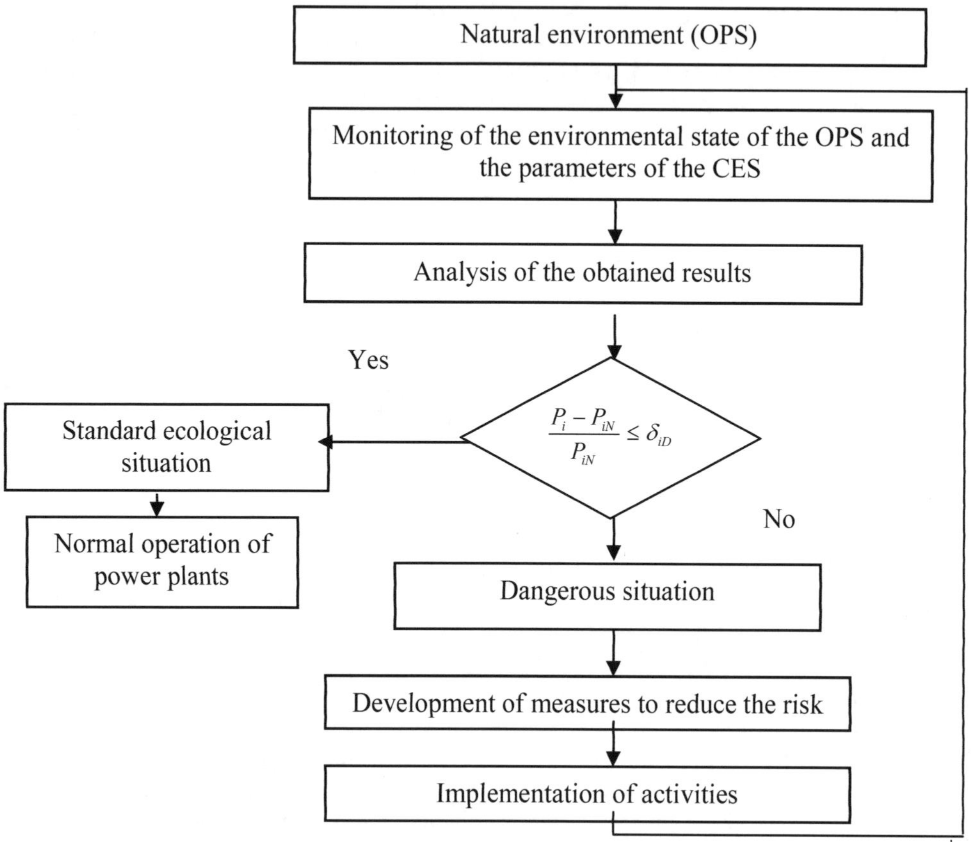

Figure 5. Algorithm of actions to prevent the dangerous impact of environmental risks on generating sets using RES.

References

[1] Bezrukikh P P, Arbuzov Yu D and Borisov G A 2002 *Resources and efficiency of using renewable energy sources in Russia* (St. Petersburg: Nauka) 313

[2] *Alternative energy sources and energy saving* 2014 (Moscow: Science and technology) 320

[3] Bezrukikh P P 2015 Renewable Energy Efficiency. Myths and facts. *Bulletin of Agrarian Science of the Don* **29** 5-17

[4] Petrov Yu S, Sakhanskiy Yu V, Zorina I Yu and Ilikoev G V 2014 Wind power plant. *Bulletin* **35** 148781

[5] Petrov Yu S, Sakhansky Yu V, Zorina I Yu and Ilikoev G V 2016 Autonomous electric power plant. *Bulletin* **2** 158761
[6] Alborov I D and Zorina I Yu 2017 Solar collector. *Bulletin* **11** 170208
[7] Petrov Yu S, Khadikov M K, Sokolov A A and Muzaev A K 2019 Wind turbine plant. *Bulletin* **11** 188444
[8] Petrov Yu S, Khadikov M K and Muzaev A K 2020 Device for autonomous lighting of a road separation barrier. *Bulletin* **6** 196315
[9] Vel'kin V I 2015 *Methodology for calculating complex RES systems for use at autonomous facilities: monograph* (Ekaterinburg: UrFU Publishing House) 226
[10] Kuachning Volker 2013 Renewable energy systems (Astana) 432
[11] Malyshenko S P and Popel O S 2010 *Renewable energy sources. Physical and technical foundations: textbook* (Dolgoprudny: Intellect Publishing House; Moscow: Publishing house MEI) 704

IOP Conf. Series: Materials Science and Engineering **976** (2020) 012003 doi:10.1088/1757-899X/976/1/012003

Influence of capacity, distance from arcs to walls on the efficiency of arcs and electricity consumption in arc steel-making furnaces

A N Makarov

Tver State Technical University, 22, Afanasy Nikitin Embankment, Tver, 170026, Russia

E-mail: tgtu_kafedra_ese@mail.ru

Abstract. The influence of the distance from the arc to the walls in the entire range of arc steel-making furnaces with a capacity from 0.5 to 120 tons on the efficiency of arcs and specific power consumption in the furnaces was investigated. In low-tonnage arc steel-making furnaces, the arcs are close to the walls and the efficiency of the arcs is small and amounts to 30-35%, and the specific power consumption for melting is 650 kWh/t. In heavy-duty steel-arc furnaces, the arcs are located far from the walls, the efficiency of the arcs is 46-49% in the absence of a slag layer, and the specific energy consumption for melting is 375-385 kWh/t.

1. Introduction

In [1], the results of the calculation and analysis of heat transfer in electric arc furnaces (EAF) of small and large capacity are presented, however, there is no explanation for the significantly large specific energy consumption in EAF of small capacity compared to EAF of large capacity. It is necessary to investigate the effect of the distance from the arcs to the walls in the entire range of arc steel-making furnaces from 0.5 to 120 t on the efficiency of arcs and specific power consumption in arc steel-making furnaces, as well as establish a proportional relationship between the efficiency of arcs and specific electricity consumption by furnaces.

2. Heat transfer and efficiency of arcs in EAF

Electric arcs are the main sources of thermal energy in electric arc furnaces, they account for 55-65% of the thermal energy supplied to a modern high-power EAF [2]. From gas-oxygen burners, exothermic reactions of oxidation of iron and other elements of the charge with oxygen, reactions of oxidation of coke and electrodes, 35-45% of energy is supplied to EAF [1-2]. All electrical energy in the arcs of the chipboard is converted into heat energy, heat flux. According to the results of numerous experimental studies, the heat flux of arcs is 90-96% formed by thermal radiation [3-7]. The share of thermal conductivity and convection accounts for 4-10% of the arc power.

The power released in the taxiway arc is distributed as follows: most of the power is released in the arc column P_{st}, part of the power is released in the anodic-cathode spots R_{ak}, the height of which is several microns, part of the power is taken by convection by the furnace gases P_k washing the arc (1):

$$P_D = P_{ak} + P_k + P_{st} \quad (1)$$

Based on the research results V.Pashkis, N.V. Okorov [4], A.V. Egorov [5], L.Ye. Nikolsky, V.D. Smolyarenko, N.L. Kuznetsov [3], A.N. Makarov [1] assume that 90% of the arc power is released in the arc column in the arcs of high-power heavy-duty steel-making arc furnaces in the form of a heat radiation flux. Anode-cathode spots and gases washing the arc account for 10% of the arc power:

$$P_D = P_{ak} + P_k + P_{st} = 0{,}1P_D + 0{,}9P_D \qquad (2)$$

The ratio of the useful power of the arc P_{pol}, going for heating, melting of metal and slag, to the power of the RD arc characterizes the efficiency of the arc heat or the efficiency of the arc:

$$\eta_D = \frac{P_{pol}}{P_D} \qquad (3)$$

Since the reference anode-cathode spot of the arc on the bath by thermal conductivity transfers all the power released in it to the metal, the reference spot on the graphite electrode, intensely radiating to the metal bath, and the gas washing the arc, transfer less than 80% of the power released in them to the metal bath, the expression for calculating the efficiency of the arc of heavy-duty EAF has the form [1]:

$$\eta_D = \frac{P_{pol}}{P_D} = \frac{(0{,}8\cdot 0{,}1)P_D + 0{,}9P_D\varphi_{DM}}{P_D} = 0{,}08 + 0{,}9\varphi_{DM} \qquad (4)$$

where φ_{DM} is the average angular emissivity of the arc column to the metal, showing the fraction of the power radiated by the arc column to the metal bath and metal charge.

In expression (4), which was used to determine the efficiency of arcs of heavy-duty 100 t EAF in the late 1990s and early 2000s, the share of power released in the arc column was taken equal to 0.9, in the anode-cathode regions 0.1. In the 2000s, the power of electric furnace transformers feeding 100 tons of chipboard was increased, since that time they began to use long arcs for metal melting. When using long arcs, the share of power released in the arc column has increased and it is necessary to correct the calculated expression (4) to calculate the efficiency of the arcs. In furnaces with a capacity of 3-50 tons, the fraction of the arc power released in the arc column is slightly less than 90%, therefore, expression (4) for furnaces with a capacity of 3-50 tons should be corrected.

3. Investigation of the efficiency of EAF arcs

The share of power released in the arc column was determined by the expression:

$$\frac{P_{st}}{P_D} = \frac{I_D U_{st}}{I_D U_D} = \frac{U_{st}}{U_D} \qquad (5)$$

where I_D, U_{CT}-respectively, arc current and voltage across the arc column.

The share of power released in the anode-cathode spots was determined by the expression:

$$\frac{P_{ak}}{P_D} = \frac{I_D U_{ak}}{I_D U_D} = \frac{U_{ak}}{U_D} \qquad (6)$$

ude U_{ak} is the anode-cathode voltage drop, $U_{ak} = 20V$ [1].

The results of calculating the distribution of the arc power in the column and anode-cathodic regions for the end of the melting period of the charge are shown in table 1.

From the results of the calculation and analysis of the calculated data in table 1, it follows that the share of power released in the arc column varies from 0.82 in the DSP-0.5 furnaces to 0.94 in the DSP-120 furnaces. The share of power released in the anodic-cathode spots of the arc varies from 0.18 in DSP-0.5 to 0.06 in DSP-120.

Considering the above, the analytical expression (4) for calculating the efficiency of EAF arcs for the entire range of furnace capacities takes the form:

$$\eta_D = \frac{P_{pol}}{P_D} = \frac{0{,}8\left(\frac{P_{pol}}{P_D}\right)P_D + \left(\frac{U_{st}}{U_D}\right)P_D\varphi_D}{P_D} = 0{,}8\left(\frac{U_{ak}}{U_D}\right) + \left(\frac{U_{st}}{U_D}\right)\varphi_{DM} \qquad (7)$$

Table 1. Distribution of voltage drop in the column and anode-cathode regions of the arc of arc steel-making furnaces of the entire range of capacities.

Parameter	DSP-0,5	DSP-3	DSP-6	DSP-25	DSP-50	DSP-100 short arcs	DSP-100 long arcs	DSP-120
U_D, V	111	128	140	163	185	200	280	320
U_{st}, V	91	108	120	143	165	180	260	300
U_{ak}/U_D	0,18	0,16	0,14	0,12	0,11	0,10	0,07	0,06
U_{st}/U_D	0,82	0,84	0,88	0,88	0,89	0,90	0,93	0,94

The influence of the design of the furnaces was investigated, in particular, the effect of the diameter of decay and electrodes and the distance from the arcs to the walls of the furnaces on the efficiency of arcs for the entire range of capacities from 0.5 to 120 tons (figure 1). Figure 1 shows a part of the working space of a chipboard, a wall, a metal bath, electrodes, electric arcs. Lining of the hearth, slopes, water-cooled wall panels are not shown. Figure 1 is a combined drawing for DSP-0.5-120 ovens. The arrangement of electrodes, arcs, and walls of the furnaces in Figure 1 on a scale repeats their location in the furnaces DSP-0,5-DSP-120. The technical data of arc steel-making furnaces are given in table 2. Figure 1 shows an example of finding an arc of a DSP-100 furnace under a thick layer of slag.

Table 2. Technical data of electric arc furnaces.

Parameter	G, t	S, MVA	D_v, m	d_r, m	r_{st}, m	G_e, kWh/t	U_{21}, V	I_D, kA
DSP-0,5	0,5	0,63	1,1	0,48	0,31	650	98-216	1,7
DSP-1,5	1,5	1,25	1,5	0,52	0,49	540	103-225	3,2
DSP-3	3	2,5	1,8	0,7	0,55	520	140-245	5
DSP-6	6	5,0	2,27	0,9	0,69	475	130-291	9
DSP-12	12	9,0	2,7	1,0	0,85	450	115-318	17
DSP-25	25	15,0	3,54	1,25	1,15	440	130-390	23
DSP-50	50	40,0	4,56	1,6	1,48	420	133-417	50
DSP-100	100	80	5,4	1,4	2,0	385	300-829	80
DSP-120	120	100	6,8	1,75	2,53	375	600-1100	75

Symbols used in table 2: G - capacity, t; S - power of the transformer, MVA; D_v - bath diameter at the level of the slopes, m; d_r - diameter of disintegration of electrodes, m; r_{st} - distance from walls to arcs, m; G_e - specific energy consumption for melting, kWh/t; U_{21} - secondary line voltage of the electric furnace transformer, V; I_D - arc current, kA.

In order to exclude the influence of the height of the slag layer on the efficiency of the arcs and to reveal the influence of the design features of the furnaces, the distance from the arcs to the walls on the efficiency of the arcs, the calculation of the efficiency of the arcs was carried out to end the melting period in the absence of a slag layer h_{sh} on the metal bath in all furnaces. The arc length was determined by the expression [1]:

$$l_D = \frac{U_D - U_{ak}}{gradU_{st}} \quad (8)$$

where $gradU_{st}$ = 0.8V/mm is the voltage gradient in the arc column.

The height of the arc penetration h_3 in the absence of slag is equal to the height of the arc penetration into the metal hm:

$$h_z = h_M,\ h_z = k_z I_D \quad (9)$$

where k_z is the coefficient of arc penetration into the metal bath, is determined from the graphs [1] and is equal to k_z=0.8 mm/kA for currents up to 10 kA, k_z=1.25÷1.40 for currents, respectively, 60-80 kA.

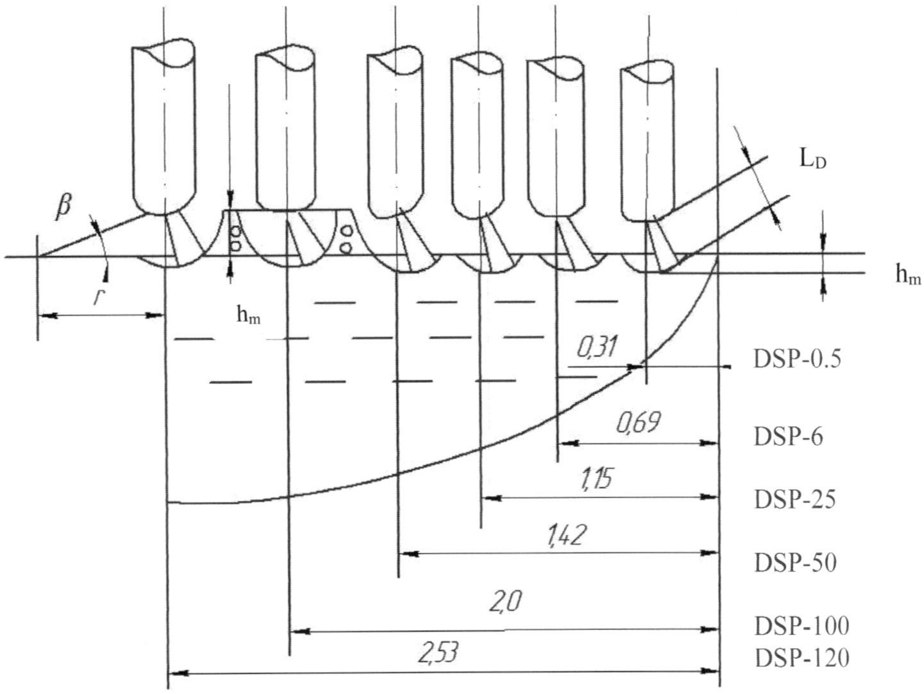

Figure 1. Arrangement of arcs in furnaces of small and large capacity DSP-0.5-DSP-120 and the distance from arches to walls.

The calculation of the efficiency of arc furnaces DSP-0.5, DSP-100, DSP-120 was carried out according to the expression (7). For calculations, we used the data in table 1 and table 2. The average angular emissivity of the arc per metal bath was determined by the method described in [1], according to which the surface of the metal bath is divided into several hundred calculated areas of the same area F_k, in this case, the calculation is 216 sites (figure 2) for each of which, according to the formulas set forth in [1], the fraction of the arc column radiation is determined taking into account its penetration into the metal bath, that is, the local angular emissivity of the arc column to the calculated area is determined. To calculate the local and average angular emissivity of the arc on the metal bath, the software package Microsoft, Excel, MathCad, Visual Fortran 6.0 was used.

4. Results of calculating the efficiency of arcs in a chipboard

Figure 2 shows the results of calculating the average angular coefficients of arc radiation on a metal bath φ_{DM}, average angular coefficients of arc radiation on the walls and roof of the furnace φ_{DS}, efficiency of arcs η_D. As can be seen from figure 2, the closer the arcs are to the walls of the chipboard, the lower the average angular emissivity of the arc to the metal bath and the efficiency of the arc, and the greater the average angular emissivity of the arc to the walls and roof of the furnace and the specific power consumption for melting.

In the DSP-0.5 furnace, the arcs are very close to the walls, r_{st} = 0.31 m, therefore, the share of the thermal radiation of the arc on the walls, the roof is large and is φ_{DC} = 0.82, the proportion of the useful thermal radiation of the arc on the metal bath is small, φ_{DM} = 0.18, therefore, the efficiency of EAF-0.5 arcs is very low, η_D = 0.31, and the specific power consumption is high and amounts to G_e = 650 kWh/t for melting the charge. In the chipboard furnace - 6, the distance from the arc to the walls $r_{st} = 0.69$ m is greater than in the chipboard furnace - 0.5, respectively, the fraction of the thermal radiation of the arc on the walls and roof is less, $\varphi_{DC} = 0.75$, more for the metal bath $\varphi_{DM} = 0.25$ and more arc efficiency, $\eta_D = 0.35$, less specific power consumption for melting $G_e = 475$ kWh/t in comparison with an EAF furnace - 0.5.

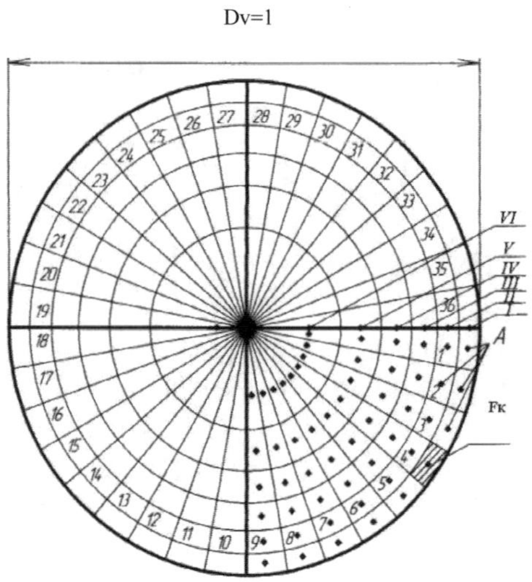

Figure 2. Dividing the metal bath into 216 elementary computational sites F_k.

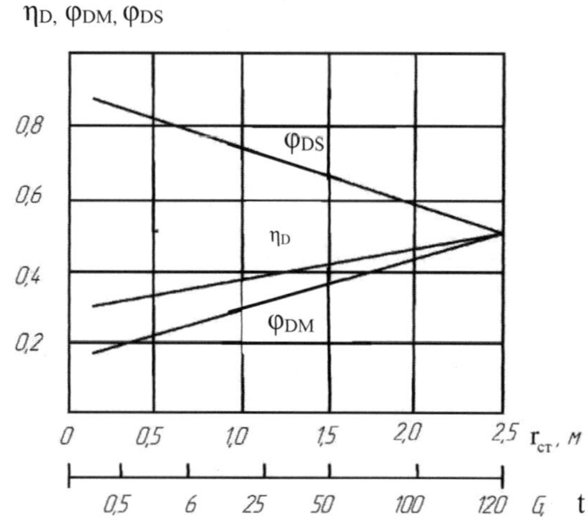

Figure 3. Dependence of the efficiency of arcs, average angular coefficients of radiation of the column of arcs on the metal bath, walls, and vault on the distance from arcs to walls and furnace capacity.

In furnaces chipboard - 100, chipboard - 120, the diameter of the bath of which, respectively, 5.4 and 6.8 m, the ratio $d_r/D_v = 0.26$, which corresponds to the recommended, the arcs are close to the center of the bath and are removed from the walls at a distance, respectively, 2.0 and 2.53 m, which resulted in an increase in the efficiency of arcs to 0.46 and 0.49, respectively, and a decrease in the specific consumption of electricity for melting to 385 and 375 kWh/t.

5. Conclusion

Analytical studies of the efficiency of arcs have established the following. In furnaces of small capacity, the diameter of the casing, the metal bath is small, the arcs are close to the walls, and the fraction of the thermal radiation of the arcs on the walls and the roof is 4 times greater than the fraction of the thermal radiation of the arcs on the metal bath. The efficiency of small-capacity furnace arcs is small, and the specific power consumption is high. In large-capacity furnaces, the diameter of the casing and the metal bath is 5-7 m, the arcs are removed from the walls, the share of thermal radiation of the arcs per metal bath, the efficiency of the arcs increase compared to small-capacity furnaces, and the specific power consumption decreases. The results of calculating the efficiency of arcs are confirmed by the practice of operating an EAF. Specific power consumption for melting is 650-700 kWh/t in small-capacity furnaces, 375-385 kWh/t in large-capacity furnaces.

Thus, theoretical studies, the results of calculating the angular coefficients of thermal radiation of arcs on the bath, walls, vault, efficiency of arcs and the practice of operating arc steel-making furnaces are in harmonious agreement with each other, the results of the calculation are confirmed by the practice of operating furnaces: the more efficiency of arcs in EAF obtained by calculation, the lower the specific power consumption for melting the charge, obtained as a result of the practice of operating the furnaces. Since practice is a criterion for the truth of theoretical developments, in this case, the practice of operating the furnaces confirms the correctness of the methodology and the performed calculations of the average angular coefficients of radiation of arcs per metal bath, walls, roof, efficiency of arcs of arc steel-making furnaces of the entire range of capacities from 0.5 to 120 t.

References

[1] Makarov A N 2014 *Heat exchange in electric arc and flare metallurgical furnaces and power plants* (St. Petersburg: Lan) 384
[2] Evstratov V G, Kiselev A D and Zinurov I Yu 2012 Features of the thermal operation of the EAF – 120 Consteel electric furnace at the Ashinskiy metallurgical plant. *Electrometallurgy* **8** 2-6
[3] Nikolsky L E, Smolyarenko V D and Kuznetsov L N 1981 *Thermal operation of arc steel-making furnaces* (Moscow: Metallurgy) 344
[4] Okorokov N V 1971 *Electric arc furnaces* (Moscow: Metallurgy) 347
[5] Egorov A V 1985 *Electric melting furnaces of ferrous metallurgy* (Moscow: Metallurgy) 280
[6] Soifer V M 2009 *Steel smelting in acid electric furnaces* (Moscow: Mechanical Engineering) 480
[7] Samokhvalov G V and Chernyi G I 1984 *Electric furnaces of ferrous metallurgy* (Moscow: Metallurgy) 232
[8] Leskov G I 1970 *Electric welding arc* (Moscow: Mashinostroenie) 335
[9] Tikhodeev G M 1961 *Energy properties of an electric welding arc* (Moscow: Publishing house of the Academy of Sciences of the USSR) 236
[10] Kuznetsov M S, Yakushev E V and Kulagin SA 2010 Influence of the charge weight and slag thickness on the technology of steelmaking in an arc furnace. *Electrometallurgy* **2** 2-6

IOP Conf. Series: Materials Science and Engineering **976** (2020) 012004 doi:10.1088/1757-899X/976/1/012004

Production automation system for company manufacturing oil country tubular goods

S Y Krotova[1], E N Ovchinnikova[1], T V Sarapulova[1] and I V Remizova[2]

[1] Saint Petersburg mining University, 2, 21th line of Vasilievsky island, Saint Petersburg, 199106, Russia
[2] Saint Petersburg state University of industrial technologies and design, 18, Bolshaya Morskaya str., Saint Petersburg, 191186, Russia

E-mail: su18@mail.ru

Abstract. This work presents a developments of a single integrated solution for the process automation of oilfield service companies that allows organizing production and management systems in a single information space. The analysis of the main factors and causes that affect the occurrence of production losses, and their consequences. In the scope of this work, algorithms for solving individual automation problems that are interconnected into certain algorithm modules have been developed. The results of the integration analysis of modern ERP automation systems are presented, in particular, their level of adaptation to the stages of the product life cycle, compliance with standards and safety requirements, and the complexity of the transition to an automated system. Possible risks of the implementation of the algorithms were also assessed, describing an applied importance of the development of a single production control algorithm. Further, an evaluation of main advantages and economic effect of the introduction of process control automation system at the company manufacturing oil country tubular goods is presented.

1. Introduction

Currently available automated systems used in oil and gas extracting, processing and storing companies are not suitable for the technological process of oil country tubular goods manufacturing or organization and management accounting at oilfield service companies. Therefore, the development of a single integrated solution for the process automation of oilfield service companies that allows organizing production and management systems in a single information space, optimize information flows, and reduce production costs, is an urgent task. This would allow improving financial results through increasing the controllability of the company and rational distribution of all types of resources thus leading to the reduction of production and services cost.

Currently, the development of an automated production cycle management system for oil and gas enterprises gained a special interest in scientific research [1]. Most of the works are focused on the systems for the companies that are extracting, processing, storing, transporting and selling oil and gas.

There is a yearly increase in oil and gas patch with exploration of new fields and installation of new pipelines. Successful drilling and service of wells requires quality tubular products.

2. Materials and methods

In the course of this research, General scientific and theoretical-empirical methods were used: analysis and synthesis, generalization, comparison, systematization, modeling.

In particular, the analysis of modern ERP-class automation systems is based on the study of the integration area, the level of adaptation to the stages of the product life cycle, compliance with the requirements of Russian standards and safety, and the complexity of the transition to an automation system.

The following types of calculations and analyses were used to evaluate the economic efficiency of designing and implementing software modules of the unified algorithm for organizing and planning a technological line for manufacturing pump and compressor pipes: quantitative calculation of the economic effect, calculation of capital costs for design and implementation, calculation of savings due to increased labor productivity, as well as project risk analysis when implementing the algorithm for organizing and planning a technological line [2].

3. Results

There are two rather independent ways of industrial company management automation.

A modern automated company management system consists of several levels, one of which is the ERP (Enterprise Resources Planning) level. A versatility of production tasks arising at the oilfield service companies impose different requirements on functionality, scalability and technology. Therefore, a reasonable choice of an ERP automation system plays a paramount role in the further successful work of the whole company [3].

According to the specifics of the production process at oilfield service companies, the main criteria for choosing an automated ERP system were identified: integrability; the possibility to transfer initial balances and other data from programs that were previously used at the company; system flexibility; adaptability of the information system to new conditions, new needs of the company; functions and possibilities of the ERP system; data exchange and subsequent automatic data export / import in ERP systems; adaptability of the system to the stages of the product life cycle at an the oil and gas company; the integrity of quality management processes at all stages of the product life cycle; preparation of integrated reports; adaptation of the system to the requirements of Russian standards.

There are 10 automated ERP systems that to a given extent meet main criteria for automated management systems of oilfield service companies: SAP R/3 (SAP AG – Germane), Oracle Applications (Oracle – USA), IFS Application (IFS – Sweden), Baan ERP (SSA Global – USA), iRenaissance (Ross Systems – USA), MBS Axapta Navision (Microsoft – USA), J.D.Edwards OneWorld (J.D.Edwards – USA), Galaktika (Galaktika – Russia), 1C: Enterprise (1C – Russia), BOSS-Corporation ("IT" Company – Russia) [4-6].

These ERP products were analyzed from the point of view of the main criteria required by the oil-well tubing manufacture process at oilfield service companies. The analysis was carried out according to the following criteria: Integrability; System flexibility; Data exchange; Adaptability to the stages of the product life cycle; Integrity ofquality management processes; Preparation of integrated reports.

SSAP R/3:

- Electronic document management is possible with significant modifications to the software.
- Meets all the standard business processes within the entire company, but the modernization of the program code is complicated.
- Its implementation requires a preparation of a complete information base and recording of operations online.
- Needs refinement, depending on the accounting specifics of the end-user company.
- Each module performs specific business tasks independently, but is closely connected to other modules.
- A production planning system that allows taking into account a large number of technological process details and perform very detailed calculations.

Oracle Applications:

- Used in projects for the implementation of information systems based on Oracle products, including interfacing with software products and software and hardware platforms of other manufacturers.
- A large number of open interfaces, the composition of the database is described in detail and comes with open source code.
- Extensive protocol support standards deployment: RosettaNet, Electronic Data Interchange - EDI, Applicability Statement 2 - AS2.
- Needs refinement, depending on the accounting specifics of the end-user company.
- One of the most fully covering the main areas of company management. The set of ERP applications includes more than 90 modules.
- Very flexible in reporting according to the requirements.

IFS Application:

- Initial setup is done by an automation company.
- With the help of additional levels in the configuration module of the system, the cost of ownership is reduced, and programmers are developing specific business tasks instead of re-adjusting the current modification of the system.
- Electronic information exchange with other software.
- Needs refinement, depending on the accounting specifics of the end-user company.
- The entire product life cycle is managed using PLM programs.
- Automated business processes, 3LM management concept that works simultaneously with fixed assets, customer service and products.

Baan ERP:

- Implemented in a three-tier client-server architecture, can interact with other software (supports CORBA, OLE technologies). The "Enterprise Modeling" package provides the possibility of a phased analysis of the company, allowing obtaining the complete structure of the system. The package is compatible with many information databases and has a fairly easy pairing with systems already installed at a customer.
- Written in 4GL - 4 Generation Language - that allows working with events. Allows using existing scripts when programming.
- Allows connecting with third-party industry solutions.
- Requires refinement; various relationships can be established between units of measurement, the conversion of units of measurements can be specified.
- Has the following features: adaptability, scalability, versatility of functions, covers a large number of management tasks.
- Provides grounds for the distribution of both complex functional modules and a variety of analytical reports.

IRenaissance:

- Configured to receive data directly from the field automation controllers (pressure and temperature sensors, weighing equipment), the enterprise is controlled at the process management level.
- It is possible to add the necessary functions to the system using the built-in CASE-tools and 4GL.

- The following DBMSs are used: Oracle Database, MSSQL-Server, Sybase SQL Server, or RDB.
- Needs refinement, depending on the accounting specifics of the end-user company.
- The system has the following modules: manufacturing; finance; personnel accounting; payroll preparation.
- The system stores 1400 ready-made business – procedures. A single database contains all corporate information. Analysis of the accumulated information is possible using decision support tools (DSS): DSSuite, ROSS Systems proprietary system.

MBS Axapta, Navision:

- The built-in COM/DCOM mechanism allows that internal objects work with external applications. To do this, a special Axapta COM Connector library is used, which allows an external application to see Axapta as a COM object and provides a possibility to directly access Axapta data, manage transactions and use the system's business logic to process them through the COM/DCOM interface. This ensures that Axapta is fully integrated into the company's document management technology.
- Has an integrated development environment MorphX. To change the standard business logic, as well as to create new functionality, the system provides an object-oriented language X ++, as well as access to SQL data.
- The COM / DCOM mechanism allows the system to interact with third-party applications, the interaction is carried out using COM objects created in the system.
- Needs refinement, depending on the accounting specifics of the end-user company.
- A multifunctional ERPII solution covers the business of the company as a whole, including production and distribution, supply chain management (SCM) and project management, financial management and business analysis tools, customer relationship management (CRM), and personnel management.
- Axapta - an integrated automated manufacturing company management system, which provides company's managers and investors with full information about the company's activities. Reports are divided into operational (to which the requirements of business processes apply) and statistical (processing a large amount of information).

Galaktika:

- Possible integration options: documents export and import (using standard system interfaces; data export and import in generally accepted formats; a possibility to organize direct interaction of third-party development tools with physical tables through the SQL query mechanism.
- ERP and BI systems are free of charge. Only support and implementation are charged. Free software options have limited functionality compared to the paid versions. "Galaktika express" contains the same source code as the paid version of the ERP system, but part of it has become open.
- The system is not integrated. A large number of modules are not connected. Documents in the finance module require manual input that significantly.
- Needs refinement, depending on the accounting specifics of the end-user company.
- The system does not implement algorithms for generating optimal production and supply requests depending on the demand, plans, forecasts, or their combination.
- The toolkit allows building stand-alone reports based on a simple database query language (SQL). Reports can be saved for reuse and exported to MS Excel, HTML, GIF, JPG, PNG, BMP, CSV, TXT, PDF formats.

1C: Enterprise:

- Supports the most common types of operating systems and databases: Linux, Windows, Mac OS, PostgreSQL, MS SQL Server, IBM DB2, Oracle Database.
- An open system that enables the exchange of data with many external applications and technical equipment, thanks to open standards and data transfer protocols.
- The system is written in its own internal programming language, which provides access to data and a possibility to interact with other programs.
- Needs refinement, depending on the accounting specifics of the end-user company.
- The system does not implement algorithms for generating optimal production and supply requests depending on the demand, plans, forecasts, or their combination.
- The toolkit allows building stand-alone reports based on a simple database query language (SQL). Reports can be saved for reuse and exported to MS Excel, HTML, GIF, JPG, PNG, BMP, CSV, TXT, PDF formats.

BOSS-Corporation:

- The architecture of the system has an integrated and modular structure, which allows phased automation of a company.
- Is an open client-server system that allows increasing functionality and access data through any programming tools.
- Allows data exchange with other expert and decision-making systems.
- Needs refinement, depending on the accounting specifics of the end-user company.
- Does not have a product quality management function at all stages of product life cycle.
- The system has a unique graphical tool for configuring the business logic of the application, which governs all possible stages of reporting.

The analysis of the automated EPR systems and to which extent they meet main needs of manufacture process at oilfield service companies has demonstrated that none of the evaluated systems fully solves all the problems of the company. Therefore, an additional module with necessary functionality shall be purchased or developed and consequently, integrated. Besides, the statement that the ERP-system is already integrated is rather questionable, since an installation of a new version of any module included in the ERP-system requires updating of other modules. Therefore, suppliers should provide a possibility to implement various versions of their applications, which also requires integration [7].

None of the analyzed ERP systems is adapted to the oil pipes production specifics and cannot receive reports at the stages of the product life cycle at oilfield service companies.

BAAN and AXAPTA systems have shown the largest integration possibilities, since they have mechanisms allowing that the external applications work with internal system object.

The lowest integration was observed in the SAP R/3 system, as it requires significant software improvements, entailing high financial investments. Most Russian automation systems are not aimed at the idea of quality management processes integrity at all stages of the product life cycle [8].

Thus, in the automation of oilfield service companies a development of a single algorithm for organizing and planning of oil-well tubing manufacture process that would optimize the operation of production, management and information flows, as well as reduce downtime and overproduction, is highly relevant.

The developed automation system of an oil and gas industry company at the process control level includes an algorithm for organizing and planning of oil-well tubing manufacture process divided into algorithm modules according to individual tasks (functions). The tasks of the system are interconnected in such a way that individual modules of the general algorithm work in specific production processes [9].

The algorithmic support for an automated process control system is based on the progressive principle of modularity, according to which software can be decomposed into a set (or synthesized from a set) of autonomous parts of algorithmic tasks [10].

Automation system of an oilfield service company at the process control level developed on a basis of a single algorithm for organizing and planning of oil-well tubing manufacture process was implemented in an active service company that manufactures and sells tubular goods along with a wide scope of maintenance services.

4. Discussion

The result of the present study is the development of a single algorithm for solving the main tasks of the automation system of an oil and gas industry company at the process control level.

The main economic effect of the implementation of the process control automation system at the company manufacturing oil country tubular goods is an optimization of accounting processes, which results in increased economic indicators.

The direct benefits of introducing the algorithm for organizing and planning of oil-well tubing manufacture process include employees' time saving, saving on raw materials, etc.

Indirect benefits from the implementation of the algorithm include transparency of management, balanced production processes, control of the workload of the production line, tracking of defects and planning of raw materials procurement. Savings due to increased productivity are calculated based on the user-performed tasks, such as:

- Information input - input of the information about a new order (order number, oil-well pipe type and geometry, and deadlines) into the database, entering into the database the information from the reference documentation and standards (order number and technological process operation), entering into the database; information about the defects in the oil-well pipe (order number and operation of the process, at which a defect was discovered).
- Calculations - calculation of the amount of raw materials needed to meed customer requirements and assessment of the possibility of launching a new order in the production cycle.
- Preparation and printing of reports - generation of reports on the workload of the production line and tracking of each order at the stages of the product life cycle, including deadlines.
- Data analysis and sampling - defect detection in operations of a production line. Defect/operation ratio calculation.

The assessment of the economic efficiency of designing and implementing of program modules of single algorithm for organizing and planning of oil-well tubing manufacture process demonstrates mid- and long-term benefits [11]. An expected economic efficiency is about $ 25 000.

5. Conclusion

Based on the algorithms of the tasks analyzed within the project, an oil and gas industry company automation system software was developed. The tasks are solved using forms.

The implementation of the algorithm for solving the tasks "Launch to the plant" and "Plant load" allows: determination of the logical sequence of the process from receiving of raw materials to finished products; showing the stages of coordination of technological and technical communication between the main operations of the production process; visualization of the workload of the production line; assessment of the possibility of fulfilling a new production order in accordance with the technical specifications before deadline and of timely material preparation for manufacturing.

A balance between the loading of the production line and the timely execution of orders without delay is essential allows avoiding overproduction and save on storage of finished products.

Solving of the task "Defect detection" allows systematization of the factors affecting the production losses. A table allows displaying the ratio of defects to production line operations over a given period. The developed module for solving the "Procurement planning" task, included into the algorithm for organizing and planning of oil-well tubing manufacture process allows increasing the conversion accuracy from tons to meters and minimizes production losses associated with irrational raw materials procurement planning.

According to the economic efficiency assessment verified at an active company, the costs of implementation of program modules of single algorithm for organizing and planning of oil-well tubing manufacture process are minimal, and the estimated project payback is around one month.

Application of the developed procedure allows adjustment of the business process itself, as the tasks are completed faster. Employees can process larger amounts of information during their working hours, which can be used either to reduce staff expenditures or for quick business growth at the same number of employees.

The technical contribution of the research project lies within the fact that the developed algorithm for organizing and planning of manufacture process can be integrated into any production accounting system, because each algorithm module allows downloading data in xls format.

The results of the project can be applied in the companies producing oil country tubular goods for oil and gas industry.

References

[1] Linnik Y and Kiryukhin M 2019 Digital technologies in the oil and gas industry. *Vestnik Universiteta* **1(7)** 37-40

[2] Polyakov R K 2007 Development of an information and logical model of the risk management algorithm in the enterprise management system. *Russian economic online journal* **1** 90

[3] Makhovikov A B 2016 Principles of corporate communications management system development for mineral resources companies. *Journal of Mining Institute* **217** 24-8

[4] Cosenz F and Not G A 2018 Dynamic business modelling approach to design and experiment new business venture strategies. *Long Range Planning* **51(1)** 127-70

[5] Ageronok V A, Gabets A P and Kozyrev D V 2012 *Professional development in 1C system: Enterprise* (Moscow: 1C-Publishing) 690

[6] Malyutin A G 2011 *Corporate electronic systems* (Omsk: Omsk State Transport University) 38

[7] Islamov Sh R, Bondarenko A V, Korobov G Y and Podoprigora D G 2019 Complex Algorithm for Developing Effective Kill Fluids for Oil and Gas Condensate. *Reservoirs International Journal of Civil Engineering and Technology* **10(1)** 2697-713

[8] Artamonova T E, Ovsyannikova A V, Vorobyeva A V and Popovich A E 2016 ERPsystems. Efficiency and implementation problems of ERP systems. *Natural and technical sciences* **4(94)** 173-174

[9] Remizova I V, Krugovaya T S, Krotova S Yu and Ovchinnikova E N 2020 Automation system for oil and gas service enterprises. *IOP Conference Series: Materials Science and Engineering* **905(1)** 012041

[10] Krugovaya T S 2019 Automation of the activities of enterprises producing pipes for the oil and gas sector at the process control level. *Proceedings of materials of the XI regional scientific and practical conference of students, undergraduates, graduate students St. Petersburg* 19-20

[11] Rozhkov A A and Karpenko M S 2015 Economic and mathematical models and algorithms for optimizing financial costs in the management of energy-saving innovations in mining enterprises. *Mountain Information and Analytical Newsletter* **6** 307-13

IOP Conf. Series: Materials Science and Engineering **976** (2020) 012005 doi:10.1088/1757-899X/976/1/012005

Ultrasonic multipath monitoring of turbulent gas flows: pulse signals correlation processing

I Y Bychkova[1], A V Bychkov[1], L A Slavutskii[1], V G Grigoriev[1] and A I Kitaev[2]

[1] Chuvash State University, 15, Moskovsky prospekt, Cheboksary, 428015, Russia
[2] Chuvash State Pedagogical University, 38, Karl Marx street, Cheboksary, 428000, Russia

E-mail: bav.xlab@gmail.com

Abstract. The article proposes digital phase modulation and correlation processing of pulsed ultrasonic signals for the turbulent gas flows monitoring. Estimates of errors in measuring the relative time delay of ultrasonic pulses during multipath propagation and their interference in the receiver were made. The data obtained from the experiment show that the relative pulse delay measured in a turbulent gas flow has a resolution exceeding tenths of the carrier frequency period. As an example the possibilities of turbulence's quantitative control the air flow boundaries of a fan are shown. The perspectives to use the developed software and hardware in flow meters and other ultrasonic control and measuring devices are discussed.

1. Introduction

Ultrasonic (US) control methods are widely used in power engineering, in particular for flow measurement [1-4]. To improve the liquid and gas flows velocity measurement's accuracy, multipath ultrasonic measurement schemes are most often used [5-6]. A gas flows feature is a high level of turbulence, which leads to increased variability of the ultrasonic signals delay [7-11]. In these conditions, it is necessary to develop methods of digital shaping and processing of US signals to increase the resolution of the measuring equipment [12-14]. In this work, to control gas flows with a high turbulence, software and hardware based on digital phase modulation and correlation processing of ultrasonic signals are proposed.

2. Materials and methods

At multipath propagation, ultrasonic signals in the receiver are shaped as a result of the interference of pulses that arrive to the receiver along different paths. For two-rays control, as a particular case of multipath control, in the cross-correlation function (CCF) of the dual and reference signals two peaks appear, the time position difference of which corresponds to the pulses relative delay. To increase the resolution of the relative position of signals digital phase coding can be used according to the Barker sequence [15-17], which, in comparison with unmodulated signals use, increases the time resolution of US devices several times (figure 1).

Measurement of small relative delays ΔT_D between signals in the receiver (figure 2, *b*) in the work is proposed for the analysis of the CCF's power W asymmetry [12], i.e. by its weighted average time position (indicated by a vertical dashed line in figure 2).

$$\tau_{CoG} = \frac{\int \tau W(\tau)\, d\tau}{\int W(\tau)\, d\tau}. \qquad (1)$$

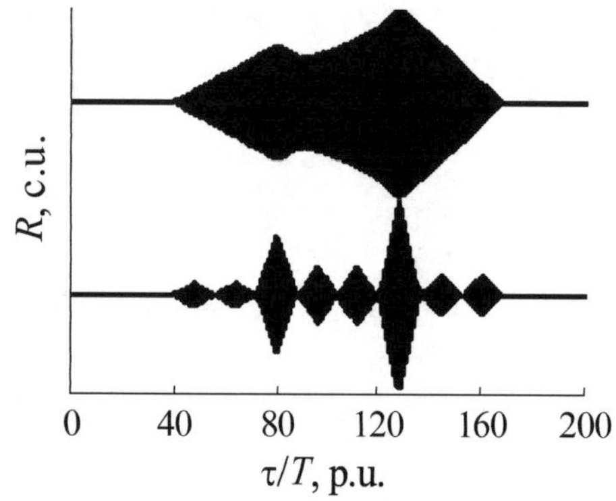

Figure 1. Comparison of the CCF of the dual and reference signals without modulation and using phase-modulation according to a Barker code.

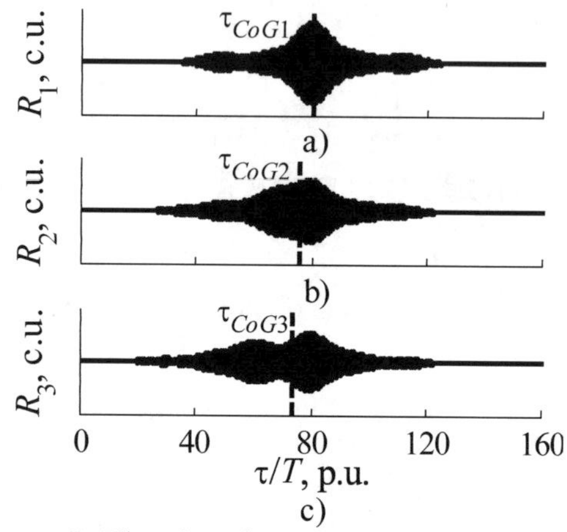

Figure 2. The changing of CCF's shape with relative pulse delay changing (T is an acoustic wave period).

At the same time, the symmetric CCF is assumed at a zero value of delay (figure 2, *a*). $\Delta T_D(\tau_{CoG})$ dependence changes linearly with the superposition of significant quasi-regular oscillations [12]. Such oscillations are smoothed out in real experimental measurements with long-term averaging due to phase fluctuations of the signals. The proposed approach does not impose any restrictions on the lower bound of the resolution of pulses relative time position. The theoretical resolution is about 15 ns at 40 kHz ultrasound carrier frequency and 1 ms pulse duration.

FPGA-based functional blocks of a pulsed ultrasonic device [12] are enclosed in a receiver–transmitter device (RTD) (figure 3). The shape of the pulse packet can be adjusted in the device (ordinary or phase-modulated according to a Barker code [15]). Ultrasonic transducers (UST) have a relatively wide radiation pattern (about 60°). In the general case, the signal at the receiver input is a superposition of two signals: direct and reflected. According to the radiation patterns, the radiation power of the side rays differs from the power of the direct ray by about 5 dB. To increase the system sensitivity to the signal reflected from the flow the USTs were turned about 45° in the experiment (figure 3). For the laboratory experiment, an industrial fan with a power of 170 W, an orifice diameter of 0.15 m, a rated velocity of 2430 rpm and an output of 1200 m^3/h was used. Using the digital phototachometer (P) the fan velocity was calibrated. It provides contactless measurement with a resolution of 0.1 rpm and 1.0 rpm in the range of less than 1000 rpm and more than 1000 rpm, respectively.

By measurement of the spatial and temporal flow velocity changes using an X-Line AeroTemp anemometer (A) the structure of the air flow was experimentally controlled. Velocity values were monitored for 20 s with a periodicity of 0.25 s at different points of the flow: in 11 points along the flow and in 5 points across the flow. The fan velocity was varied with a controller (C) (figure 3).

The invariant embedding method for the theoretical estimates was used to solve the problem of acoustic wave scattering from a randomly inhomogeneous layer of the medium. This method allows to transform problem with boundary conditions into problem with initial conditions [18]. The process is described by a first order nonlinear differential equation. Its solution is the reflection coefficient of a harmonic wave with ω frequency [19]:

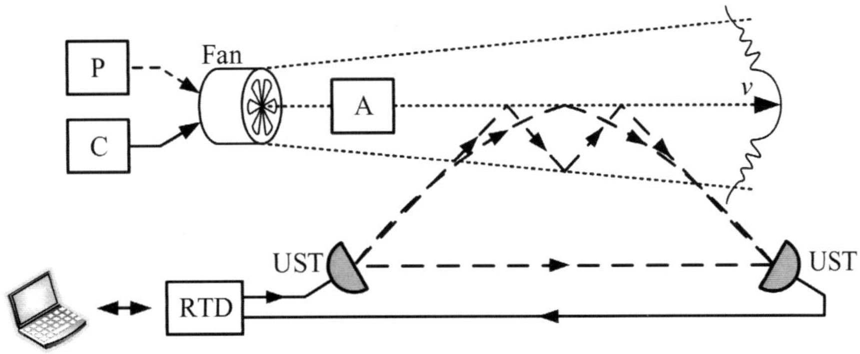

Figure 3. Functional scheme of the experiment.

$$\frac{dR_L}{dL} = 2ikR_L + i\frac{k}{2}\varepsilon(L)(1+R_L)^2, \tag{2}$$

where $k = \omega/c$ – acoustic wave number in a homogeneous medium; c – sound speed; $\varepsilon(L)$ – random layer inhomogeneity function; L – layer thickness (integration parameter). The frequency response $R_L(\omega)$ of the reflection coefficient can be obtained by solving (2) for different frequencies. The shape of the reflected pulse finally can be get by $R_L(\omega)$ inverse Fourier transform.

3. Results

Based on model calculations figure 4 shows an example of the layer inhomogeneity function expressed in terms of the sound speed. It consists of two components: random and regular $\varepsilon(x) = \varepsilon_0(x) + \tilde{\varepsilon}(x)$. The random component was specified as a function with a Gauss distribution and variance corresponding to the characteristic in homogeneities of the medium. The value of random inhomogeneity is comparable to the acoustic wavelength. Figure 5 shows as the example pulse waveform of an ultrasonic resonant transducer and form of distorted pulse after reflection.

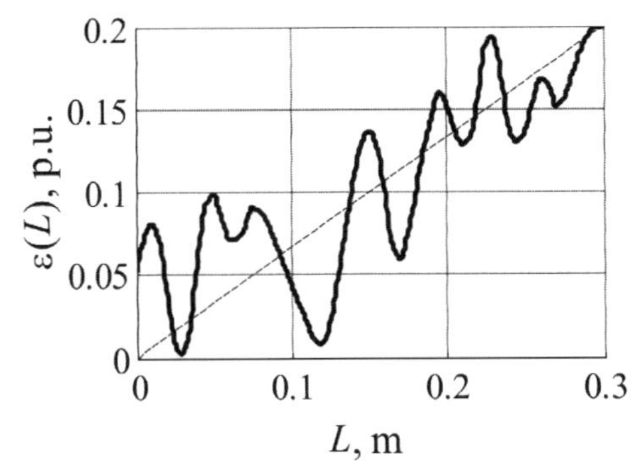

Figure 4. Sound speed profile.

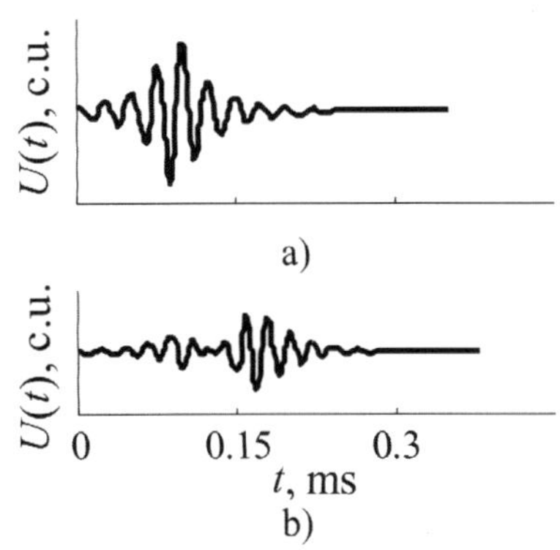

Figure 5. Input and output pulses.

According to the simulation results, we can say that an increase in the amplitude of the random component $\tilde{\varepsilon}(x)$ changes not only the amplitude and phase of the reflected signal, but also its duration. As a whole this affects the time delay of the signal.

A comparison of modeling results of the measured delay's error using the proposed algorithm based on estimating CCF's asymmetry and the trigger method with a cutoff level of 0.5 is carried out. From the histograms of the delay's statistical distributions plotted by 1000 pulses (figure 6, *a*), the dependences of the average, standard deviation and asymmetry on the noise level σ_N in the range 1 – 9 % of pulses amplitude were obtained (at $\sigma_N > 10\%$ when measured by the trigger method, abnormal spikes appear in the delay histogram) (figure 6, *b*).

When calculating the delay based on the weighted average position of the CCF, statistical distribution's width (by about an order of magnitude) and asymmetry decrease. So in comparison with the traditional measurement method, the proposed algorithm makes it possible to reduce random and regular errors, thereby increasing the time resolution of pulsed ultrasonic devices.

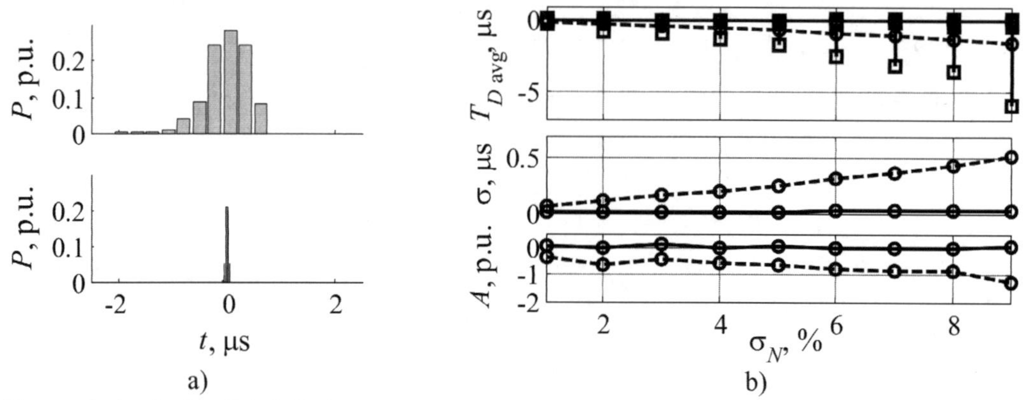

Figure 6. Analysis of statistical error
a) histograms of the delay distribution by triggering (top) and by the weighted average position of the CCF's power (bottom) with noise variance $\sigma_N = 7\%$;
b) characteristics of distributions depending on σ_N (average $T_{D\,\text{avg}}$, standard deviation σ and asymmetry A) when measuring the delay by triggering (dashed line) and by integral estimate of the CCF's shape (solid line).

Figure 7 shows the distribution of the flow velocity's fluctuations Δv (the degree of turbulence in the nominal fan operation) relative to its average value v_{avg}

$$e = \left(\frac{|\Delta v|}{v_{\text{avg}}} \right) \cdot 100\%. \tag{3}$$

As can be seen from the figure, at the edges of the flow, the degree of turbulence reaches 10-15 %.

The US pulses relative delay was measured at different values of the flow velocity by processing and averaging two hundred pulses. The averaging time can vary depending on the measurement conditions and the required accuracy. Figure 8 shows the dependence of the CCF's weighted average time position on the fan velocity. Thus, with flow velocity growth the relative time delay between the direct and the reflected from the flow signals monotonically increases. According to the results of experimental studies, the most important is the fact that when the transmitter and the receiver positions are swapped corresponding dependences of the delay on the flow velocity differ very little. In other words, the delay increases with the propagation of ultrasound both along and against the flow.

IOP Conf. Series: Materials Science and Engineering **976** (2020) 012005 doi:10.1088/1757-899X/976/1/012005

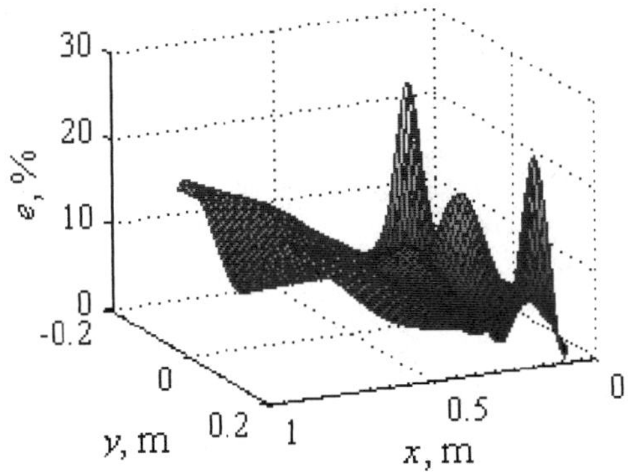

Figure 7. Distribution of relative fluctuations of flow velocity.

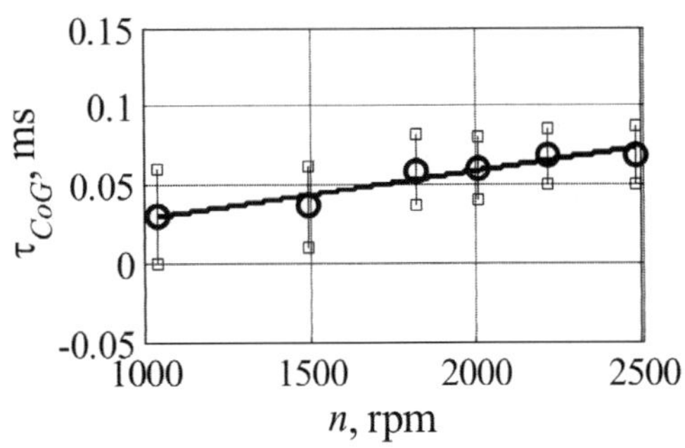

Figure 8. Experimental dependence τ_{CoG} on the fan velocity. Circles – average value, vertical lines – value of standard deviation.

4. Discussion

Under re-reflections of the acoustic wave between random inhomogeneities of the layer the additional temporal delay according to the experimental results is appeared . The wave is "captured" by random stratification and propagates in a randomly inhomogeneous layer like in waveguide. This phenomenon is called the "localization effect" [19]. At a flow velocity of about 5 m/s on the fan's longitudinal axis the relative time delay between the direct and the reflected signals differs about 3-5 μs for forward (downstream) and backward (upstream) ultrasonic wave propagation. With flow velocity's increase the change range of the delay is about 25-30 μs. (figure 8). When calculating refraction using ray approximation, the delay is about few microseconds. In other words, the ultrasonic wave's propagation in a flow is associated not only with refraction, but also with the scattering, the occurrence of effects typical of waveguide propagation and localization effects when reflected from turbulent inhomogeneities [20]. Thus, experimental measurements based on the correlation processing of phase-modulated signals allow, due to the high temporal resolution in multipath propagation of ultrasound, to evaluate the mechanisms of acoustic wave scattering in a turbulent air flow. Besides the ray approximation the small-angle approximation (parabolic equation) can't be used too at modeling due to the scattering angles of ultrasound are at least 15-20 degrees according to the theoretical estimates and experimental results.

5. Conclusion

The measurement scheme presented in the article is an example of using the proposed technique for the digital forming and processing of acoustic signals. The technique makes it possible to quantify the parameters of the turbulent layer at the gas flow boundary. The approach can be applied even in the classical scheme of an ultrasonic flow meter based on measuring the relative delay of pulses that have passed along and against the flow direction. Similar signal processing can be used to increase resolution for a wide range of pulsed ultrasonic control and measuring equipment.

References

[1] Kupnik M, Schröder A, O'Leary P, Benes E and Gröschl M 2006 Adaptive Pulse Repetition Frequency Technique for an Ultrasonic Transit-Time Gas Flowmeter for Hot Pulsating Gases. *IEEE Sensors journal* **4** 906-15

[2] Yadarova O N, Okhotkin G P, Slavutskii L A 2017 Fan drive control system based on Doppler ultrasound control of the air flow. *Russian Electrical Engineering* **7** 27-30

[3] Kakolaki J B, Henry R M 2001 Ultrasonic flow measurement using two asymmetric 'sing-around' paths. *Measurement + Control* **34** 306-11

[4] Raine A B, Aslam N, Underwood C P, Danaher S 2015 Development of an Ultrasonic Airflow Measurement Device for Ducted Air. *Sensors* **15** 10705-22

[5] Yang L, Feng X, Zhang J and Shu X 2019 Multi-Ray Modeling of Ultrasonic Sensors and Application for Micro-UAV Localization in Indoor Environments. *Sensors* **19** 1-18
[6] Thompson E 2011 Fundamentals of Multipath Ultrasonic flow meters for Gas Measurement. *American School of Gas Measurement Technology* 1-23
[7] Bui G T, Jiang Y T and Pang D C 2016 Two Capacitive Micro-Machined Ultrasonic Transducers for Wind Speed Measurement. *Sensors* **6** 814-22
[8] Cruette D, Marillier A, Dufresne J L and Grandpeix J Y 1999 Fast Temperature and True Airspeed Measurements with the Airborne Ultrasonic Anemometer-Thermometer. *Journal of atmospheric and oceanic technology* **17** 1020-39
[9] Arens E, Ghahramani A, Przybyla R, Andersen M, Min S, Peffer T, Raftery P, Zhu M, Luu V and Zhang H 2020 Measuring 3D indoor air velocity via an inexpensive low-power ultrasonic anemometer. *Energy and Buildings* 1-26
[10] Dinchev Z, Gorbunov Y and Kostadinova N 2018 Velocity field visualization, measured with 3D ultrasonic anemometer. *Journal of mining and geological sciences* **61** 39-44
[11] Yakunin A G 2017 3D Ultrasonic Anemometer with tetrahedral arrangement of sensors. *IOP Conf. Series: Journal of Physics: Conf. Series* **881** 1-7
[12] Bychkova I Y, Bychkov A V and Slavutskii L A 2018 Digital phase modulation and correlation processing of ultrasound signals for pulsed measurements in an inhomogeneous medium. *Instruments and Experimental Techniques* **61** 421-6
[13] Aldawi F J, Longstaff A P, Fletcher S, Mather P and Myers A 2007 A high accuracy ultrasound distance measurement system using binary frequency shift-keyed signal and phase detection. *Proceedings of Computing and Engineering Annual Researchers* 1-7
[14] Elmer H and Schweinzer H 2003 Effect of frequency dependent radiation of ultrasonic transducers to correlation based distance measurement systems. *Intelligent Components and Instruments for Control Applications* 273-278
[15] Barker R H 1953 Group synchronizing of binary digital sequences. *Communication theory* (London: Butterworth Publ.) 273-287
[16] Fan Z, Rudlin J, Asfis G and Meng H 2019 Convolution of Barker and Golay Codes for Low Voltage Ultrasonic Testing. *Technologies* **7** 1-16
[17] Stedtnitz W and Schott T 2017 Super Deltaflex – Advanced development of transit time acoustic flow measurement. *River Sedimentation* 436-442
[18] Ishimaru A 1978 *Wave Propagation and Scattering in Random Media* (New-York: Academic Press) 272
[19] Klyatskin V I 2015 *Stochastic Equations: Theory and Applications in Acoustics, Hydrodynamics, Magnetohydrodynamics, and Radiophysics. Volume 2. Coherent phenomena in stochastic dynamic systems* (Switzerland: Springer International Publ.) 508
[20] Blokhintsev D I 1981 *Acoustics of inhomogeneous moving medium* (Moscow: Nauka Publ.) 208

Heat supply efficiency improvement in the Arctic regions with an increased wind potency

A V Bezhan

Northern Energetics Research Centre Kola Science Centre of the Russian Academy of Sciences, 14, Fersman street, Apatity, Murmansk Region, 184209, Russian Federation

E-mail: abezh@rambler.ru

Abstract. Arctic regions are known to be characterized by a situation when the cash costs associated with the heat power production for heat supply overshoot the heat power tariff. Therefore, part of the financial burden placed on the government by way of subsidizing the fossil fuel purchase and delivery to the settlements of the Arctic regions. Through the example the Tsypnavolok settlement, which is located in one of such regions, wind power plants using for heat supply are shown to be considered as a measure to reduce the government financial burdens. For Tsypnavolok the cost of power produced by the wind power plant revealed by calculations is amount to 67.02 USD/Gcal. This is not so many as the average heat tariff (67.9 USD/Gcal). At such heat supply indexes, wind power plants are not only to reduce the government financial burdens, but also to bring an extra income.

1. Introduction

Many countries view their future development in close liaison with the Arctic regions. One of the main concerns of the successful exploration and development of these areas is efficient and reliable operation of heat supply systems, which is being complicated by the severe climatic conditions and heating period duration. Large amounts of fossil fuel herewith are required for the uninterrupted operation of heat supply systems in the Arctic regions. This entails high cash costs for the fuel purchase and delivery to settlements. This problem is particular interest in hard-to-reach and remote settlements of the Arctic regions [1-2]. Therefore, solving the problems associated both with the fossil fuel delivery to the settlements of the Arctic regions and with the efficiency improvement of their heat supply is especially important for such regions.

The use of local renewable energy sources in heat supply systems may be one of the options aimed at solving the problems. The experience of Iceland gives a good example where geothermal energy is successfully used for heating needs [3].

2. Materials and methods

In the Arctic regions with an increased wind potency, there is a possibility of using wind power plants (WPPs) as an extra source of energy [4], including for heat supply [5-7]. In this case, it is important to assess the efficiency that can be, using WPPs in the heat supply systems of the Arctic regions, which is the main objective of this research.

In the regions mentioned, there are many settlements that have a similar organization of heat supply and the related problems described above. Therefore, the chosen object of the study was the heat supply system for the Tsypnavolok, which is one of such settlements.

IOP Conf. Series: Materials Science and Engineering **976** (2020) 012006 doi:10.1088/1757-899X/976/1/012006

The research presented in this paper was carried out in two main phases. The first phase included the collection and analysis of data on the heat supply state in Tsypnavolok. The data was obtained from reports on the heat supply companies operations and from regional government regulations. The review and analysis of the documents revealed a number of specific problems related to the heat supply management. The next phase of the research exemplified by Tsypnavolok showed that the use of WPPs for heat supply could be one of the problem-oriented options.

Various scientific approaches and methods applied in the research resulted in author's achievement the research objective.

3. Results

Tsypnavolok is located in the Murmansk region in northwestern Russia. This inhabited locality is part of the Pechenga municipal settlement (Pechenga MS). As of 2020, Pechenga's heat supply organizations, providing coverage of Tsypnavolok heating demand as well, imposed the tariff for heating to be at 49.4 USD/Gcal for the inhabitants and also approximately 71-102 USD/Gcal for other consumers (table 1). Given this, the average heating tariff for the Pechenga MS is possible to be approximately calculated. It is around 67.9 USD/Gcal. But this tariff is not enough to cover all the heat production costs. Therefore, the government has to subsidize this heat production, namely, purchasing and delivering of fossil fuel to the Pechenga MS inhabited localities.

Table 1. Heat tariffs for some heat supply organizations in the Pechenga MS.

Organization name	Unit of Measurement	Customer Class	Period	
			from 01.01.2020 to 30.06.2020	from 01.07.2020 to 31.12.2020
OOO "Teplostroy Plus" (Limited Liability Company)	USD/Gcal	Inhabitants	49.4	49.4
		Other	61.7	70.9
OOO "PromVoenStroy" (Limited Liability Company)	USD/Gcal	Inhabitants	49.4	49.4
		Other	81.1	102

Table 2 shows data on the amount of subsidies, provided to some heat supply organizations in the Pechenga MS in 2019. These organizations has been annually provided by the subsidies in order to compensate for shortfall in revenue associated with the fossil fuel price hike. The amount of subsidies is not constant from year to year and depends on the real cash costs for the fossil fuel purchase and delivery to the settlements of the Arctic regions.

Table 2. Subsidies to heat supply organizations in the Pechenga MS in 2019.

Organization name	Amount, USD
OOO "Teplostroy Plus"	22180
OOO "Teplostroy Plus"	389430

As discussed above, to improve the heat supply efficiency as an extra source of energy, the use WPPs is available. Such WPPs can operate jointly with fossil fuel boiler rooms. This option of using WPPs reduces the boiler room involvement into heat supply and thereby save fossil fuel. The use of WPPs herewith is justified when the power cost produced by the WPP proved to be either less than or equal to the heating tariff. As exemplified by Tsypnavolok, the calculation results for such option of WPP using are further presented.

Tsypnavolok's number of inhabitants is about 35 people. Given this, the 0.2 Gcal/h boiler room is required to meet industrial and municipal needs. The research [8] found that the WPPs optimal power operating jointly with a boiler room may amount to 0.6-0.8 of the boiler room power. 150 kW Bonus WPP can be recommended for Tsypnavolok (65% of the boiler room power) with the 23.8 m wind wheel diameter and the 30 m tower height.

To calculate the WPP annual power output, the probility function of the wind speed at the tower height and the WPP performance curve are both to be known. The Tsypnavolok average annual wind speed at the tower height is of 8.2 m/s. Hence, the probility function of the wind speed at the tower height looks as shown in figure 1. The WPP performance curve is shown in figure 2. Calculations showed that the WPP annual power output was of 570 thousand kWh per year.

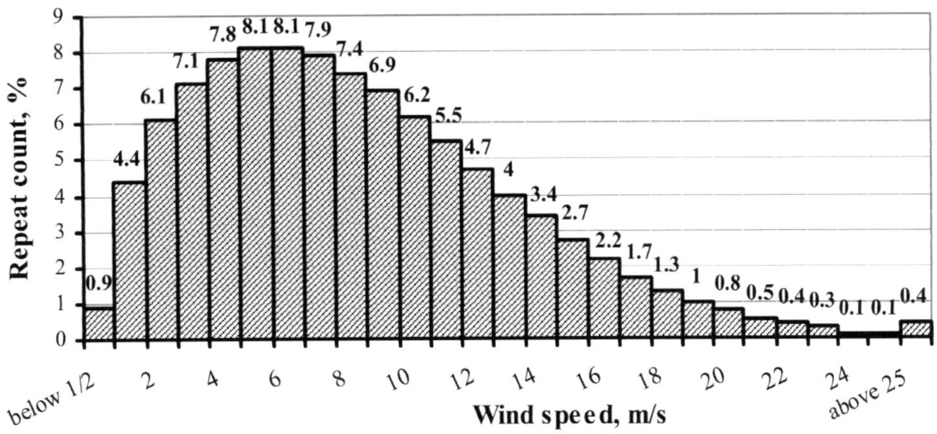

Figure 1. Probility function of the wind speed at an average annual wind speed of 8.2 m/s.

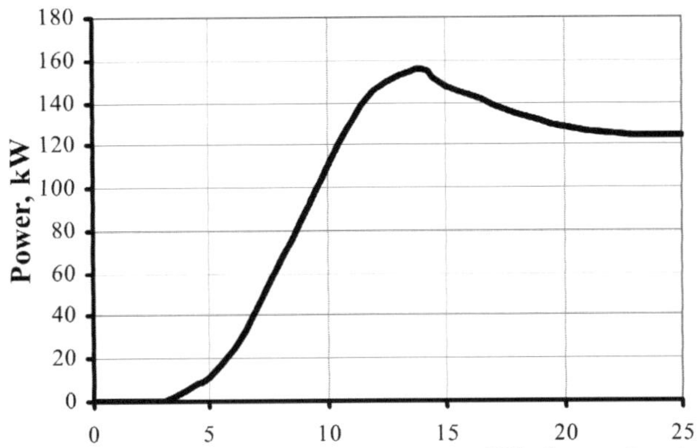

Figure 2. 150 kW Bonus WPP performance curve.

The cost of energy produced by the WPP is possible to be determined, using the data presented. Relying on world prices for WPPs [9-10], unit costs on investments in WPPs amount to about 1,700 USD/kW. They include all costs associated with construction and commissioning of the WPP.

In some Arctic regions, there has already been an experience in the WPP construction. For example, one of the promising projects is the construction of the 201 MW "Kola Wind Farm" wind farm, whose unit costs on investments are estimated at 1.360 EUR/kW [11].

The power cost produced by a WPP depends on the annual depreciation charges, wage-and-salary disbursements and other expenses. If the WPP payback period is set equal to 15 years, then the power cost produced by the WPP is amount to 67.02 USD/Gcal. This is not so many as the average heat tariff (67.9 USD/Gcal).

This indicates that under such conditions, in the Arctic regions with an increased wind potency, the joint use of WPPs and boiler rooms for heat supply is more economically justified than using the boiler rooms only. By this token, this option for WPPs using provides an opportunity to reduce the boiler room involvement into heat supply and thereby save fossil fuel. Therefore this reduces government subsidies for such fuel to be purchased and delivered to the settlements of the Arctic regions.

Thus, the costing of power produced by the WPP and its comparison with the heat supply indexes (tariff and subsidies) of boiler rooms enabled the efficiency assessment of WPPs using in the heat supply systems of the Arctic regions.

4. Discussion

High cash costs for heat supply to settlements in the Arctic regions, funded in subsidies by the government for the fossil fuel purchase and delivery to these settlements, along with the environmental and public health degradation declined by fossil fuel boiler rooms, enhance the attractiveness of WPPs using for heat supply, providing the fossil fuel heat power replacement with power produced by WPPs.

The author points out that the possibility of WPPs operating for heat supply jointly with boiler rooms is of particular importance for the Arctic regions with an increased wind potency, since their use in heat supply systems is to minimize the problems of fossil fuel delivery to settlements in the Arctic regions and to improve the efficiency of their heat supply. At the same time, the need for subsidies in the share of power generated by WPP for heat supply completely disappears.

Thus, WPPs joint operation with boiler rooms for heat supply in the Arctic regions with an increased wind potency can be regarded as a measure to reduce the government financial burdens, which is of particular importance with the prevailing harsh economic environment in the country.

5. Conclusion

In order to solve the problems associated with high cash costs for settlements heat supply in the Arctic regions, funded in subsidies by the government for the fossil fuel purchase and delivery to these settlements, there is an available joint operation of WPPs with boiler rooms for heat supply. This option of using WPPs reduces the boiler room involvement into heat supply and thereby save fossil fuel. Therefore, this reduces government subsidies for such fuel to be purchased and delivered to the settlements of the Arctic regions.

Tsypnavolok, located in the Murmansk region in northwestern Russia, exemplifies that if the power cost produced by the WPP proved to be either less than or equal to the heating tariff, then joint operation of WPPs with boiler rooms for heat supply can be regarded as a measure to reduce the government financial burdens.

References

[1] Mortensen L, Hansen A M and Shestakov A 2017 How three key factors are driving and challenging implementation of renewable energy systems in remote Arctic communities. *Polar Geography* **40(3)** 163-85

[2] Hossain Y, Loring P A and Marsik T 2016 Defining energy security in the rural North—Historical and contemporary perspectives from Alaska. *Energy Research & Social Science* **16** 89-97

[3] Bustaffa E, Cori L, Manzella A, Nuvolone D, Minichilli F, Bianchi F and Gorini F 2020 The health of communities living in proximity of geothermal plants generating heat and electricity: A review. *Science of The Total Environment* **706** 135998

[4] Rafique S F, Shen P, Wang Z, Rafique R, Iqbal T, Ijaz S and Javaid U 2018 Global power grid interconnection for sustainable growth: concept, project and research direction. *IET*

Generation, Transmission & Distribution **12(13)** 3114-23

[5] Li G, Zhang R, Jiang T, Chen H, Bai L, Cui H and Li X 2017 Optimal dispatch strategy for integrated energy systems with CCHP and wind power. *Applied Energy* **192** 408-19

[6] Zheng J, Zhou Zh, Zhao J and Wang J 2018 Integrated heat and power dispatch truly utilizing thermal inertia of district heating network for wind power integration. *Applied Energy* **211** 865-74

[7] Minin V A and Furtaev A I 2019 Prospects for the Development of Wind Energy Resources in the Western Sector of the Arctic Zone of Russia. International Scientific Multi-Conference on Industrial Engineering and Modern Technologies (FarEastCon) 8602694

[8] Zubarev V V, Minin V A and Stepanov I R 1989 *Wind energy using in Nothern regions* (Leningrad: Science) 208

[9] Zore Z, Cucek L, Sirovnik D, Novak Pintaric Z and Kravanja Z 2018 Maximizing the sustainability net present value of renewable energy supply networks. *Chemical Engineering Research and Design* **131** 245-65

[10] Duc Luong N 2015 A critical review on potential and current status of wind energy in Vietnam. *Renewable and Sustainable Energy Reviews* **43** 440-8

[11] Bezhan A V 2020 Experience and prospects for the wind power plants constructing in the north western part of the Euro-Arctic region of Russia. *IOP Conference Series: Earth and Environmental Science* **539** 012145

Development of a control system for a rectification column with a neural network adjustment of the flag liquid flow and upper temperature

E A Muravyova

Ufa State Petroleum Technological University, Branch of the University in the City of Sterlitamak, 453118, Russian

E-mail: muraveva_ea@mail.ru

Abstract. The gas and oil industries are high-tech industries that are based on modern advances in science and technology. This contributed to the development of the control system and the increase in the level of automation of the oil refining process. At a refinery, which is complex production, a process control system plays a key role. There is a need to provide continuous and error-free control, safe and stable operation of the system. One of the perspectives in control systems is neural networks. They can implement any desired process of a non-linear control algorithm with an incomplete, inaccurate description of the control object, and also provide easily implemented adaptation with unstable statistics. This article presents the development of a control system with a neural network adjustment of the reflux fluid flow rate and the top temperature of the column and a comparison with the number of deviations from the norm of the process parameters under the control of the neuroregulator and under the control of the PID controller .Block diagrams of a control system with a neural network controller and a PID controller are compiled. The expediency of using a neural network controller to control the object is substantiated. A mathematical model of the control object was developed taking into account its inherent internal relationships between the parameters of the technological mode and taking into account the influence of external disturbing factors. Processing of the research results was carried out using the software package "MatLab".

1. Introduction

The temperature in the upper part of the columns ensures the production of a distillate, for example, gasoline of a certain fractional composition according to the temperature of its boiling end, respectively, by changing its temperature or flow rate, the temperature of the upper part of the column and, therefore, the quality of the distillate can also be controlled. To solve this problem, it is necessary to develop a parameter control system for production with an accurate fractional composition and to improve the quality of the distillate.

Standard PID controllers cannot always be used to solve this problem, since not all parameters are linear, and have a mutual influence on each other. In this regard, it is advisable to use other types of regulators.

Today, the most promising is the use of neural network regulators to control the flow rate and temperature of the distillation column. The creation of such a system will allow: to improve the adaptive capabilities of the regulator, to improve the quality of management and, therefore, the quality indicators of the technological process [1-3].

2. Description of the process

The task of automation of all types of oil treatment plants is to provide operational automated control of the quality of the oil produced, computer control and management of all stages of oil preparation, control and management of technological equipment and processes.

Figure 1 shows the automation circuit of a distillation column.

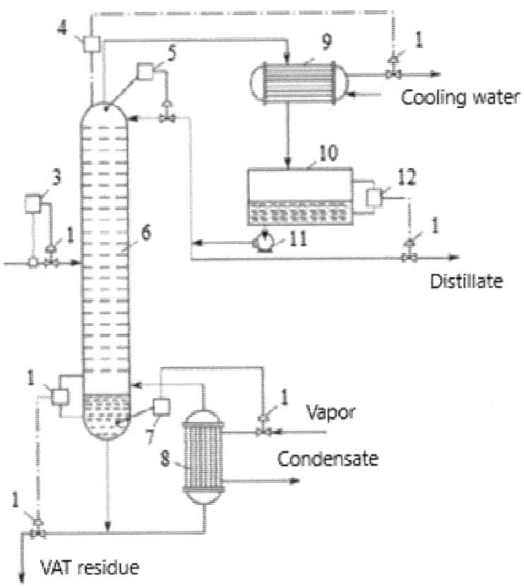

Figure1. Diagram of the automation of a distillation column.
1 – Diaphragm valve;
2,12 – Level sensors;
3 – Flow meter;
4 – Pressure meter;
5,7 – Temperature sensors;
8 – Evaporator;
9 – Capacitor;
10 – Capacity;
11 – Pump.

The base mix heated up to a feed temperature in the vapor, vapor-liquid or liquid phase, inflows the column as feed. The area in which the power is supplied is called evaporation (from the English evaporation - vaporization), as far as there is a process of evaporation–one-time separation of steam from the liquid.

To create an upward flow of vapors in the bottom (lower, splitter) part of the distillation column, part of the bottom liquid is sent to the interchanger, the formed vapors are fed back under the bottom plate of the column.

Vapors rise to the top of the column, cool, condensates in the condenser-refrigerator.

3. Development of a neural network

It is necessary to build a distillation column control system with neural network construction of parameters.

ANN development was carried out in the Matlab environment. When solving the problem, the Matlab Neural Network package and the Simulink graphic tool were used.

The process of constructing a neural network model can be divided into 4 main stages [4].

The first step is to construct an ANN.

The second stage is related to network training, which can be carried out because of a constructive or destructive approach.

In the third stage, a neural network model is created in the Simulink graphical environment.

At the fourth stage, the obtained ANN model is tested on an independent sample of examples [5].

To solve the problem with the help of a neural network, it is necessary to collect data for training (table 1). Input parameters: temperature of the cube of the column Tc, temperature of the top of the column Tt. Output parameters: transfer function channel flow rate of superheated steam W_1 for the PID controller and neuroregulator, transfer function channel flow rate of reflux fluid W_2 for the PID controller and neuroregulator, transfer function channel concentration of the feed column W_3 for PID controller and neuroregulator.

IOP Conf. Series: Materials Science and Engineering **976** (2020) 012007 doi:10.1088/1757-899X/976/1/012007

Table 1. The training dataset

No	Input		Output		
	T_c, °C	T_t, °C	W_1, PID/Neuroregulator	W_2, PID/Neuroregulator	W_3, PID/Neuroregulator
1	263	70	70,9/88,06	153,8/77,83	1,051/29,55
2	217	90	129,1/112,9	190,4/93,46	1,881/16,16
3	285	74	66,77/72,68	109,2/78,2	0,984/31,46
…	…	…	…	…	…
998	275	76	59,44/64,59	100,3/80,1	0,897/29,55
999	281	73	75,7/85,57	147,5/77,69	1,101/12,16
1000	248	84	101,5/75,57	143,3/75,12	1,432/14,33

The transfer function is one of the methods of mathematical description of a dynamic system. It is used mainly in control theory, communications and digital signal processing [6]. It is a differential operator expressing the relationship between the input and output of a linear stationary system. Knowing the input signal of the system and the transfer function, you can restore the output signal.

In control theory, the transfer function of a continuous system is the ratio of the Laplace transform of the output signal to the Laplace transform of the input signal under zero initial conditions [7].

Since the transfer function of the system completely determines its dynamic properties, the initial task of calculating the automatic control system is reduced to determining its transfer function [8]. When calculating the settings of regulators, rather simple dynamic models of industrial control objects are widely used. The transfer function is a fractional rational function of a complex variable for different systems [9,10]. Measurements will be carried out using the three transfer functions W_1 – channel flow rate of superheated steam, W_2 – channel flow rate of reflux fluid, W_3 – channel concentration of the feed column (concentration of the reaction mixture in the feed column).

For calculation, we choose a dynamic transfer function. We begin the construction of the first block diagram using the transfer function W_1 for a distillation column and select parameters for it (superheated steam flow channel W_1 and column temperature Tc):

$$W_1 = \frac{5{,}12}{15{,}349s^3 + 10{,}812s^2 + 4{,}785s + 1} \tag{1}$$

The following models are used in the flowcharts: Random Reference (generating a signal with a uniform distribution); Neural Network Predictive Controller (neuroregulator); Plant (Column) – control object; Graph (graph); Clock (generates a signal whose value at each calculation step is equal to the current simulation time); PID Controller (represents the sum of the input signal, the integral of the input signal and the derivative of the input signal); Display (illustrates the application of this source and the measurement of its input signal); Sum (summing element); Integrator (displays the value of the integral of its input signal relative to time).

Let's build a model in a Simulink graphical environment with a neural network controller (figure 2).

After creating a block diagram of the control environment of the Simulink environment, we use the Plant Identification tool to plot the values and calculate the deviations from the norm of the parameters.

From the analysis of the data obtained, it follows that the reaction of the system to step effects with a random amplitude is quite satisfactory, has an oscillatory character with a fairly fast attenuation; at an interval of 15 s, all impacts are effectively worked out.

Next, we move on to constructing a block diagram for the first transfer function, but instead of the neuroregulator we use the PID controller.

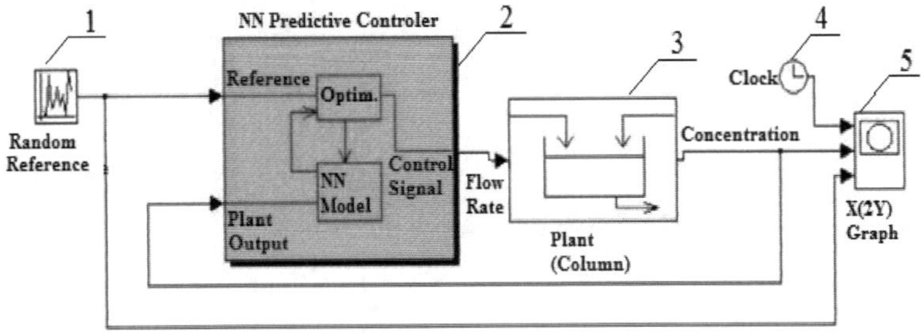

Figure 2. Block diagram of a control system with a neural network controller.
1 – Evenly distributed random signal; 2 – Neuroregulator; 3 – Control object (column apparatus); 4 – Signal, the value of which at each step of the calculation is equal to the current simulation time; 5 – Graph.

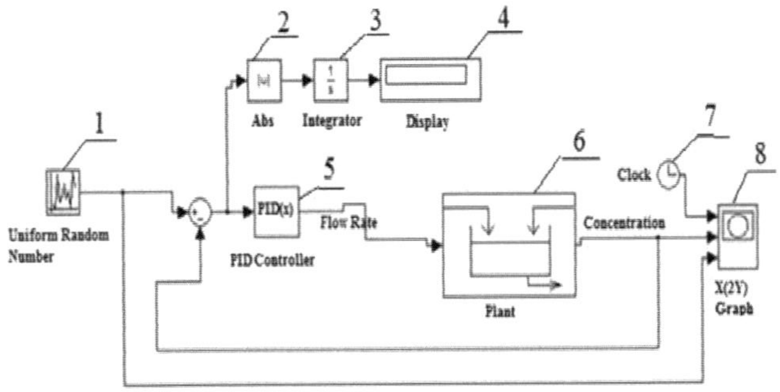

Figure 3. Scheme with additional blocks to obtain a numerical error value.
1 – Evenly distributed random signal;
2 – Abs;
3 – Integrator;
4 – Display;
5 – PID Controller;
6 – Control object (column apparatus);
7 – Signal, the value of which at each step is equal to the current simulation time;
8 – Graph.

We change the transfer function to W_2 to measure deviations from the norm on the second and third transfer functions:

$$W_2 = \frac{1{,}663}{5{,}083s^3 + 4{,}343s^2 + 1{,}698s + 1} \tag{2}$$

We consider the flow channels of the reflux liquid W_2 and the temperature of the top of the column Tt.

For the transfer function W_2, we construct mathematical control models with a neuroregulator and a PID controller.

Using the Plant Identification tool, we train the model and display the graphs of the input Input value, the output value of the Plant Output control object and the NN Output neural network, as well as the errors of the second transfer function Error (figure 4), these parameters indicate the learning results for the column, it follows from the graphs that the system reaction on step effects with a random amplitude is quite satisfactory, has an oscillatory character with a fairly rapid attenuation; at an interval of 20 s, all impacts are effectively worked out.

Deviations from the norm of a control system with a neuroregulator are shown in figure 5, with a PID controller in figure 6.

On the graph (figure 5) we see that the deviations in the second transfer function (W_2, NN Controller) are included in the norm of parameters (Optimal parameters), since the average values of the minimum and maximum points are close to the norm of the parameters of the neuroregulator.

On the graphs (figure 6) we see that the deviations in the second transfer function (W_2, PID Controller) of the model with the PID controller are not included in the optimal parameters norm(Optimal parameters), since the deviation values have a big difference with the norm of the parameters.

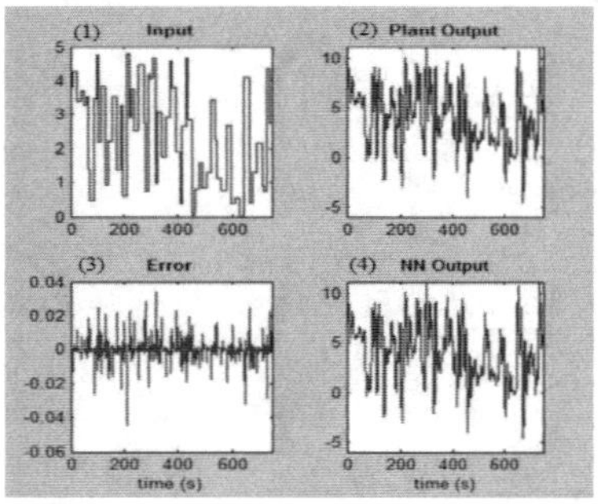

Figure 4. Graphs of parameters for the second transfer function with a neuroregulatory.
1 – input value; 2 – output value of the control object; 3 – transfer function errors;
4 – neuroregulator output value

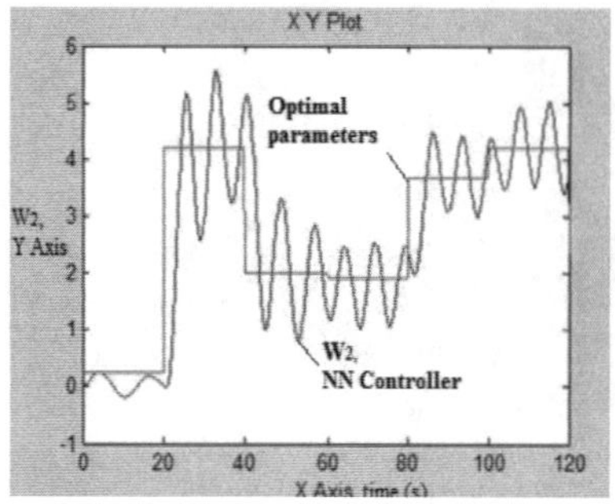

Figure 5. Deviation from the model when using the neuroregulator W_2.

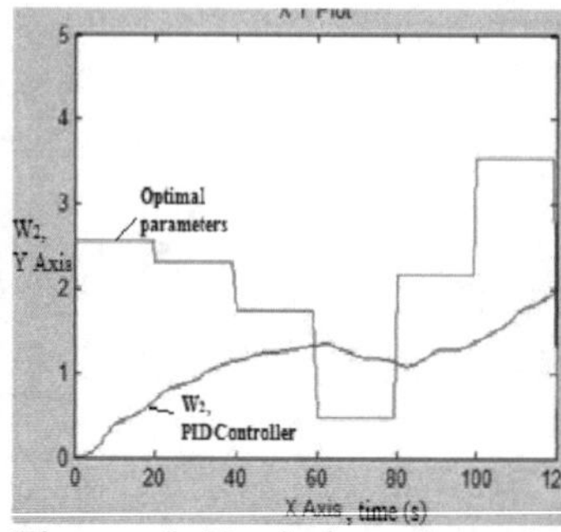

Figure 6. Deviations from the normal model with a PID controller.

The following transfer function W_3 with the parameters of the channels of concentration of the power supply column W_3 and the temperature of the top of the column Tt:

$$W_3 = \frac{0{,}765s^2+1{,}816s+1{,}72}{5{,}083s^3+4{,}343s^2+1{,}698s+1} \quad (3)$$

The deviations in the third transfer function (W3, PID Controller) are included in the optimal parameters norm (Optimal parameters), since the graph also covers the average norm values of the PID controller parameters.

The deviations in the third transfer function (W3, NN Controller) are included in the norm of parameters (Optimal parameters), since the average values of the minimum and maximum points are close to the norm of the parameters of the neuroregulator.

The numerical values (deviations from the norm) obtained during the study were summarized in table 2. The average value for each transfer function was also derived.

Table 2. Numerical values of the errors of the neuroregulator and PID controller

No measurement	First transfer function, W_1		Second transfer function, W_2		Third transfer function, W_3	
	PID Controller	*NN Predictive Controller*	*PID Controller*	*NN Predictive Controller*	*PID Controller*	*NN Predictive Controller*
1	70,90	88,06	153,80	77,83	1,051	29,55
2	129,00	112,90	190,40	93,46	1,881	16,16
3	66,77	72,68	109,20	78,2	0,984	31,46
4	80,52	93,89	168,80	83,87	1,199	17,13
5	103,70	87,78	158,60	83,26	1,517	11,51
6	104,70	108,8	153,90	93,30	1,542	18,48
7	136,90	131,00	205,30	99,84	2,006	12,89
8	60,96	84,45	168,90	74,23	0,916	27,56
9	108,60	93,83	177,40	95,77	1,615	27,64
10	92,20	100,70	188,00	74,54	1,369	19,11
Mean value	95,43	97,41	167,43	85,43	1,41	21,15

4. Neural network testing

According to the first transfer function, the difference in deviations between the studied control system with the neural network controller and the PID controller is insignificant, but the PID controller shows a lower error value.

In the second case, the situation is radically different, the control system with a neuroregulator shows the result 2 times better than the PID controller, therefore it is rational to choose a neural network controller to control the reflux fluid flow channel and the top temperature of the column.

To control the feed concentration channel of the column, a PID controller is suitable, for which the deviation from the norm is only 1.41, while the deviation of the neuroregulator is 21.15.

5. Conclusion

During the study, the following main results were obtained: the feasibility of developing a distillation column control system with a neural network adjustment of parameters using mathematical modeling methods was substantiated, and the efficiency of the distillation column control system was evaluated. The study also compares deviations from the norms of the PID controller and the neural network controller. Conclusions are drawn from the numerical data.

After comparing the control system with the PID controller and the neural network controller in three cases, we can conclude that it is more rational to use a neural network to control the reflux fluid flow channels and the top temperature of the columns, and the PID controller to control the column supply concentration channel.

References

[1] Muravyova E A 2011 Autoregressive neural network for model predictive control of multivariable cracking catalyst calcinatory. *Optical Memory & Neural Networks* 213-6
[2] Yahyaeva G E 2006 *Fuzzy sets and neural networks* (Moscow: Internet University of Information Technologies) 316
[3] Nielsen M 2015 *Neural networks and deep learning Publishing house* (Moscow: Determination Press) 224
[4] Rashid T 2016 *Make your own neural network* (Moscow: CreateSpace) 222
[5] Kriesel D 2014 *A brief introduction to neural networks* (Moscow: Autoedicin) 244
[6] Samarasinghe S 2016 *Neural networks for applied sciences and engineering* (Moscow: CRC Press) 570
[7] Evmenov V P 2009 *Intelligent Control Systems* (Moscow: Book House Librocom) 290
[8] Khaikin S 2006 *Neural networks: the full course* (Moscow: Williams) 1104
[9] Callan R 2003 *Basic concepts of neural networks* (Moscow: Williams) 291
[10] Nikolenko S, Kadurin A and Arkhangelskaya E 2018 *Deeplearning. Immersion in the world of neural networks* (St. Petersburg: Peter) 477

Sintering in a microwave field is a promising ecological technology for producing nanostructured ceramic composites from natural raw materials

I A Zhenzhurist

Kazan State Energetics University, 51, Krasnoselskaya, Kazan, 420066, Russia

E-mail: ir.jenjur@yandex.ru

Abstract. The article provides literature data on sintering inorganic substances in an electromagnetic field, shows the prospects for research in this direction. The results of sintering samples from masses based on natural compositions: marl, low-melting loam and diatomite with carbonate additives in a microwave oven are considered. The possibility of sintering in a microwave field of natural raw materials, which is traditional for the production of many fired ceramic products, has been established. The phase compositions of the fired materials have been determined. The composition of sintered compositions, consisting of nanoscale phases, has been established. The effect on the sintering process of the composition of the raw material composition from the carbonate component, diatomite and the addition of NaCl salt is shown. Samples from a mass based on marl, with the addition of diatomite and NaCl salt show high quality and increased strength of the samples, the presence of a large percentage of the amorphous phase in the fired sample, and a uniform microstructure of the material. Specimens from masses based on clay with the addition of chalk and NaCl salt after firing in the microwave field showed a significant increase in strength.

1. Introduction

Compliance with environmental, energy and resource-saving requirements is the priority aspects of any promising material production technology. For fired materials, heat treatment is the most energy-intensive and environmentally problematic process due to the combustion of hydrocarbon fuels. In addition, the duration of the thermal process does not contribute to the formation of a fine-grained structure, on which the strength of the material depends. This is especially important with a simplified scheme for processing raw materials, which does not allow obtaining a high degree of dispersion of the components of the molding mass. The aforementioned problems of traditional firing technology and the need for materials with special, often extreme properties create the need to search for alternative solutions. One of these options is the development of a technology for the synthesis of a material under the action of a microwave electromagnetic field. The study of the sintering process at a high heating rate in an electric field showed the possibility of obtaining a material with an ultrafine, with increased density and strength of a material with a microcrystalline structure, which has increased viscous fracture rates [1-3]. This is especially important for ceramic, traditionally fragile materials. Differences are found in the structure of crystalline materials obtained by heat treatment in an electromagnetic field and in the process of convective heating [4].

For clay compositions, it was found that the microstructure of the samples and its main phases are formed at low firing temperatures and include nanoscale formations [5-6].

In the technology of firing materials, sintering is the main stage of heat treatment, during which the final structure of the material is formed. In recent years, the possibility of obtaining materials with special properties is associated with the formation of a nanoscale structure in them - this is the level of passage of the main interfacial processes during sintering. This level of interaction is possible in compositions with nanosized dispersion of components and is achieved at a high degree of grinding, for example, during slip preparation of the molding mass. According to the traditional firing technology, with prolonged heating and holding of the material at a final temperature, the developing recrystallization processes prevent the formation of its nanoscale structure.

The intense development of nanotechnology in recent years is associated with the discovery of new materials with extraordinary mechanical and physical properties. The study of the structural features of such materials showed that the appearance of special properties in them is caused by the achievement of a certain size of the structure elements of the material in each case [7-12]. A decrease in the grain size below the critical level and structural defects causes a significant increase in strength, and the fracture toughness of traditionally brittle ceramics increases. This fact was recorded in a number of works [13-18]. The effect of the structure, dispersion of raw materials, and the conditions for forming the product on the final properties of the material was shown [19-23]. Most of the studies were carried out on simple compounds or mixtures with a small amount of components. Natural polymineral compositions are multicomponent mixtures of minerals of various chemical and grain composition. This complicates the technological process of obtaining the material.

Sintering of natural mineral compositions, mostly consisting of hydroaluminosilicates of various chemical compositions, occurs after their decomposition and begins between the grains in the interfacial region. The study of heating in the electromagnetic field of dielectrics, which include the components of most natural minerals, indicates the beginning of the process in the interphase region, continuing throughout the entire volume due to diffusion [24].

The results of studies of the behavior of compositions from charges of various polymineral composition showed a significant difference in the sintering mechanism in the microwave field and in a muffle furnace. Qualitatively, without cracks and ruptures, samples made from a charge, the main clay-forming mineral of which decomposes upon heating with a slight energy effect (polymineral mixtures based on montmorillonite - bentonite clay) and samples modified with NaCl salt, were sintered in the microwave field. Specimens from compositions on refractory clay cracked during firing in a microwave field [18-19].

The results of testing a large number of compositions showed an increased strength of the samples, the charge of which was activated or the samples were fired in the microwave field, in comparison with the samples fired by the convective method in a muffle furnace [18-19].

The study of the features of microwave sintering of polymineral composites is of great importance for the development of the fundamental principles of the technology for obtaining materials with special properties and the technology for producing ceramic composites.

The aim of the work was to study the features of heat treatment in the microwave field of compositions from natural polymineral raw materials with impurities of carbonate rocks.

2. Materials and methods

Investigations of the effect of the microwave field on natural aluminosilicates containing Ca and Mg carbonates have been carried out. The final structure of the fired material in the microwave field and the structure after convective firing in a muffle furnace is compared.

Natural polymineral raw materials were investigated: Maksimkovsky marl composition in mass %: 33.2% SiO_2, 11.4% Al_2O_3, 26.1% $CaCO_3$ + $MgCO_3$, 3.6% Fe_2O_3, 3.6% Na_2O + K_2O; Kalininsky loam (42% montmorillonite, 5% mica) with the content of basic oxides (wt%): 68.1% SiO_2, 12.6% Al_2O_3 + TiO_2, 6.4% Fe_2O_3, 3.3% CaO + MgO, 3, 2% Na_2O + K_2O; Inza diatomite (98.7% SiO_2). As a carbonate additive, chalk was used with the composition: $CaCO_3$ and $MgCO_3$ 98.2 - 98.4, R_2O_3 0.001, Fe_2O_3 0.15. As a mineralizer salt NaCl (extra following Russian Federation State Standard GOST 51574-20000) due to its ability to form fusible mixtures with silicates.

Molding mixtures were prepared from raw materials, ground and sieved through a sieve with a hole diameter of 1 mm. The charge components were mixed in a ball mill for an hour. The mass was moistened until a plastic dough was obtained, and specimens with a size of 20 × 20 × 20 mm were molded. Samples without preliminary drying were fired in a microwave oven at 1000 °C with holding at a maximum temperature of 5 min.

Microwave treatment was carried out in a microwave oven (Samsung m 1711 NR) with an output radiation power of 800 W at an operating frequency of 2.45 GHz. The magnetic field is created by current of the industrial frequency of 50 Hz, which proceeds in a furnace power supply system. For sintering, a muffle made of mullite-silica plates is installed inside the furnace. The temperature was monitored by a thermocouple with a radiation-protected junction coating installed near the sample.

X-ray phase analysis of the fired samples was carried out on a Shimadzu XRD 6000 diffractometer in CuK_{α} -radiation (PDF 4+ base, POWDERCELL 2.4 full-profile analysis program), sampling of sample breaks - on a system with an electron and focused ion beam (Quanta 200 3D) in the Tomsk Regional center of collective use of NU TSU.

3. Results

All the samples under study, which included a carbonate additive and NaCl salt, sintered without destruction and showed a significant increase in strength.

The greatest increase in strength was obtained from the composition based on marl with the addition of diatomite and mineralizing salt NaCl (figure 1). Samples without defects with increased strength were obtained from diatomite and clay with the addition of chalk and NaCl salt.

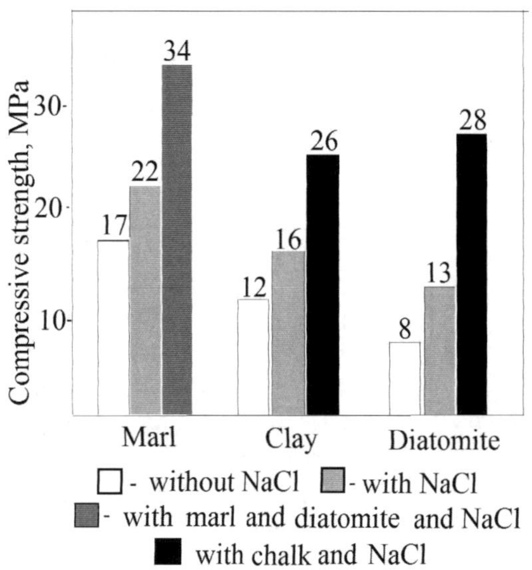

Figure 1. Durability on compression of the samples burned at 1000 °C from: marl, clay and diatomite with additives.

The decomposition of the constituent components of the charge for all investigated compositions occurs in the same temperature range (600– 900 °C). Salt NaCl contributes to the formation in this temperature range of the liquid phase and the formation of high-temperature phases, which, in the presence of a liquid phase, are formed in the low-temperature range and with lower energy costs [5, 6]. Samples from all structures were received free of defects.

The results of X-ray phase analysis of the fired samples, which showed the highest strength, are shown in table 1.

Table 1. Phase composition of the samples after firing.

Composition	Weight percent	Maintenance of phases, mass%	Crystallite size, nm
	$Ca_2MgSi_2O_7$	21,6	12,36
	SiO_2	17,6	51,84
Marl, diatomite, NaCl	$CaMgSi_2O_6$	24,6	15,15
	Amorphous phase	< 40	
	$Ca_2Mg0{,}08Al{,}84Si1{,}08O_7$		
	SiO_2	36,7	92
	$Al_2Si_2O_5$	8.7	24.9
	Fe_2O_3	1.1	40.8
Diatomite, chalk, NaCl	$Na_{0.7}K_{0.3}Al_{1.02}Si_{2.98}O_8$	5.5	16.9
	Amorphous phase, phase traces $CaCO_3$	43	
	SiO_2	48.9	139
	Al_2SiO_5	12.2	29.6
Clay, chalk, NaCl	γ-Fe_2O_3	3.4	>100
	$(Na_{0,7}K_{0,3})(Al_{1,02}Si_{2,98}O_8)$	35.4	34

The X-ray diffraction pattern of the composition of the marl sample with the addition of diatomite and salt is shown in figure 2.

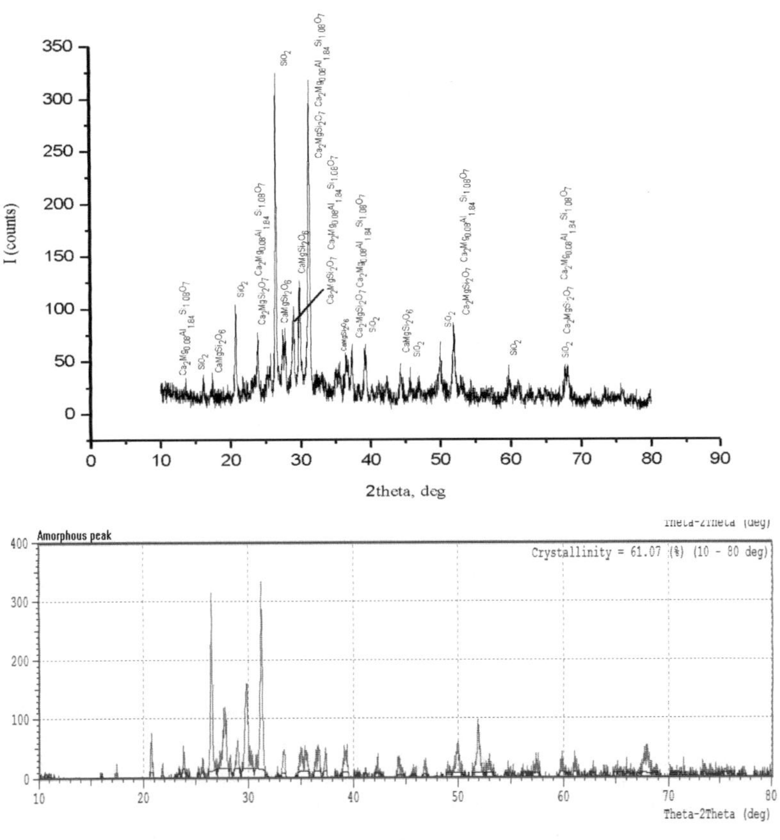

a

b

Figure 2. Results of the X-ray phase analysis of a sample from composition: marl-diatomite-NaCl with interpretation on phases (a) and allocation of an amorphous phase (b).

X-ray phase analysis of the samples showed the presence of nanoscale phases in all compositions, and in the compositions with marl, diatomite, and salt, a significant amount of an amorphous phase. The presence of an amorphous phase in fired samples can be associated with the formation of a large amount of a liquid phase during sintering and the formation of a material structure, which is responsible for the high strength of the sample.

In the composition of the sample based on marl, diatomite, and salt, during sintering, phases were formed that are the basis of clinker and cement binder. This feature creates a prospect for studying the possibility of obtaining cement clinker by sintering in a microwave field and is important for further research.

The microstructure of an alloy based on marl, diatomite and salt is shown in figure 3.

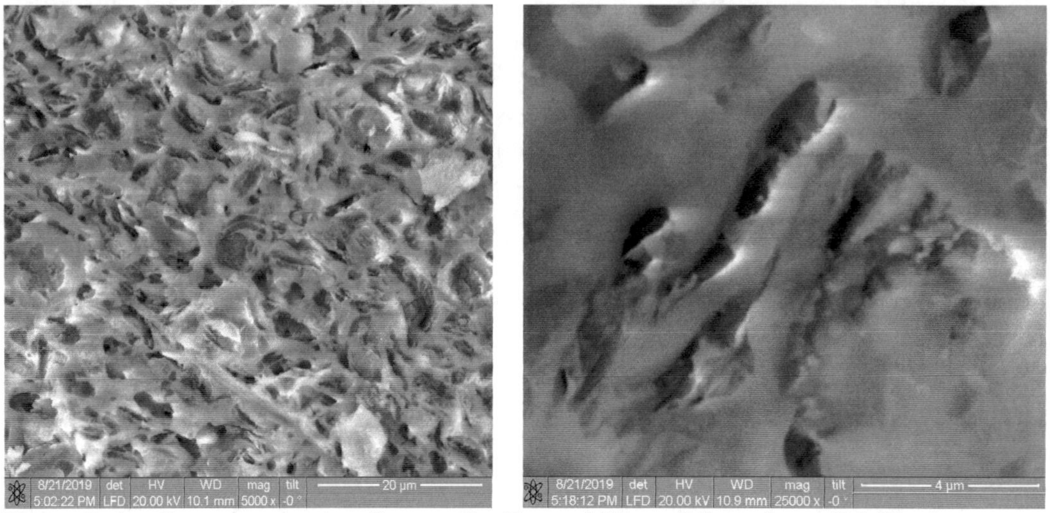

Figure 3. An alloy microstructure from marl with an additive of diatomite and NaCl salt.

The dense lattice structure of the composition with vitrified ribs is visible. Such a uniform structure must have high strength, which is confirmed by test results.

4. Discussion

The paper presents an analysis of the literature data on a possible solution to the ecological problem of traditional firing technologies. Listed are studies on the synthesis of materials in an electromagnetic field, which are distinguished by special, often extreme properties. The study of the sintering process at a high heating rate in an electric field has shown the possibility of obtaining a material with an ultrafine, with increased density and strength of a material with a microcrystalline structure, which has increased viscous fracture rates, which is especially important for strengthening traditionally brittle ceramics.

The research results showed the differences in the structure. formed under the conditions of traditional convective firing and high-speed sintering under the influence of the microwave electromagnetic field. It was noted that under the conditions of sintering the composition in a microwave field, a microstructure with inclusions of nanosized phases is formed. A number of studies have shown the influence of characteristics, primarily the fineness of raw materials, molding conditions on the final properties of the material, the role of interfacial processes during sintering in an electromagnetic field is noted.

Materials with extraordinary mechanical and physical properties have been obtained, primarily based on low-component mixtures. The prospects of research in this direction in materials science and the importance of research for the development of the fundamental principles of the technology for obtaining materials with special properties from polymineral raw materials are shown.

The presented work is devoted to the study of the possibility of using polymineral, containing carbonates of natural raw materials to obtain materials with high strength indicators under conditions of firing in an electromagnetic field of microwave frequency, to the study of the features of the sintering process.

5. Conclusion

The result of the study was the process of sintering samples from natural polymineral raw materials with carbonate inclusions in the microwave electromagnetic field.

As a result of the research it was found:

- The fundamental possibility of sintering a polymineral composition from natural raw materials with carbonate inclusions in a microwave oven;
- The dependence of the sintering process on the composition of the carbonate-silica component of the composition;
- The presence of a fluxing additive (NaCl) in the composition of the carbonate-silica polymineral composition contributes to the production of defect-free samples of increased strength;
- Formation of a nanoscale phase during sintering in a microwave electromagnetic field, the composition of which depends on the composition of the raw material composition;
- In compositions with a carbonate-diatomite component and NaCl salt, a glass phase is formed during sintering in a microwave field and a uniform microstructure of the material is formed.

The results of the study on sintering in the microwave field of polymineral compositions with carbonate inclusions showed the prospects for the technology of obtaining many firing materials.

The results obtained confirmed some of the conclusions made earlier on the sintering of low-component mixtures and compositions based on carbonate and silica raw materials. An increase in the strength of the material is associated with the formation of a special structure of the material, the presence of nanoscale phases.

The firing of samples without defects from the polymineral composition was possible due to the introduction of the NaCl salt into the composition, which forms low-melting compositions.

References

[1] Mattia Biesuz 2017 *Flash Sintering of Alumina-based Ceramics* (Trento: University of Trento Departament of Industrial Engineering) 179

[2] Jaehun Cho, Qiang Li, Han Wang, Zhe Fan, Jin Li, Sichuang Xue, Vikrant, Haiyan Wang, Troy B, Amiya K, García R Edwin and Xinghang Zhang 2018 High temperature deformability of ductile flash-sintered ceramics via in-situ compression. *Nature communications* **9** 1-9

[3] Dancer C E J 2016 Flash sintering of ceramic materials. *Materials Research Express* **3(10)** 102001

[4] Tompsett G A, Conner W C, Yngvesson K S 2006 Microwave Synthesis of Nanoporous Materials. *ChemPhysChem* **7(2)** 296-319

[5] Zhenzhurist I A 2018 Effect of Microwave Energy on Phase Transformations of Aluminosilicates and Properties of Related Materials. *Inorganic Materials* **54(9)** 103-7

[6] Zhenzhurist I A 2017 Effect of Microwave Fields on the Properties of Silicates with an Aluminum-Sodium-Chloride Modifier. *Glass and Ceramics* **73(11–12)** 423-6

[7] Naglieri V, Palmer P, Montanaro L and Chevalier J 2013 Elaboration of alumina-zirconia composites: role of the zirconia content on the microstructure and mechanical properties. *Materials* **6(5)** 2090-2102

[8] Mohamed F A and Li Y 2001 Greep and superplasticity in nanocrystalline materials: current understanding and future prospects. *Materials. Science and Engineering* **A 298** 1-15

[9] Han B Q, Lavernia E J and Mohamed F A 2005 Mechanical properties of nanostructured

materials. *Reviers on Advance Materials Science* **9(1)** 1-16
[10] Nam K W 2010 Crack-healing behavior and bending strength of Al2O3/SiC composite ceramics according to the amount of added Y2O3. *Journal of Ceramic Processing Research* **11(4)** 471-474
[11] Anderson T L 2005 Fracture mechanics: fundamentals and application 3rd edn (CRC Press Francis Group) 357
[12] Balakrishnan A, Pizette P, Martin C L, Joshi S V and Saha B P 2010 Effect of Particle Size in Aggregated and Agglomerated Ceramic Powders. *Acta Materialia***58(3)** 802-12
[13] Milligan W W 2003 In: Comprehensive structural integrity, ed. by I. Milnc, R. O. Ritchie, and B. Karihaloo (Amsterdam: Elsevier Pergamon) **8** 529
[14] Camargo P H C, Satyanarayana K G and Wypych F 2009 Nanocomposites: synthesis, structure, properties and new application opportunities. *Materials Research* **12(1)** 1-39
[15] Khalaf M M, Ibrahimov H G and Ismailov E H 2012 Nanostructured materials: importance, synthesis and characterization - a review. *Chemistry Journal* **2(3)** 118-25
[16] Cain M and Morrell R 2001 Nanostructured ceramics: a review of their potential. *Applied Organometallic Chemistry* **15** 321-30
[17] Fan X, Yang F, Rong Z, Cai X and Li G Characterization and thermoelectric properties of Bi0.4Sb1.6Te3 nanostructured bulk prepared by mechanical alloying and microwave activate. d hot pressing. *Ceramics international* **41(5)** 6817-23
[18] Hadjipanayis G C and Prinz G A 1991 *Science and Technology of Nanostructured Magnetic Materials* (New York: Springer) 342
[19] Tellkamp V L, Melmed A and Lavernia E J 2001 Mechanical behavior and microstructure of a thermally stable bulk nanostructured. *Al alloy Metallurgical and Materials Transctions* **A 32** 2335
[20] Parchoviansky M, Galusek D, Sedlacek J, Svancarek P, Kasiarova M, Dusza J and Sajgalik P 2013 Microstructure and mechanical properties of hot pressed Al2O3/SiC nanocomposites. *Journal of the European Ceramic Society* **33(12)** 2291-8
[21] Sun X K, Cong H T, Sun M and Yang M C 2000 Preparation and mechanical properties of highly densified nanocrystalline. *Al Metallurgical and Materials Transctions* **A 31** 1017-24
[22] Sedlacek J, Galusek D, Svancarek P, Riedel R, Atkinson A and Wang X 2008 Abrasive wear of Al2O3–SiC and Al2O3–(SiC)–C composites with micrometer- and submicrometer-sized alumina matrix grains. *Journal of the European. Ceramic Society* **28(15)** 2983-93
[23] Trunec M and Maca K 2007 Compaction and pressureless sintering of zirconia nanocomposites. *Journal of the American Ceramic Society* **90(9)** 2735-40
[24] Anenkov Y M and Ivashutenko A S 2005 Fizicheskaya model of agglomeration and modifying of ceramics in high-frequency and superhigh-frequency fields. *News of Tomsk Polytechnic University* **308(7)** 30-5

IOP Conf. Series: Materials Science and Engineering **976** (2020) 012009 doi:10.1088/1757-899X/976/1/012009

Review of tritium in Bangladesh before commissioning of NPP Rooppur

Ramit Kumar Sadhukhan and B I Synzynys

Obninsk Institute for Nuclear Power Engineering of NRNU «MEPhI», 249040, Obninsk, Kaluga Region, Studgorodok-1, IATE, Russian Federation

E-mail: ramitkumar777@gmail.com

Abstract. The comprehensive review article was carried out to present an overview of the tritium radioactivity concentrations in different components of observed freshwater ecosystems, primarily underground water (for drinking) and precipitation in People's Republic of Bangladesh. Most of the reported data have been obtained using either ^{3}He in-growth technique which is an indirect method or using direct method with liquid scintillation counter. The results of the radio-ecological monitoring of environment in Bangladesh show that the tritium content in groundwater, precipitation and river water vary from 0-25 TU, 5-10 TU and 5-10 TU respectively. Around proposed Rooppur Nuclear Power Plant site, tritium radionuclide concentration of 6.7-17.6 TU, <25.2 TU, <25.2 TU and 3.4-10.1 TU were obtained for surface water of Padma, bottom sediments, aquatic flora and drinking water respectively. Globally, the natural and technogenic background of tritium is assumed to be 2.2±0.7 Bq/L (18.5±5.9 TU) and 5 Bq/L (42.0 TU) respectively. Henceforth, the investigation authenticated that the tritium content in all samples is below cut-off point that requires intervention. The results obtained during the radio-ecological studies allow for subsequent evaluation of radiation doses and radiological health hazards that organisms/people experience due to the anthropogenic changes of background radiation in the studied regions.

1. Introduction

Tritium (^{3}H) is one of the radioactive isotopes of hydrogen existing in the environment either as tritiated water (HTO) or as organically bound tritium (OBT) [1]. Being a weak beta emitter, it emits an electron with median energy of 5.7 keV, maximum energy of 18.6 keV, and has a physical half-life of 12.3 years [1-2]. Natural source of tritium is through the action of cosmic rays in the upper atmosphere while technogenic origin is through scheduled releases by almost all nuclear facilities (nuclear power plants, defense facilities, fusion facilities, fuel processing plants, etc) and through thermonuclear explosions in atmosphere during 1950s and early 1960s. When ingested with food, water, or inhaled air, as well as through damaged body coverings, tritium can become a source of risk if it enters the cells and tissues of the body. OBT through substituting hydrogen in the DNA molecule can instigate considerable damage to genetic material and instill high risk of deep-rooted reverberations of radiation [1; 3].

2. Materials and methods

The research paper set forth a synopsis of the tritium radioactivity concentrations in different components of observed freshwater ecosystems, underground water for drinking purpose and precipitation in Bangladesh (BD). Tritium radioactivity corresponding to the site of the first Nuclear

Power Plant (NPP) in Bangladesh, that is Rooppur has also been included in the present study. In the referenced articles, the ^{3}H analysis was performed either using ^{3}He in-growth technique [4-7] which is an indirect method or using direct method with a liquid scintillation counter [8-10].

The results obtained during the radio-ecological studies allow for the obtaining data about the content of tritium radionuclide in the environment in Bangladesh which also include area in the vicinity of Rooppur NPP and subsequent evaluation of radiation doses and radiological risks that organisms/people experience due to the anthropogenic changes of tritium background radiation in these studied localities.

3. Results

The results of the radio-ecological surveillance of habitat in Bangladesh is presented in table 1 while outcomes of study carried out around Rooppur Nuclear Power Plant is represented in table 2. In the aforementioned tables, 1 Bq/kg of tritium activity corresponds to 8.390 ± 0.015 TU [11; 13], making use of the new value for the tritium half-life [12] and the CODATA 1998 value for the Avogadro constant [14]. One tritium unit (TU) is commensurate with one atom of tritium in 10^{18} atoms of hydrogen. Additionally, figure 1 represents the distribution of ^{3}H content (2006-2008) in southern site shallow wells, central site shallow wells and intermediate wells [8] while in figure 2 the spatial distribution of ^{3}H content (2006-2008) in shallow groundwater and intermediate groundwater [8] is emphasized.

4. Discussion

The meticulous analysis of the different works set forth in the preceding section in order to determine tritium concentration and subsequent deduction of radiation doses and radiological risks involved due to the background presence of tritium in Bangladesh is carried out in the current segment.

4.1. Bengal Delta Study.

In Bangladesh, about 97% of the populace depends on groundwater as a source of innocuous potable water [15]. Therefore, an exhaustive study [8] covering about 85% land area of Bangladesh excluding only the eastern Tertiary Hill Range was carried out to assess groundwater using isotopes technique during the period 2006-2008. Tritium concentration was measured applying liquid scintillation counter after electrolytic enrichment and was found to fluctuate between 0-5.0 TU (shown in table 1) in 328 groundwater samples that were analysed. Globally, the natural and anthropogenic background of tritium is assumed to be 2.2±0.7 Bq/L (18.5±5.9 TU) and 5 Bq/L (42.0 TU) respectively [11]. So the results obtained in the Bengal Delta study is below the mean global value and several orders of magnitude below the level demanding intervention, according to the Russian radiation safety legislation.

Two well-defined clusters of shallow groundwater can be seen in figure 1 [8] having ^{3}H content <1.1 TU and >1.1 TU. And as for five intermediate depth ground waters, four samples show ^{3}H values even at greater depths ranging from 76 to 168 m while ^{3}H is absent for the sample collected at a depth of 152 m.

Tritium was obtained in majority of the samples that were analysed as is shown in the spatial distribution in figure 2 [8]. This is a proof that the active recharge nourished Bengal Delta aquifers all over the country. Active recharge of 0.5-0.7 TU nourished groundwater of intermediate depths of 88-134 m at west-central setting. But higher tritium concentrations were obtained at the north and central sites as shown in figure 2. Tritium concentrations of 1.1-5.0 TU in groundwater of these regions signal that considerable percentage of water is post-bomb in nature. Conversely, sites in the southern segment demonstrate low tritium concentrations at shallow depth with maximum concentration of 1.1 TU. This signifies the fact that these ground waters are pre-bomb in origin. Somewhat large times are required for water to travel to the aquifer in these unsaturated regions. Consequently, tritium concentration decreases due to radioactive decay during the long travel time.

Table 1. ^{3}H concentrations in groundwater, river water and precipitation across different areas of Bangladesh (BD).

Component	Location	Year	Depth (m)	Tritium Level (TU)	Tritium Level (Bq/l)	Reference
Groundwater	9 south-western districts	1999	10-335	0-7.2	0-0.86	[4]
	4 western districts	2000	10-90	0-7.2	0-0.86	[4]
	Eastern, south-eastern	1999-2000	0-280	0-25	0-3.0	[5]
	Araihazar, Central	2003	<20	0.4-5.4	0.05-0.64	[6]
	Munshiganj, Southern	2004	9	5.02± 0.16	0.60±0.02	[7]
			14	9.23±0.27	1.10±0.03	
			18	18.5±0.5	2.21±0.06	
			24	12.4±0.4	1.48±0.05	
			30	2.22±0.13	0.26±0.02	
			38	0.12±0.08	0.01±0.01	
			46-107	0-0.04±0.09	0-0.01±0.01	
	Bengal Delta (85 % land area of BD except eastern Tertiary Hill range)	2006-2008	7-152	0-5.0	0-0.60	[8]
Precipitation	Dhaka, Central	1953	NA	5.9	0.70	[9]
		1955		9.7	1.16	
		1958		111.2	13.25	
		1962		224.2	26.72	
		1963		597.2	71.18	
		1964		383.2	45.67	
		1965		199.4	23.77	
		1966		114.1	13.60	
		1985-1995		5-10	0.60-1.19	
River water	9 south-western districts and 4 western districts	1999, 2000	Surface water	5-10	0.60-1.19	[4]

Table 2. The content of tritium in the components of freshwater ecosystems in the vicinity of Rooppur NPP [10].

Component	Tritium Unit (TU)	Tritium Level (Bq/kg (L))
Surface water of Padma	6.7-17.6	0.8-2.1
Bottom sediments	<25.2	<3
Aquatic flora	<25.2	<3
Drinking water	3.4-10.1	0.4 – 1.2

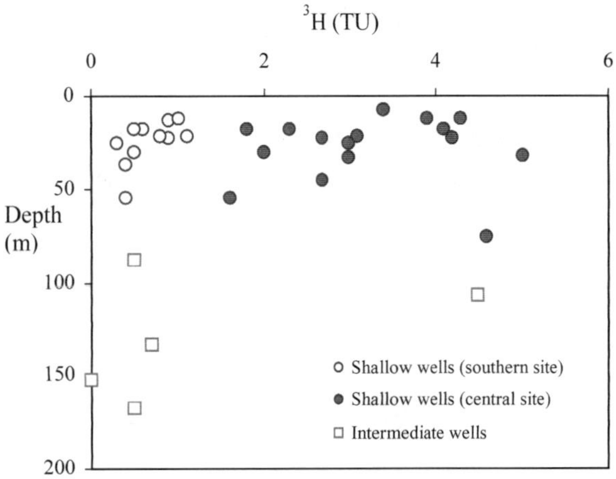

Figure 1. Distribution of ^3H content (2006-2008) in southern site shallow wells, central site shallow wells and intermediate wells [8].

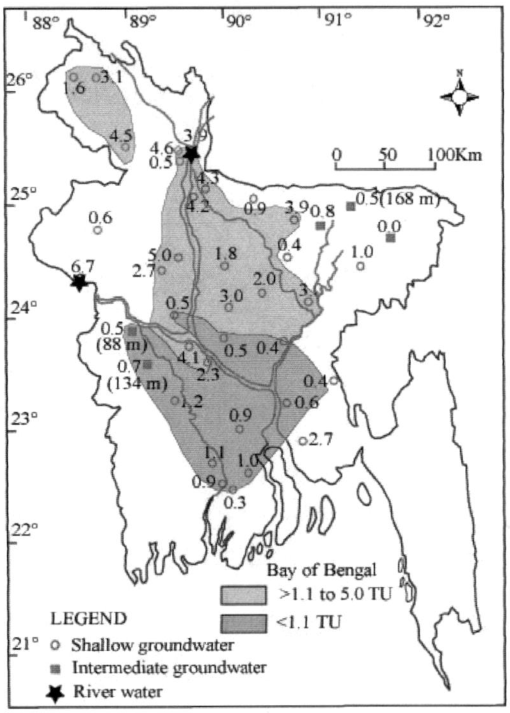

Figure 2. Spatial distribution of ^3H content (2006-2008) in shallow groundwater and intermediate groundwater [8].

4.2. Some Other Groundwater Studies.

In [7], groundwater dynamics were studied using noble gases and tritium in the rural Munshiganj district as representative of southern region of Bangladesh. In the investigation carried out in January 2004 by Klump et.al., the samples of ^3H were analysed using the ^3He in-growth method using a high-sensitivity compressor-source noble gas mass spectrometer. From table 1, it can be seen that the ^3H content is positively correlated with depth and peaks at a value of 18.5 TU at a depth of approximately 20 m. At depths below 20 m, ^3H concentration decreases, and below approximately 35 m, the groundwater is virtually free of ^3H. In conclusion, the environmental tracer data suggest the presence of a young, shallow groundwater body that contains bomb-derived ^3H, indicating ages less than roughly 55 years, and of an old, deep groundwater body, which is virtually ^3H free. The peak concentration of 18.5 TU is relatively low and below the average global value indicating negligible danger to populace due to tritium radiation.

Groundwater in the central part of Bangladesh was investigated by Stute et. al. [6] using indirect method of ^{3}He in-growth technique. Araihazar Upazila was selected as representative of central region of Bangladesh. Like in other studies, tritium concentration of 0.4-5.4 TU in groundwater samples at depths below 20 m collected from this region is below world mean value of 25 TU (2.2±0.7 Bq/L) and hence tritium radioactivity in this region is relatively low.

Tritium concentration was also found out in ground water and in river water [4]. In May 1999, during the first sampling campaign conducted in nine south-western regions of Bangladesh, 36 water samples were collected from shallow and deep tube wells, ranging in depth from 10 to 335 m. During the second mission organized in January 2000 in four western parts of Bangladesh, 20 samples from shallow and deep tube wells, ranging in depth from 10 to 90 m were collected. The concentration of tritium in underground water from both south-western and western districts was relatively low and measured 0-7.2 TU, showing that danger from tritium radioactivity is negligible. River water in the region also has a similar trivial tritium content. The low tritium content of shallow groundwater is consistent with recharge from local rain and floodwaters. The absence of tritium in many of the shallow samples indicates relatively large travel times in the unsaturated zone (several years to tens of years), resulting in the loss of tritium by radioactive decay before reaching the aquifer.

4.3. Study of Precipitation.

Research [9] reconstructed annual mean ^{3}H concentration in precipitation utilizing International Atomic Energy Agency (IAEA)/ World Meteorological Organization (WMO) database for Dhaka, Bangladesh, for 1953-1995 in order to apply the ^{3}H and ^{3}H/^{3}He methods to constrain the groundwater flow regime in Bangladesh.

Tritium concentration in precipitation before 1953 was about 5 tritium units (TU). In the 1950s and early 1960s atmospheric nuclear tests introduced much higher amounts of tritium into the atmosphere. Consequently, in 1963/64, the concentration in precipitation increased by three orders of magnitude in the northern hemisphere as compared to that arising from the cosmic rays [16]. The peak concentration of tritium in precipitation occurred in 1963 which in case of Dhaka, Bangladesh amounted to 597.2 TU. Since then, tritium concentration has decreased substantially to present levels of about 5-10 TU in the Northern Hemisphere which is same as the pre-nuclear tests level. Present day precipitation in Bangladesh is estimated to have a tritium content of around 5-10 TU which is below the threshold point that requires intervention.

4.4. Research in the Vicinity of Rooppur Nuclear Power Plant.

The site around the first NPP in Bangladesh, i.e. Rooppur NPP was investigated [10-11] and tritium concentration assessed using liquid scintillation counter. The results are illustrated in table 2. Panov et. al. in their research [10] measured tritium radionuclide concentration of 0.8-2.1 Bq/L, <3 Bq/kg, <3 Bq/kg and 0.4-1.2 Bq/L for surface water of Padma, bottom sediments, aquatic flora and drinking water respectively. Henceforth, the investigation authenticated that the tritium content in the surface waters of the Padma River together with the components of freshwater ecosystem in the vicinity of the Rooppur NPP is below the world average value and possesses no radiological threat to the population residing in the locality.

5. Conclusion

The extensive review article presents an overview of the tritium radioactivity concentrations in different components of observed freshwater ecosystems, underground water for drinking purpose and precipitation in Bangladesh with special attention given to the site of the first Rooppur NPP in Bangladesh. The tritium content in groundwater, precipitation and river water fluctuates between 0-25 TU, 5-10 TU and 5-10 TU respectively as obtained from the radio-ecological monitoring of environment in Bangladesh. The tritium radionuclide concentration of 6.7-17.6 TU, <25.2 TU, <25.2 TU and 3.4-10.1 TU were detected around Rooppur NPP for surface water of Padma, bottom sediments, aquatic flora and drinking water respectively.

On the basis of these results, one can state that tritium concentration in groundwater, precipitation and different components of observed freshwater ecosystem is relatively low (0 to 25 TU) all over

Bangladesh at this moment in time and is below the cut-off point that demands intervention. This comprehensive review study provides a baseline of tritium concentration in Bangladesh including the proposed Rooppur NPP site area and will allow the observation of changes and the evaluation of radiation doses and radiological health hazards that organisms/people experience due to the technogenic changes of radiation on the environmental situation in these studied regions.

References

[1] Bridges B A *et al.* 2007 *Review of risks from tritium.* Report of the independent Advisory Group on Ionising Radiation. Documents of the Health Protection Agency Radiation, Chemical and Environmental Hazards (United Kingdom) 90

[2] *Transfer of tritium in the environment after accidental releases from nuclear facilities* 2014 (Vienna: IAEA) 1738

[3] Chao TC, Wang CC, Li J, Li C and Tung CJ 2012 Cellular- and micro-dosimetry of heterogeneously distributed tritium. *Int J Radiat Biol.* **88(1-2)** 151-7

[4] Aggarwal P K *et al.* 2000 A report on isotope hydrology of groundwater in Bangladesh: implications for characterization and mitigation of arsenic in groundwater (Vienna: IAEA) 648

[5] Zheng Y, Stute M, Geen A V, Gavrieli I, Dhar R, Simpson H J, Schlosser P and Ahmed K M 2004 Redox control of arsenic mobilization in Bangladesh groundwater. *Applied Geochemistry* **19(2)** 201-14

[6] Stute M *et al.* 2007 Hydrological control of As concentrations in Bangladesh groundwater. *Water Resour. Res.* **43** W09417

[7] Klump S, Kipfer R, Cirpka O A, Harvey C F, Brennwald M S, Ashfaque N S, Badruzzaman A B M, Hug S J and Imboden D M 2006 Groundwater dynamics and arsenic mobilization in Bangladesh assessed using noble gases and tritium. *Environmental Science & Technology* **40(1)** 243-50

[8] Majumder R K, Halim M A, Saha B B, Ikawa R, Nakamura T, Kagabu M and Shimada J 2011 Groundwater flow system in Bengal Delta, Bangladesh revealed by environmental isotopes. *Environmental Earth Sciences* **64** 1343-52

[9] Stute M 2001 Tritium in precipitation in Bangladesh. *Proc. of the KTH-Dhaka University Seminar, University of Dhaka, Bangladesh KTH Special Publication TRITA-AMI Report* **3084** 109-17

[10] Panov A V, Isamov N N, Kuznetsov V K and Kurbakov D N 2019 Developing of the radioecological monitoring system of atmospheric air, terrestrial and freshwater ecosystems in the vicinity of Rooppur NPP (People's Republic of Bangladesh). *ENVIRA 2019 Proceedings* 69-73

[11] Mikailova R, Panov A V and Kurbakov D N 2019 The programme and results of the radioecological monitoring of freshwater ecosystems in the vicinity of Rooppur NPP (People's Republic of Bangladesh). *RAP Conference Proceedings* **4** 108-12

[12] Lucas L L and Unterweger M P 2000 Comprehensive review and critical evaluation of the half-life of tritium. *J Res Natl Inst Stand Technol* **105(4)** 541-9

[13] Rozanski K and Gröning M 1998 Tritium assay in water samples using electrolytic enrichment and liquid scintillation spectrometry. *Proceedings of a Consultants Meeting on Quantifying Uncertainty in Nuclear Analytical Measurements* 195-217

[14] Mohr P J and Taylor B N 2000 CODATA recommended values of the fundamental physical constants: 1998. *Rev. Mod. Phys.* **72**(2) 351-495

[15] Hasan M A, Ahmed K M, Sracek O, Bhattacharya P, von Bromssen M, Broms S, Fogelstrom J, Mazumder M L and Jacks G 2007 Arsenic in shallow groundwater of Bangladesh: investigations from three different physiographic settings. *Hydrol J* **15** 1507–22

[16] Cook P G and Herezeg A L 2000 *Environmental Tracers in Subsurface Hydrology* (Boston: Kluwer Academic Publishers) 529

IOP Conf. Series: Materials Science and Engineering **976** (2020) 012010 doi:10.1088/1757-899X/976/1/012010

Features of the development of a mathematical model of an electric multipole with memresistive branches for nanoelectronic components of quantum computing systems

A V Bondarev[1] and V N Efanov[2]

[1]FSBEI HE "Orenburg State University", Kumertau Branch OSU, 3b, 2 Soviet lane, Kumertau, 453300, Russia
[2]FSBEI HE "Ufa State Aviation Technical University", 12, C. Marx str., Ufa, 450000, Russia

E-mail: bondarevav@kfosu.edu.ru

Abstract. The article discusses the features of constructing a mathematical model of a memristor, one of the components of quantum computing systems, in the finite increments of its parameters. A modification of the basic set of elements for the design of nanoelectronic circuits with the addition of a memristor as one of the components is described. As a result of the modification, a mathematical model of an electric multipole with memresistive branches in the finite increments of its parameters is constructed.

1. Introduction

The mathematical model of an electric multipole is formed for a certain basic set of elements. The universal set includes: linear and nonlinear resistor, linear and nonlinear capacitance, linear and nonlinear inductance, independent current and voltage sources, dependent (controlled) sources - a current or voltage controlled current source, and a voltage or current controlled voltage source. Thus, this set contains 12 elements, the parametric equations of which are known in finite increments [1-16].

In 1971, the American researcher Chua L.O. suggested the possibility of creating in the future a fourth basic element of electrical circuits. In fact, at the tip of his pen, he predicted the presence of a physical dependence linking 4 main variables of electrical circuits: current I, voltage U, charge Q and magnetic flux F. Five out of six possible relationships connecting these parameters are well known. For the sixth ratio, Chua suggested that the magnetic flux and electric charge be linked through a new characteristic - memresistivity M. And the "absent" element was called memristor:

$$dF = M \cdot dQ \tag{1}$$

Note that flux in this case should be understood as an integral of voltage over time Chua showed that in the general case the memristivity should depend on q [5; 13].

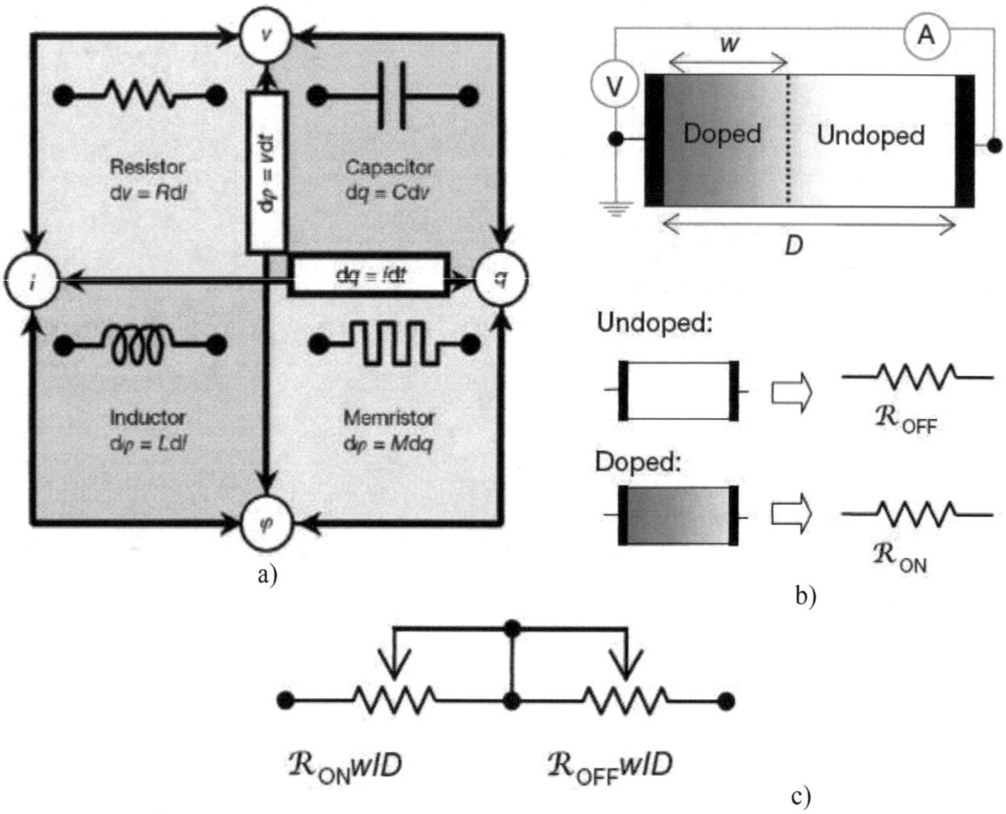

Figure 1. a - basic elements of electrical circuits: resistor, capacitor, inductance and memristor; b - memristor structure; c - equivalent circuit of memristor [13]

In theory, the memristor model can be represented as follows [13]:

$$u(t) = (R_{ON}\frac{w(t)}{D} + R_{OFF}\left(1 - \frac{w(t)}{D}\right)) \cdot i(t) \quad (2)$$

In this case, the border shifts according to the law:

$$\frac{dw(t)}{dt} = \mu_u \cdot \frac{R_{ON}}{D} \cdot i(t) \quad (3)$$

where μ_u is the average ion mobility.
Integration gives the formula for *w(t)*:

$$w(t) = \mu_u \cdot \frac{R_{ON}}{D} \cdot q(t) \quad (4)$$

Considering that $R_{ON} << R_{OFF}$, we get an expression for memristivity [13]:

$$M(q) = R_{OFF}\left(1 - \frac{\mu_u \cdot R_{ON}}{D^2}\right) \cdot q(t) \quad (5)$$

For more than a quarter of a century, the memristor remained a hypothetical circuit element with no material implementation. But in 2008, a team of HP researchers led by Stanley Williams finally created a real memristor. Its properties corresponded to the model proposed by Chua. The memristor, developed by Williams' group, was a thin layer of semiconductor material sandwiched between two metal contacts. On one side of the layer, there is a dopant (positive ions) - in the region of width w [13; 14].

IOP Conf. Series: Materials Science and Engineering **976** (2020) 012010 doi:10.1088/1757-899X/976/1/012010

2. Materials and methods

I-V characteristic of the memristor. Due to the fact that the memristor has moved from the category of hypothetical components to the field of practical implementation and has been used by leading manufacturers of nanoelectronics for a considerable time, it makes sense to take a fresh look at the mathematical modeling of multipoles. Since the characteristics of nanoelectronic components largely depend on the intensity of the magnetic field, for an adequate mathematical description of such systems, it is simply necessary to introduce a linear and nonlinear memristor into the basic set of elements, which is responsible for the interconnection of the electric charge and magnetic field in the structure of nanoelectronic devices.

Obviously, we have to talk about not one, but two new elements of the basic set. This follows from the well-known experimental dependences of the I-V characteristic obtained by Western manufacturers. They showed that if an alternating sinusoidal voltage of a certain frequency is applied to the memristor, its current-voltage characteristic takes a form resembling a Lissajous figure centered at the origin. Therefore, a memristor, unlike a resistor, has a hysteresis. With increasing voltage frequency, the hysteresis curve degenerates into a straight line. This circumstance makes a memristor related to resistance, only in the reverse analogy (at high frequencies, the I–V characteristic of the memristor becomes linear). That is why we are talking about two new elements of the basic set: linear and nonlinear memristors.

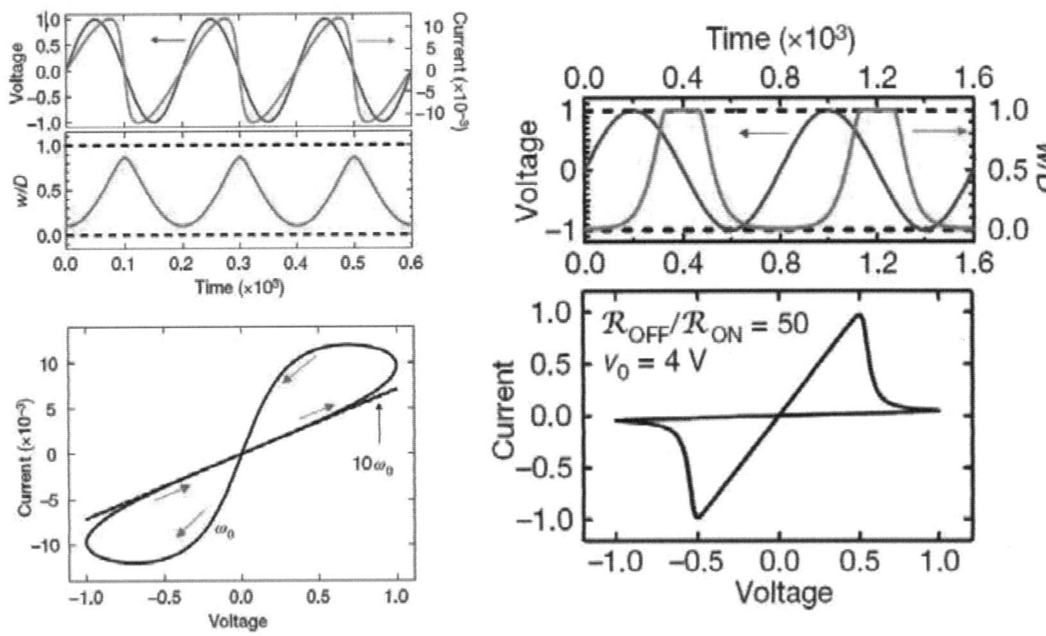

a) b)

Figure 2. Results of modeling the behavior of the memristor: a - based on the model described by equations (2-5); b - when multiplying the right side of equation (3) by the window function w(1-w)/D_2. The axes plotted dimensionless values of voltage, current and time, normalized to v_0 = 1 V (Fig. 2, a) and 4V (Fig. 2, b), i_0 = 10 mA, t_0 = 10 ms, respectively [13].

Topological equations of a multipole. Earlier, a mathematical model of an electric multipole was obtained in a hybrid basis of topological and parametric equations in finite increments of standard components of the base set [12]. With the introduction of a memristor into this set of elements, it becomes necessary to build a new mathematical model that takes into account not only internal connections and processes in current and voltage, but also in magnetic components.

The generalized mathematical model of a multipole network will consist of n topological and n parametric equations. Indeed, topological equations in finite increments for the previous base set [12]:

$$\begin{cases} D \cdot \Delta I = 0 \\ B \cdot \Delta U = 0 \end{cases} \tag{6}$$

where D is the matrix of principal sections of *(k-1)* equations; B - matrix of main contours of *(n-k + 1)* equations; n is the number of branches and k is the number of nodes; ΔI, ΔU - increments of currents and voltages in the branches of the directed graph of the circuit.

Decomposed form for vectors of currents and voltages increments:

$$\begin{cases} \Delta U = [\Delta U_L^X, \Delta U_H^X, \Delta U_R^X, \Delta U_M^X, \Delta U_C^P, \Delta U_H^P, \Delta U_R^P, \Delta U_M^P] \\ \Delta I = [\Delta I_L^X, \Delta I_H^X, \Delta I_R^X, \Delta I_M^X, \Delta I_C^P, \Delta I_H^P, \Delta I_R^P, \Delta I_M^P] \end{cases} \tag{7}$$

where ΔU_L^X, ΔI_L^X - increments of voltages and currents on inductive chords; ΔU_C^P, ΔI_C^P - voltage and current increments on the capacitive edges; ΔU_R^X, ΔU_R^P, ΔI_R^P, ΔI_R^X - voltage and current increments on resistive edges and chords; ΔU_H^X, ΔU_H^P, ΔI_H^P, ΔI_H^X - voltage and current increments on nonlinear edges and chords; ΔU_M^X, ΔU_M^P, ΔI_M^X, ΔI_M^P - voltage and current increments on memresistive edges and chords.

The system of topological equations for currents will take the form:

$$\begin{bmatrix} \Delta I_C^P \\ \Delta I_H^P \\ \Delta I_R^P \\ \Delta I_M^P \end{bmatrix} = - \begin{bmatrix} D_I & D_{II} & D_{III} & D_{IV} \\ D_V & D_{VI} & D_{VII} & D_{VIII} \\ D_{IX} & D_X & D_{XI} & D_{XII} \\ D_{XIII} & D_{XIV} & D_{XV} & D_{XVI} \end{bmatrix} \cdot \begin{bmatrix} \Delta I_L^X \\ \Delta I_H^X \\ \Delta I_R^X \\ \Delta I_M^X \end{bmatrix} \tag{8}$$

Concerning voltages:

$$\begin{bmatrix} \Delta U_L^X \\ \Delta U_H^X \\ \Delta U_R^X \\ \Delta U_M^X \end{bmatrix} = - \begin{bmatrix} B_I & B_{II} & B_{III} & B_{IV} \\ B_V & B_{VI} & B_{VII} & B_{VIII} \\ B_{IX} & B_X & B_{XI} & B_{XII} \\ B_{XIII} & B_{XIV} & B_{XV} & B_{XVI} \end{bmatrix} \cdot \begin{bmatrix} \Delta U_C^P \\ \Delta U_H^P \\ \Delta U_R^P \\ \Delta U_M^P \end{bmatrix} \tag{9}$$

Taking into account the above, the set of topological equations in finite increments in a hybrid basis of currents and voltages, charges and magnetic fluxes will consist of 4 groups:

$$\begin{cases}
\begin{bmatrix}\Delta U_L^X\\ \Delta I_C^P\end{bmatrix}=\begin{bmatrix}-B_I & 0\\ 0 & -D_I\end{bmatrix}\cdot\begin{bmatrix}\Delta U_C^P\\ \Delta I_L^X\end{bmatrix}+\begin{bmatrix}-B_{II} & 0\\ 0 & -D_{II}\end{bmatrix}\cdot\begin{bmatrix}\Delta U_H^P\\ \Delta I_H^X\end{bmatrix}+\begin{bmatrix}-B_{III} & 0\\ 0 & -D_{III}\end{bmatrix}\cdot\begin{bmatrix}\Delta U_R^P\\ \Delta I_R^X\end{bmatrix}+\\
\qquad+\begin{bmatrix}-B_{IV} & 0\\ 0 & -D_{IV}\end{bmatrix}\cdot\begin{bmatrix}\Delta U_M^P\\ \Delta I_M^X\end{bmatrix}\\
\begin{bmatrix}\Delta U_H^X\\ \Delta I_H^P\end{bmatrix}=\begin{bmatrix}-B_V & 0\\ 0 & -D_V\end{bmatrix}\cdot\begin{bmatrix}\Delta U_C^P\\ \Delta I_L^X\end{bmatrix}+\begin{bmatrix}-B_{VI} & 0\\ 0 & -D_{VI}\end{bmatrix}\cdot\begin{bmatrix}\Delta U_H^P\\ \Delta I_H^X\end{bmatrix}+\begin{bmatrix}-B_{VII} & 0\\ 0 & -D_{VII}\end{bmatrix}\cdot\begin{bmatrix}\Delta U_R^P\\ \Delta I_R^X\end{bmatrix}+\\
\qquad+\begin{bmatrix}-B_{VIII} & 0\\ 0 & -D_{VIII}\end{bmatrix}\cdot\begin{bmatrix}\Delta U_M^P\\ \Delta I_M^X\end{bmatrix}\\
\begin{bmatrix}\Delta U_R^X\\ \Delta I_R^P\end{bmatrix}=\begin{bmatrix}-B_{IX} & 0\\ 0 & -D_{IX}\end{bmatrix}\cdot\begin{bmatrix}\Delta U_C^P\\ \Delta I_L^X\end{bmatrix}+\begin{bmatrix}-B_X & 0\\ 0 & -D_X\end{bmatrix}\cdot\begin{bmatrix}\Delta U_H^P\\ \Delta I_H^X\end{bmatrix}+\begin{bmatrix}-B_{XI} & 0\\ 0 & -D_{XI}\end{bmatrix}\cdot\begin{bmatrix}\Delta U_R^P\\ \Delta I_R^X\end{bmatrix}+\\
\qquad+\begin{bmatrix}-B_{XII} & 0\\ 0 & -D_{XII}\end{bmatrix}\cdot\begin{bmatrix}\Delta U_M^P\\ \Delta I_M^X\end{bmatrix}\\
\begin{bmatrix}\Delta U_M^X\\ \Delta I_M^P\end{bmatrix}=\begin{bmatrix}-B_{XIII} & 0\\ 0 & -D_{XIII}\end{bmatrix}\cdot\begin{bmatrix}\Delta U_C^P\\ \Delta I_L^X\end{bmatrix}+\begin{bmatrix}-B_{XIV} & 0\\ 0 & -D_{XIV}\end{bmatrix}\cdot\begin{bmatrix}\Delta U_H^P\\ \Delta I_H^X\end{bmatrix}+\begin{bmatrix}-B_{XV} & 0\\ 0 & -D_{XV}\end{bmatrix}\cdot\begin{bmatrix}\Delta U_R^P\\ \Delta I_R^X\end{bmatrix}+\\
\qquad+\begin{bmatrix}-B_{XVI} & 0\\ 0 & -D_{XVI}\end{bmatrix}\cdot\begin{bmatrix}\Delta U_M^P\\ \Delta I_M^X\end{bmatrix}
\end{cases}\tag{10}$$

3. Results

Parametric memristor equations. We now turn to consideration of parametric equations. Expressions for $\Delta U_L^X, \Delta U_H^X, \Delta U_R^X, \Delta U_C^P, \Delta U_H^P, \Delta U_R^P$ and $\Delta I_L^X, \Delta I_H^X, \Delta I_R^X, \Delta I_C^P, \Delta I_H^P, \Delta I_R^P$ were obtained earlier and are known [12]. The unknown are the expressions for $\Delta U_M^X, \Delta U_M^P, \Delta I_M^X, \Delta I_M^P$. Consider the parametric equation of a linear memristor. Equation (1) has an analogue according to Ohm's law [17]:

$$u(t) = M(q) \cdot i(t) \tag{11}$$

Due to the fact that only electrical parameters voltage and current are considered, then equation (11) in the case of increments can be written in the form:

$$\Delta u(t) = E_M^N + M_E \cdot \Delta i(t) \tag{12}$$

where $E_M^N = i(t) \cdot \Delta M(q)$ - interval specified independent source of EMF; $M_E \in [M(q) + |\Delta M(q)|; M(q) - |\Delta M(q)|]$ is the equivalent interval specified value of memresistivity for the disturbed mode.

Most researchers discard the last term of equations similar to Eq. (12), considering the product of two increments to be negligible [12, 15, 16]. However, to study the robustness of electronic circuits in real operating conditions, it is advisable to take this term into account. Thus, we obtain the equivalent circuit of a linear memristor, shown in figure 3.

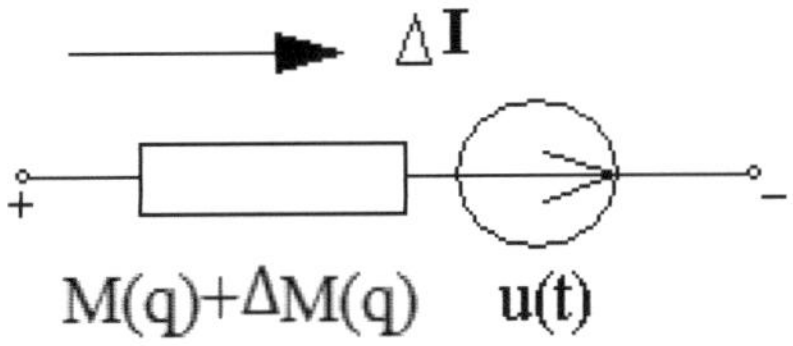

Figure 3. Equivalent circuit of a linear memristor in finite increments.

Next, consider a nonlinear memristor with an I-V characteristic similar to the Lissajous figure (see figure 2, a). In the nominal mode, the nonlinear memristor will be described by the following system of equations:

$$\begin{cases} u(t) = \alpha \cdot \sin \omega t \\ i(t) = \beta \cdot \sin(\omega t + \varphi)' \end{cases} \tag{13}$$

where α and β are constant coefficients of the loop width of the I-V characteristic; ω - frequency of current and voltage; φ is the angle of inclination of the I-V characteristic figure. Using the standard expansion of the sine of the sum, substituting the first equation of system (13) into the second and expressing the cosine of the function in terms of the sine, we obtain an expression for the nonlinear memristor in the nominal mode:

$$i(t) = u(t) \cdot \frac{\beta}{\alpha} \cdot \cos \varphi - u(t) \cdot \frac{\beta}{\alpha} \cdot \sin \varphi \cdot \sqrt{1 - \frac{u^2(t)}{\alpha^2}} \tag{14}$$

In perturbed mode:

$$i(t) + \Delta i(t) = u(t) \cdot \frac{\beta}{\alpha} \cdot \cos \varphi + \Delta u(t) \cdot \frac{\beta}{\alpha} \cdot \cos \varphi - u(t) \cdot \frac{\beta}{\alpha} \cdot \sin \varphi \cdot \sqrt{1 - \frac{(u + \Delta u)^2(t)}{\alpha^2}} - \\ -\Delta u(t) \cdot \frac{\beta}{\alpha} \cdot \sin \varphi \cdot \sqrt{1 - \frac{(u + \Delta u)^2(t)}{\alpha^2}} \tag{15}$$

The equation in finite increments for a nonlinear memristor takes the form:

$$\Delta i(t) = \Delta M^{-1}(q) \cdot \Delta u(t) - J_{\text{H}}^{Z} - J_{\text{H}}^{N} \tag{16}$$

where $\Delta M^{-1}(q) = \frac{\beta}{\alpha} \cdot \cos \varphi$ is the increment of inverse memresistivity (by analogy with conductivity); $J_{\text{H}}^{Z} = \Delta u(t) \cdot \frac{\beta}{\alpha} \cdot \sin \varphi \cdot \sqrt{1 - \frac{(u+\Delta u)^2(t)}{\alpha^2}}$ - dependent current source; $J_{\text{H}}^{N} = u(t) \cdot \frac{\beta}{\alpha} \cdot \sin \varphi \cdot \sqrt{1 - \frac{(u+\Delta u)^2(t)}{\alpha^2}}$ - independent current source.

In this case, the equivalent circuit of the nonlinear memristor will take the form shown in figure 4.

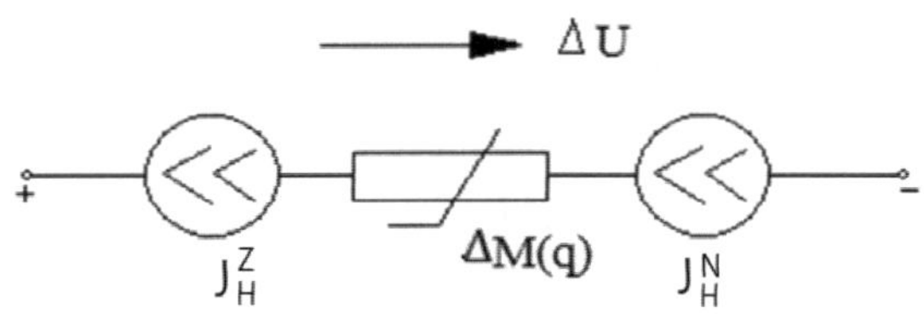

Figure 4. Equivalent circuit of a nonlinear memristor in finite increments.

In connection with the above, the system of hybrid equations for linear memresistive branches takes the following form:

$$\begin{bmatrix} \Delta U_M^X \\ \Delta I_M^P \end{bmatrix} = \begin{bmatrix} 0 & \widehat{M} \\ \widehat{M}^{-1} & 0 \end{bmatrix} \cdot \begin{bmatrix} \Delta U_M^P \\ \Delta I_M^X \end{bmatrix} + \begin{bmatrix} E_N^M \\ J_N^M \end{bmatrix} \tag{17}$$

where $\widehat{M} = M + \Delta M$ is the matrix of the branch memresistivities, $\widehat{M}^{-1} = M^{-1} + \Delta M^{-1}$ is the matrix of inverse memresistivities of the branch; E_N^M, J_N^M - equivalent vectors of independent voltage and current sources on memresistive branches.

System of hybrid equations for nonlinear memresistive branches:

$$\begin{bmatrix} \Delta U_{MH}^{X} \\ \Delta I_{MH}^{P} \end{bmatrix} = \begin{bmatrix} 0 & \widehat{M} \\ \widehat{M}^{-1} & 0 \end{bmatrix} \cdot \begin{bmatrix} \Delta U_{MH}^{P} \\ \Delta I_{MH}^{X} \end{bmatrix} - \begin{bmatrix} E_{N}^{M} \\ J_{N}^{M} \end{bmatrix} - \begin{bmatrix} E_{Z}^{M} \\ J_{Z}^{M} \end{bmatrix} \quad (18)$$

4. Discussion

Model synthesis. To construct a mathematical model of an electric multipole in a hybrid basis of topological and parametric equations, we will use the previously obtained expressions [12]. The introduction of linear and nonlinear memristors into the basic set of elements will not affect the properties of the reactive and resistive branches of the multipole. Therefore, the expressions for them will remain unchanged.

For reactive branches:

$$\begin{bmatrix} \Delta U_{L}^{X} \\ \Delta I_{C}^{P} \end{bmatrix} = \begin{bmatrix} 0 & \hat{L} \\ \hat{C} & 0 \end{bmatrix} \cdot \frac{d}{dt} \begin{bmatrix} \Delta U_{C}^{P} \\ \Delta I_{L}^{X} \end{bmatrix} + \begin{bmatrix} 0 & Z_{E} \\ V_{E} & 0 \end{bmatrix} \cdot \begin{bmatrix} \Delta U_{C}^{P} \\ \Delta I_{L}^{X} \end{bmatrix} + \begin{bmatrix} E_{E}^{L} \\ J_{E}^{C} \end{bmatrix} \quad (19)$$

where $\hat{L} = \begin{bmatrix} diag\{L_k + \Delta L_k\} & 0 \\ 0 & diag\{L_i^N + L_i^Z(\Delta i)\} \end{bmatrix}$ is the matrix of inductances ($k = \overline{1, n_{\text{Л}L}^{X}}$, $l = \overline{1, n_{HL}^{X}}$, $n_{\text{Л}L}^{X} + n_{HL}^{X} = dim\Delta U_{L}^{X}$); $\hat{C} = \begin{bmatrix} diag\{C_k + \Delta C_k\} & 0 \\ 0 & diag\{C_i^N + C_i^Z(\Delta i)\} \end{bmatrix}$ is the capacity matrix ($k = \overline{1, n_{\text{Л}C}^{X}}$, $l = \overline{1, n_{HC}^{X}}$, $n_{\text{Л}C}^{X} + n_{HC}^{X} = dim\Delta I_{C}^{P}$); $Z_E = \left[[0]_{n_L^X \times n_{\text{Л}L}^X} \quad diag\{(Z_N^L + Z_Z^L(\Delta i))_l\} \right]$ - matrix of equivalent resistances and $V_E = \left[[0]_{n_C^P \times n_{\text{Л}C}^P} \quad diag\left\{ \left(G_N^C + G_Z^C(\Delta u) \right)_l \right\} \right]$ - equivalent conductivities of nonlinear inductances and capacities; $E_E^L = \left[[(E_N)_k]_{n_{\text{Л}L}^X \times 1} \quad [(E_N^L + E_Z^L(\Delta i))_k]_{n_{HL}^X \times 1} \right]^T$, $J_E^C = \left[[(J_N)_k]_{n_{\text{Л}C}^P \times 1} \quad \left[\left(J_N^C + J_Z^C(\Delta u) \right)_k \right]_{n_{HC}^P \times 1} \right]^T$ - equivalent vectors of current and voltage sources [12].

For resistive branches:

$$\begin{bmatrix} \Delta U_{R}^{X} \\ \Delta I_{R}^{P} \end{bmatrix} = \begin{bmatrix} 0 & \hat{R} \\ \hat{Y} & 0 \end{bmatrix} \cdot \begin{bmatrix} \Delta U_{R}^{P} \\ \Delta I_{R}^{X} \end{bmatrix} + \begin{bmatrix} E_{N}^{R} \\ J_{N}^{R} \end{bmatrix} \quad (20)$$

where $\hat{R} = R + \Delta R$ is the resistance matrix of the branch, $\hat{Y} = Y + \Delta Y$ is the matrix of the branch conductivity; E_N^R, J_N^R - equivalent vectors of independent voltage and current sources on resistive branches [12].

And for nonlinear branches, taking into account nonlinear memresistivities (18), the system of hybrid equations takes the form:

$$\begin{bmatrix} \Delta U_{H}^{X} \\ \Delta I_{H}^{P} \end{bmatrix} = \begin{bmatrix} \Delta M^{-1} & Z_X(\Delta I_H^X) \\ G_P(\Delta U_H^P) & \Delta M \end{bmatrix} \cdot \begin{bmatrix} \Delta U_{H}^{P} \\ \Delta I_{H}^{X} \end{bmatrix} + F^I \quad (21)$$

where $F^I = \begin{bmatrix} E & E & 0 & 0 & -E & -E & 0 & 0 \\ 0 & 0 & E & E & 0 & 0 & -E & -E \end{bmatrix} \cdot \begin{bmatrix} E_H^H \\ E_X^H \\ J_H^H \\ J_P^H \\ E_H^Z \\ E_H^N \\ J_H^Z \\ J_H^N \end{bmatrix}$; $Z_X(\Delta I_H^X)$ and $G_P(\Delta U_H^P)$ are diagonal matrices of equivalent resistances and conductivities; ΔM^{-1} and ΔM are diagonal matrices of equivalent inverse and forward memresistivities; E is the identity matrix; E_H^H, E_X^H, J_H^H, J_P^H, E_H^Z, E_H^N, J_H^Z and $J_H^N$are equivalent sources of EMF and current on nonlinear chords and edges.

Since the left sides of equation (21) and the second equation of system (10) are equal, equating their right sides, we get:

$$\begin{bmatrix}\Delta U_H^P\\ \Delta I_H^X\end{bmatrix} = A_1\begin{bmatrix}\Delta U_C^P\\ \Delta I_L^X\end{bmatrix} + A_2\begin{bmatrix}\Delta U_R^P\\ \Delta I_R^X\end{bmatrix} + A_3\begin{bmatrix}\Delta U_M^P\\ \Delta I_M^X\end{bmatrix} - H\cdot F^I \tag{22}$$

where $A_1 = \begin{bmatrix}\Delta M^{-1} & Z_X(\Delta I_H^X)\\ G_P(\Delta U_H^P) & \Delta M\end{bmatrix}\cdot\begin{bmatrix}-B_V & 0\\ 0 & -D_V\end{bmatrix}$, $A_2 = \begin{bmatrix}\Delta M^{-1} & Z_X(\Delta I_H^X)\\ G_P(\Delta U_H^P) & \Delta M\end{bmatrix}\cdot\begin{bmatrix}-B_{VII} & 0\\ 0 & -D_{VII}\end{bmatrix}$, $A_3 = \begin{bmatrix}-B_{VIII} & 0\\ 0 & -D_{VIII}\end{bmatrix}$ и $H = \begin{bmatrix}\Delta M^{-1} & Z_X(\Delta I_H^X)\\ G_P(\Delta U_H^P) & \Delta M\end{bmatrix}$.

Because the right-hand sides of the third equation of system (10) and equation (20) are equal, giving similar terms we get the following expression:

$$\begin{bmatrix}\Delta U_R^P\\ \Delta I_R^X\end{bmatrix} = G_1\begin{bmatrix}\Delta U_C^P\\ \Delta I_L^X\end{bmatrix} + G_2\begin{bmatrix}\Delta U_H^P\\ \Delta I_H^X\end{bmatrix} + G_3\begin{bmatrix}\Delta U_M^P\\ \Delta I_M^X\end{bmatrix} - W\begin{bmatrix}E_N^R\\ J_N^R\end{bmatrix} \tag{23}$$

where $G_1 = \begin{bmatrix}B_{XI} & \hat{R}\\ \hat{Y} & D_{XI}\end{bmatrix}^{-1}\cdot\begin{bmatrix}-B_{IX} & 0\\ 0 & -D_{IX}\end{bmatrix}$, $G_2 = \begin{bmatrix}B_{XI} & \hat{R}\\ \hat{Y} & D_{XI}\end{bmatrix}^{-1}\cdot\begin{bmatrix}-B_X & 0\\ 0 & -D_X\end{bmatrix}$, $G_3 = \begin{bmatrix}B_{XI} & \hat{R}\\ \hat{Y} & D_{XI}\end{bmatrix}^{-1}\cdot\begin{bmatrix}-B_{XII} & 0\\ 0 & -D_{XII}\end{bmatrix}$ и $W = \begin{bmatrix}B_{XI} & \hat{R}\\ \hat{Y} & D_{XI}\end{bmatrix}^{-1}$.

Let us compare the fourth equation of system (10) and expression (17) and, applying similar reasoning as for resistive branches, we obtain:

$$\begin{bmatrix}\Delta U_M^P\\ \Delta I_M^X\end{bmatrix} = S_I\begin{bmatrix}\Delta U_C^P\\ \Delta I_L^X\end{bmatrix} + S_2\begin{bmatrix}\Delta U_H^P\\ \Delta I_H^X\end{bmatrix} + S_3\begin{bmatrix}\Delta U_R^P\\ \Delta I_R^X\end{bmatrix} - K\begin{bmatrix}E_N^M\\ J_N^M\end{bmatrix} \tag{24}$$

where $S_I = \begin{bmatrix}-B_{XVI} & \widehat{M}\\ \widehat{M}^{-1} & -D_{XVI}\end{bmatrix}^{-1}\cdot\begin{bmatrix}-B_{XIII} & 0\\ 0 & -D_{XIII}\end{bmatrix}$, $S_2 = \begin{bmatrix}-B_{XVI} & \widehat{M}\\ \widehat{M}^{-1} & -D_{XVI}\end{bmatrix}^{-1}\cdot\begin{bmatrix}-B_{XIV} & 0\\ 0 & -D_{XIV}\end{bmatrix}$, $S_3 = \begin{bmatrix}-B_{XVI} & \widehat{M}\\ \widehat{M}^{-1} & -D_{XVI}\end{bmatrix}^{-1}\cdot\begin{bmatrix}-B_{XV} & 0\\ 0 & -D_{XV}\end{bmatrix}$ и $K = \begin{bmatrix}-B_{XVI} & \widehat{M}\\ \widehat{M}^{-1} & -D_{XVI}\end{bmatrix}^{-1}$.

Comparing the right-hand sides of the first equation of system (10) and (19), we obtain an expression in a hybrid basis for reactive branches:

$$\begin{bmatrix}0 & \hat{L}\\ \hat{C} & 0\end{bmatrix}\cdot\frac{d}{dt}\begin{bmatrix}\Delta U_C^P\\ \Delta I_L^X\end{bmatrix} = Q_1\begin{bmatrix}\Delta U_C^P\\ \Delta I_L^X\end{bmatrix} + Q_2\begin{bmatrix}\Delta U_H^P\\ \Delta I_H^X\end{bmatrix} + Q_3\begin{bmatrix}E_N^R\\ J_N^R\end{bmatrix} + Q_4\begin{bmatrix}\Delta U_M^P\\ \Delta I_M^X\end{bmatrix} - \begin{bmatrix}E_E^L\\ J_E^C\end{bmatrix} \tag{25}$$

where $Q_1 = \begin{bmatrix}-B_I & Z_E\\ V_E & -D_I\end{bmatrix} + \begin{bmatrix}-B_{IV} & 0\\ 0 & -D_{IV}\end{bmatrix}\cdot G_1$, $Q_2 = \begin{bmatrix}-B_{II} & 0\\ 0 & -D_{II}\end{bmatrix} + \begin{bmatrix}-B_{IV} & 0\\ 0 & -D_{IV}\end{bmatrix}\cdot G_2$, $Q_3 = \begin{bmatrix}-B_{II} & 0\\ 0 & -D_{II}\end{bmatrix}\cdot W$ и $Q_4 = \begin{bmatrix}-B_{III} & 0\\ 0 & -D_{III}\end{bmatrix} + \begin{bmatrix}-B_{IV} & 0\\ 0 & -D_{IV}\end{bmatrix}$.

Then we compare the second equation of system (10) and expression (21), we get:

$$P_1(\Delta I_H^X, \Delta U_H^P)\cdot\begin{bmatrix}\Delta U_H^P\\ \Delta I_H^X\end{bmatrix} = P_2\cdot\begin{bmatrix}\Delta U_C^P\\ \Delta I_L^X\end{bmatrix} + F^{II} \tag{26}$$

where $P_1 = \begin{bmatrix}B_V & Z_X(\Delta I_H^X)\\ G_P(\Delta U_H^P) & D_V\end{bmatrix} + \begin{bmatrix}-B_{VI} & 0\\ 0 & -D_{VI}\end{bmatrix}\cdot G_2$, $P_2 = \begin{bmatrix}-B_{IV} & 0\\ 0 & -D_{IV}\end{bmatrix} + \begin{bmatrix}-B_{VI} & 0\\ 0 & -D_{VI}\end{bmatrix}\cdot G_1$ и $F^{II} = \begin{bmatrix}B_{VI} & 0\\ 0 & D_{VI}\end{bmatrix}\cdot M\begin{bmatrix}E_N^R\\ J_N^R\end{bmatrix} - F_N^H$, $F_N^H = G_1\begin{bmatrix}\Delta U_C^P\\ \Delta I_L^X\end{bmatrix} + G_2\begin{bmatrix}\Delta U_H^P\\ \Delta I_H^X\end{bmatrix} - M\begin{bmatrix}E_N^R\\ J_N^R\end{bmatrix}$.

Combining expressions (22-25), we obtain a full-size mathematical model of an electric multipole with memresistive branches in a full hybrid basis using the branch decomposition method, which will look like this:

$$\begin{cases} \begin{bmatrix} 0 & \hat{L} \\ \hat{C} & 0 \end{bmatrix} \cdot \frac{d}{dt}\begin{bmatrix} \Delta U_C^P \\ \Delta I_L^X \end{bmatrix} = Q_1 \begin{bmatrix} \Delta U_C^P \\ \Delta I_L^X \end{bmatrix} + Q_2 \begin{bmatrix} \Delta U_H^P \\ \Delta I_H^X \end{bmatrix} + Q_3 \begin{bmatrix} E_N^R \\ J_N^R \end{bmatrix} + Q_4 \begin{bmatrix} \Delta U_M^P \\ \Delta I_M^X \end{bmatrix} - \begin{bmatrix} E_E^L \\ J_E^C \end{bmatrix} \\ \begin{bmatrix} \Delta U_H^P \\ \Delta I_H^X \end{bmatrix} = A_1 \begin{bmatrix} \Delta U_C^P \\ \Delta I_L^X \end{bmatrix} + A_2 \begin{bmatrix} \Delta U_R^P \\ \Delta I_R^X \end{bmatrix} + A_3 \begin{bmatrix} \Delta U_M^P \\ \Delta I_M^X \end{bmatrix} - H \cdot F^I \\ \begin{bmatrix} \Delta U_R^P \\ \Delta I_R^X \end{bmatrix} = G_1 \begin{bmatrix} \Delta U_C^P \\ \Delta I_L^X \end{bmatrix} + G_2 \begin{bmatrix} \Delta U_H^P \\ \Delta I_H^X \end{bmatrix} + G_3 \begin{bmatrix} \Delta U_M^P \\ \Delta I_M^X \end{bmatrix} - W \begin{bmatrix} E_N^R \\ J_N^R \end{bmatrix} \\ \begin{bmatrix} \Delta U_M^P \\ \Delta I_M^X \end{bmatrix} = S_I \begin{bmatrix} \Delta U_C^P \\ \Delta I_L^X \end{bmatrix} + S_2 \begin{bmatrix} \Delta U_H^P \\ \Delta I_H^X \end{bmatrix} + S_3 \begin{bmatrix} \Delta U_R^P \\ \Delta I_R^X \end{bmatrix} - K \begin{bmatrix} E_N^M \\ J_N^M \end{bmatrix} \end{cases} \quad (27)$$

5. Conclusion

Expression (27) can be applied in circuit design and analysis of circuits of nanoelectronic devices, and the element base of quantum computers. The methods for solving the obtained mathematical model are subject to further research using the apparatus of interval arithmetic.

References

[1] Bondarev A V 2015 Research problem of a robustness of electronic schemes by methods of interval calculations in the conditions of uncertainty. *Materials of the CSIT'2015 Proceedings of the 17th International Workshop on Computer Science and Information Technologies* 145-9

[2] Bondarev A V, Muravyova E A, Kadyrov R R and Rahman P A 2016 The analysis of opportunities of construction and use of avionic systems based on COTS-modules. *ARPN Journal of Engineering and Applied Sciences* **11 1** 78-92

[3] Borghetti J, Snider G S, Kuekes P J et al. 2010 Memristive switches enable stateful logic operations via material implication. *Nature letters* **464** 873-6

[4] *Bourzac K Memristor Memory Readied for Production.* Retrieved from: www.technologyreview.com/computing/25018/

[5] Chua L O 1971 Memristor – the missing circuit element. *IEEE Trans. Circuit Theory* **18** 507-19

[6] *Johnson R C End of the CPU? HP demos configurable memristor.* Retrieved from: www.eetimes.com/electronicsnews/4088557/End-of-the-CPU-HP-demos-configurablememristor

[7] Kuekes P J, Snider G S and Williams R S 2005 Crossbar nanocomputers. *Scientific American*, **293** 72-8

[8] *Markoff J HP Sees a Revolution in Memory Chip.* Retrieved from: www.nytimes.com/2010/04/08/science/08chips.html?_r=1

[9] *Memristor*. Retrieved from: en.wikipedia.org/wiki/Memristor

[10] *Merritt R HP researcher predicts memory-centric processors.* Retrieved from: www.eetimes.com/electronicsnews/4199856/HP-researcher-predicts-memory-centricprocessors

[11] Strukov D B, Snider G S, Stewart D R and Williams R S 2008 The missing memristor found. *Nature letters* **453** 80-3

[12] Bondarev A V 2011 *Decision support system for assessing the robustness of complex airborne electronic systems based on COTS products: dissertation for the degree of candidate of technical sciences* (Ufa: Ufa State Aviation Technical University) 177

[13] Eliseev N 2010 Memristors and crossbars: nanotechnology for processors. *Electronics: Science, Technology, Business* **8** 84-9

[14] Chua L O and Lin Pen-Min 1980 *Machine analysis of electronic circuits. Algorithms and computational methods* (Moscow: Energy) 640

[15] Zebrev G I 2008 *Physical foundations of silicon nanoelectronics: a textbook* (Moscow: MEPhI) 288

[16] Sigorskiy V P and Petrenko A I 1976 *Algorithms for analyzing electronic circuits* (*Moscow: Soviet radio*) 608

[17] Shary S P *Finite-dimensional interval analysis*. Retrieved from: http://www.nsc.ru/interval

Mode of "small" signal of the mathematical model of an electric multipole with memresistive branches under conditions of interval uncertainty

A V Bondarev[1] and V N Efanov[2]

[1]FSBEI HE "Orenburg State University", Kumertau Branch OSU, 3b, 2 Soviet lane, Kumertau, 453300, Russia
[2]FSBEI HE "Ufa State Aviation Technical University", 12, C. Marx str., Ufa, 450000, Russia

E-mail: bondarevav@kfosu.edu.ru

Abstract. In this article, a mathematical model of an electric multipole with memresistive branches is developed when operating under conditions of interval uncertainty in the mode of a small signal on alternating current.

1. Introduction

The emergence of a wide range of nanoelectronic components in recent years, on the one hand, certainly expands the capabilities of information and computing systems. However, the use of such elements in conditions other than laboratory ones has interval uncertainties in determining the electrical parameters of these devices, which is superimposed on the uncertainty of the entire system as a whole, which complicates the circuit analysis and the entire design process as a whole.

Despite the widespread use of information and computing technologies in all spheres of human activity, the tasks of increasing the performance of modern computing devices without increasing the cost and expanding their operational characteristics for use not only in laboratory and office conditions, but also for stable operation in an extreme environment remain urgent.

A new element of nanoelectronics, the appearance of which was predicted in 1971 [1-7], the memristor was able to change its resistance depending on the charge flowing through it. A change in the properties of a memristor is provided by chemical reactions in a thin two-layer TiO2 film, and in the classification, it belongs to the class of nanoionic devices. The additionally appearing hysteresis phenomenon makes it possible to use it as a memory cell. [5-21].

In connection with the above, it is necessary to take a fresh look at the electric multipole and its mathematical model using the memristor as one of the elements of the basic set. The introduction of such an addition to the basic set of elements also determines the appearance of new memresistive branches of the electric multipole.

In accordance with the adopted classification [21], it is customary to distinguish the following typical operating modes of the investigated electronic devices, which is quite true for the elements of nanoelectronics:

- Quasi-linear mode of a small signal when calculating with an alternating current;
- Quasi-linear small signal mode when calculating in the time domain;

- Large signal mode when calculating for direct current (static mode);
- Large signal mode when calculating in the time domain (dynamic mode).

Depending on the selected mode, the initial model of the investigated device will be an electrical multipole of one of the following types: linear resistive, linear reactive, nonlinear resistive and nonlinear reactive. At the same time, the general methodological basis for studying all the listed types of models is their description in the basis of finite deviations of currents and voltages of the branches of the corresponding multipole, since the main goal of our study is to assess the ability of a device to maintain its characteristics within specified limits in the presence of disturbing influences with undefined properties. This article analyzes the first small signal modes in the above classification for an electric multipole with memresistive branches in finite increments.

2. Materials and methods

Mathematical models of an electric multipole with memresistive branches in the nominal form and using finite increments were built earlier [1-3] using the method of decomposition of branches of the directed graph of the circuit described in [19-22].

A full-size mathematical model in a full hybrid basis in finite increments of currents and voltages will look like this [19]:

$$\begin{cases} \begin{bmatrix} 0 & \hat{L} \\ \hat{C} & 0 \end{bmatrix} \cdot \dfrac{d}{dt} \begin{bmatrix} \Delta U_C^P \\ \Delta I_L^X \end{bmatrix} = Q_1 \begin{bmatrix} \Delta U_C^P \\ \Delta I_L^X \end{bmatrix} + Q_2 \begin{bmatrix} \Delta U_H^P \\ \Delta I_H^X \end{bmatrix} + Q_3 \begin{bmatrix} E_N^R \\ J_N^R \end{bmatrix} + Q_4 \begin{bmatrix} \Delta U_M^P \\ \Delta I_M^X \end{bmatrix} - \begin{bmatrix} E_E^L \\ J_E^C \end{bmatrix} \\ \begin{bmatrix} \Delta U_H^P \\ \Delta I_H^X \end{bmatrix} = A_1 \begin{bmatrix} \Delta U_C^P \\ \Delta I_L^X \end{bmatrix} + A_2 \begin{bmatrix} \Delta U_R^P \\ \Delta I_R^X \end{bmatrix} + A_3 \begin{bmatrix} \Delta U_M^P \\ \Delta I_M^X \end{bmatrix} - H \cdot F^I \\ \begin{bmatrix} \Delta U_R^P \\ \Delta I_R^X \end{bmatrix} = G_1 \begin{bmatrix} \Delta U_C^P \\ \Delta I_L^X \end{bmatrix} + G_2 \begin{bmatrix} \Delta U_H^P \\ \Delta I_H^X \end{bmatrix} + G_3 \begin{bmatrix} \Delta U_M^P \\ \Delta I_M^X \end{bmatrix} - W \begin{bmatrix} E_N^R \\ J_N^R \end{bmatrix} \\ \begin{bmatrix} \Delta U_M^P \\ \Delta I_M^X \end{bmatrix} = S_I \begin{bmatrix} \Delta U_C^P \\ \Delta I_L^X \end{bmatrix} + S_2 \begin{bmatrix} \Delta U_H^P \\ \Delta I_H^X \end{bmatrix} + S_3 \begin{bmatrix} \Delta U_R^P \\ \Delta I_R^X \end{bmatrix} - K \begin{bmatrix} E_N^M \\ J_N^M \end{bmatrix} \end{cases} \quad (1)$$

where $\hat{L} = \begin{bmatrix} diag\{L_k + \Delta L_k\} & 0 \\ 0 & diag\{L_i^N + L_i^N(\Delta i)\} \end{bmatrix}$ is the matrix of inductances ($k = \overline{1, n_{ЛL}^X}$, $l = \overline{1, n_{HL}^X}$, $n_{ЛL}^X + n_{HL}^X = dim\Delta U_L^X$); $\hat{C} = \begin{bmatrix} diag\{C_k + \Delta C_k\} & 0 \\ 0 & diag\{C_i^N + C_i^N(\Delta i)\} \end{bmatrix}$ is the capacity matrix ($k = \overline{1, n_{ЛC}^X}$, $l = \overline{1, n_{HC}^X}$, $n_{ЛC}^X + n_{HC}^X = dim\Delta I_C^P$); ΔU_L^X, ΔI_L^X - increments of voltages and currents on inductive chords; ΔU_C^P, ΔI_C^P - voltage and current increments on the capacitive edges; ΔU_R^X, ΔU_R^P, ΔI_R^P, ΔI_R^X - voltage and current increments on resistive edges and chords; ΔU_H^X, ΔU_H^P, ΔI_H^P, ΔI_H^X - voltage and current increments on nonlinear edges and chords; ΔU_M^X, ΔU_M^P, ΔI_M^X, ΔI_M^P - voltage and current increments on memresistive edges and chords; $Q_1 = \begin{bmatrix} -B_I & Z_E \\ V_E & -D_I \end{bmatrix} + \begin{bmatrix} -B_{IV} & 0 \\ 0 & -D_{IV} \end{bmatrix} \cdot G_1$, $Q_2 = \begin{bmatrix} -B_{II} & 0 \\ 0 & -D_{II} \end{bmatrix} + \begin{bmatrix} -B_{IV} & 0 \\ 0 & -D_{IV} \end{bmatrix} \cdot G_2$, $Q_3 = \begin{bmatrix} -B_{II} & 0 \\ 0 & -D_{II} \end{bmatrix} \cdot W$ and $Q_4 = \begin{bmatrix} -B_{III} & 0 \\ 0 & -D_{III} \end{bmatrix} + \begin{bmatrix} -B_{IV} & 0 \\ 0 & -D_{IV} \end{bmatrix}$; $Z_E = [[0]_{n_L^X \times n_{ЛL}^X} \quad diag\{(Z_E^L + Z_E^L(\Delta i))_l\}]$ - matrix of equivalent resistances and $V_E = \left[[0]_{n_C^P \times n_{ЛC}^P} \quad diag\left\{\left(G_N^C + G_N^C(\Delta u)\right)_l\right\}\right]$ - equivalent conductivities of nonlinear inductances and capacities; D_i is the matrix of the main sections of the directed graph of the circuit; B_i - matrix of the main contours of the directed graph of the circuit; n is the number of branches and k is the number of nodes; $E_E^L = \left[[(E_N)_k]_{n_{ЛL}^X \times 1} \quad [(E_N^L + E_Z^L(\Delta i))_k]_{n_{HL}^X \times 1}\right]^T$, $J_E^C = \left[[(J_N)_k]_{n_{ЛC}^P \times 1} \quad \left[\left(J_N^C + J_Z^C(\Delta u)\right)_k\right]_{n_{HC}^P \times 1}\right]^T$ - equivalent vectors of current and voltage sources; where

$A_1 = \begin{bmatrix} \Delta M^{-1} & Z_X(\Delta I_H^X) \\ G_P(\Delta U_H^P) & \Delta M \end{bmatrix} \cdot \begin{bmatrix} -B_V & 0 \\ 0 & -D_V \end{bmatrix}$, $A_2 = \begin{bmatrix} \Delta M^{-1} & Z_X(\Delta I_H^X) \\ G_P(\Delta U_H^P) & \Delta M \end{bmatrix} \cdot \begin{bmatrix} -B_{VII} & 0 \\ 0 & -D_{VII} \end{bmatrix}$, $A_3 = \begin{bmatrix} -B_{VIII} & 0 \\ 0 & -D_{VIII} \end{bmatrix}$ and $H = \begin{bmatrix} \Delta M^{-1} & Z_X(\Delta I_H^X) \\ G_P(\Delta U_H^P) & \Delta M \end{bmatrix}$; $F^I = \begin{bmatrix} E & E & 0 & 0 & -E & -E & 0 & 0 \\ 0 & 0 & E & E & 0 & 0 & -E & -E \end{bmatrix} \cdot \begin{bmatrix} E_H^H \\ E_X^H \\ J_H^H \\ J_P^H \\ E_H^Z \\ E_H^N \\ J_H^Z \\ J_H^N \end{bmatrix}$; $Z_X(\Delta I_H^X)$ and $G_P(\Delta U_H^P)$ - diagonal matrices of equivalent resistances and conductivities; ΔM^{-1} and ΔM are diagonal matrices of equivalent inverse and forward memresistivities; E is the identity matrix; $E_H^H, E_X^H, J_H^H, J_P^H, E_H^Z, E_H^N, J_H^Z$ and J_H^N - equivalent sources of EMF and current on nonlinear chords and edges; $G_1 = \begin{bmatrix} B_{XI} & \hat{R} \\ \hat{Y} & D_{XI} \end{bmatrix}^{-1} \cdot \begin{bmatrix} -B_{IX} & 0 \\ 0 & -D_{IX} \end{bmatrix}$, $G_2 = \begin{bmatrix} B_{XI} & \hat{R} \\ \hat{Y} & D_{XI} \end{bmatrix}^{-1} \cdot \begin{bmatrix} -B_X & 0 \\ 0 & -D_X \end{bmatrix}$, $G_3 = \begin{bmatrix} B_{XI} & \hat{R} \\ \hat{Y} & D_{XI} \end{bmatrix}^{-1} \cdot \begin{bmatrix} -B_{XII} & 0 \\ 0 & -D_{XII} \end{bmatrix}$ and $W = \begin{bmatrix} B_{XI} & \hat{R} \\ \hat{Y} & D_{XI} \end{bmatrix}^{-1}$; $S_I = \begin{bmatrix} -B_{XVI} & \widehat{M} \\ \widehat{M}^{-1} & -D_{XVI} \end{bmatrix}^{-1} \cdot \begin{bmatrix} -B_{XIII} & 0 \\ 0 & -D_{XIII} \end{bmatrix}$, $S_2 = \begin{bmatrix} -B_{XVI} & \widehat{M} \\ \widehat{M}^{-1} & -D_{XVI} \end{bmatrix}^{-1} \cdot \begin{bmatrix} -B_{XIV} & 0 \\ 0 & -D_{XIV} \end{bmatrix}$, $S_3 = \begin{bmatrix} -B_{XVI} & \widehat{M} \\ \widehat{M}^{-1} & -D_{XVI} \end{bmatrix}^{-1} \cdot \begin{bmatrix} -B_{XV} & 0 \\ 0 & -D_{XV} \end{bmatrix}$ and $K = \begin{bmatrix} -B_{XVI} & \widehat{M} \\ \widehat{M}^{-1} & -D_{XVI} \end{bmatrix}^{-1}$; $\widehat{M} = M + \Delta M$ is the matrix of the branch memresistivities, $\widehat{M}^{-1} = M^{-1} + \Delta M^{-1}$ is the matrix of inverse memresistivities of the branch; E_N^M, J_N^M - equivalent vectors of independent voltage and current sources on memresistive branches.

Mathematical model of a multipole with memresistive branches in the "small" signal mode. The calculation of a multipole in this mode for alternating current is carried out in a certain vicinity of the nominal operating mode of the corresponding electronic devices. The same approach may well be applied to nanoelectronic components of quantum information and computing systems. In this regard, the calculation uses local linear high-frequency models of nanoelectronic devices included in their composition. Replacing the reactive elements of the equivalent circuit with complex resistances or conductivities allows us to consider the obtained equivalent circuits as linear resistive multipoles with dependent and independent sources of current and voltage.

Based on the above, the mathematical model of a multipole with memresistive branches (1) in the "small" signal mode will take the form:

$$\begin{cases} \begin{bmatrix} \Delta U_R^P \\ \Delta I_R^X \end{bmatrix} = Q^T \left(G_3 K \begin{bmatrix} E_N^M \\ J_N^M \end{bmatrix} + W \begin{bmatrix} E_N^R \\ J_N^R \end{bmatrix} \right) \\ \begin{bmatrix} \Delta U_M^P \\ \Delta I_M^X \end{bmatrix} = N^T \left(S_3 \begin{bmatrix} E_N^R \\ J_N^R \end{bmatrix} + K \begin{bmatrix} E_N^M \\ J_N^M \end{bmatrix} \right) \end{cases} \quad (2)$$

where $Q = G_3 S_3 - O$; $N = G_3 S_3 - H$; O and H are unit matrices.

Expression (2) belongs to the class of interval systems of linear algebraic equations, for the solution of which it is permissible to use Kahan's interval arithmetic and the Gauss-Seidel interval method.

The essence of Kahan's interval arithmetic is that operations with intervals containing zero have the same result as in the case of other intervals and allows you to keep the interval expansion of functions unchanged, and also guarantees, under certain conditions, not only the distributiveness of operations, but also inclusion monotony [20; 22].

The basis of the Gauss-Seidel method is an algorithm of sets of iterative procedures to refine the set of solutions. Usually this method is applied after preliminary preconditioning of an interval system of algebraic equations [6; 23].

3. Results

Computational algorithm for external estimation of the solution sets of the mathematical model of a multipole with memresistive branches in the "small" signal mode. The method involves solving a system of equations of the form $Ax = b$. Reducing the first equation of system (2) to this form, we obtain: $A = Q; b = G_3 K \begin{bmatrix} E_N^M \\ J_N^M \end{bmatrix} + W \begin{bmatrix} E_N^R \\ J_N^R \end{bmatrix}$ and $x = \begin{bmatrix} \Delta U_R^P \\ \Delta I_R^X \end{bmatrix}$. Similarly, for the second equation: $A = N; b = S_3 \begin{bmatrix} E_N^R \\ J_N^R \end{bmatrix} + K \begin{bmatrix} E_N^M \\ J_N^M \end{bmatrix}$ and $y = \begin{bmatrix} \Delta U_M^P \\ \Delta I_M^X \end{bmatrix}$. Choose the bars $x = \begin{bmatrix} \Delta U_{R1}^P, \dots, \Delta U_{Rn}^P \\ \Delta I_{R1}^X, \dots, \Delta I_{Rn}^X \end{bmatrix}^T$ and $= \begin{bmatrix} \Delta U_{M1}^P, \dots, \Delta U_{Mn}^P \\ \Delta I_{M1}^X, \dots, \Delta I_{Mn}^X \end{bmatrix}^T$ for the first and second equations of system (2) respectively.

Using the above new assigned variables, the computational algorithm can be represented as follows:

- Reduction of system (2) to the form $Ax = b$, as shown above;
- Assign the beams $x = (x_1, \dots, x_n)^T = \begin{bmatrix} \Delta U_{R1}^P, \dots, \Delta U_{Rn}^P \\ \Delta I_{R1}^X, \dots, \Delta I_{Rn}^X \end{bmatrix}^T$ and $= (y_1, \dots, y_n)^T = \begin{bmatrix} \Delta U_{M1}^P, \dots, \Delta U_{Mn}^P \\ \Delta I_{M1}^X, \dots, \Delta I_{Mn}^X \end{bmatrix}^T$ for two equations of system (2), moreover, $x, y \in \mathbb{IR}^n$, limiting the desired parts of the combined set of solutions $\Xi_x(A, b)$ and $\Xi_y(A, b)$;
- Let us introduce some constant $\varepsilon > 0$, which determines the practically significant number of possible iterations.
- $d \leftarrow +\infty$;
- $do\ while\ (d \geq \varepsilon)$:
  - $do\ for\ i = 1\ to\ n$;
  - $\tilde{x}_i \leftarrow x_i \cap \frac{\left(b_i - \sum_{j=1}^{i-1} a_{ij} x_i - \sum_{j=i+1}^{n} a_{ij} x_i\right)}{a_{ii}}$, where a_{ij} - element of matrix A;
  - $if\ \tilde{x}_i = \varnothing\ then\ stop$, signaling "the set of solutions $\Xi_x(A, b)$ does not intersect the bar x";
  - $end\ if$;
  - $end\ do$;
- $d \leftarrow$ distance between vectors x and $\tilde{x} = (x_1, \dots, x_n)^T$;
- $x \leftarrow \tilde{x}$;
- $end\ do$.

Similarly, the algorithm for solving the second equation of system (2) is implemented for the set of solutions given by the interval bar vector $y = \begin{bmatrix} \Delta U_{M1}^P, \dots, \Delta U_{Mn}^P \\ \Delta I_{M1}^X, \dots, \Delta I_{Mn}^X \end{bmatrix}^T$.

4. Discussion

The "small signal" mode considered in this article for an electric multipole with memresistive branches confirms the applicability of the chosen approach to the analysis of nanoelectronic circuits of both memory elements and the architecture of information-measuring systems as a whole.

5. Conclusion

The form of presentation of the mathematical model for each mode of operation is quite consistent with the forms required for using the methods of interval analysis. Of course, it is necessary to check the adequacy of the presented model of an electric multipole with memresistive branches both in general form and in dynamic and static modes for real nanoelectronic circuits, which is the subject of further research.

References

[1] Bondarev A and Efanov V 2019 The Principles of Forming of the Mathematical Model of Nanoelectronic Components of Quantum Computer Systems with Memresistance Branches. *Atlantis Highlights in Computer Sciences* **3** 17-22

[2] Bondarev A V 2015 Research problem of a robustness of electronic schemes by methods of interval calculations in the conditions of uncertainty. *Materials of the CSIT'2015 Proceedings of the 17th International Workshop on Computer Science and Information Technologies* 145-9

[3] Bondarev A V, Muravyova E A, Kadyrov R R and Rahman P A 2016 The analysis of opportunities of construction and use of avionic systems based on COTS-modules. *ARPN Journal of Engineering and Applied Sciences* **11 1** 78-92

[4] Borghetti J, Snider G S, Kuekes P J et al. 2010 Memristive switches enable stateful logic operations via material implication. *Nature letters* **464** 873-6

[5] *Bourzac K Memristor Memory Readied for Production.* Retrieved from: www.technologyreview.com/computing/25018/

[6] Chua L O 1971 Memristor – the missing circuit element. *IEEE Trans. Circuit Theory* **18** 507-19

[7] Alefeld G and Mayer G 2000 Interval analysis: theory and applications. *Journal of Computational Applied Mathematics* **121** 421-64

[8] *Memristor: "the missing element".* Retrieved from: http://www.nanonewsnet.ru/articles/2011/memristor-nedostayushchii-element

[9] *Strogy V V Application of memristors in pc architecture.* Retrieved from: https://eduherald.ru/ru/article/view?id=17265

[10] *Johnson R C End of the CPU? HP demos configurable memristor.* Retrieved from: www.eetimes.com/electronicsnews/4088557/End-of-the-CPU-HP-demos-configurablememristor

[11] Kuekes P J, Snider G S and Williams R S 2005 Crossbar nanocomputers. *Scientific American*, **293** 72-8

[12] *Markoff J HP Sees a Revolution in Memory Chip.* Retrieved from: www.nytimes.com/2010/04/08/science/08chips.html?_r=1

[13] *Memristor*. Retrieved from: en.wikipedia.org/wiki/Memristor

[14] *Merritt R HP researcher predicts memory-centric processors.* Retrieved from: www.eetimes.com/electronicsnews/4199856/HP-researcher-predicts-memory-centricprocessors

[15] Strukov D B, Snider G S, Stewart D R and Williams R S 2008 The missing memristor found. *Nature letters* **453** 80-3

[16] Bondarev A V 2019 A review of quantum computing algorithms. *Prospects for Science* **7(118)** 27-31

[17] Bondarev A V 2019 Review of the element base of quantum computers. *XXI century: results of the past and problems of the present plus* **8 3(47)** 96-100

[18] Bondarev A V 2019 Features of building the architecture of quantum computers. *Modern science: Series Natural and technical sciences* **6** 52-5

[19] Bondarev A and Efanov V 2020 The Principles of Forming of the Mathematical Model of Nanoelectronic Components of Quantum Computer Systems with Memresistance Branches. *Control systems and information technologies* 1(79) 4-10

[20] Bondarev A V 2011 *Decision support system for assessing the robustness of complex airborne electronic systems based on COTS products: dissertation for the degree of candidate of technical sciences* (Ufa: Ufa State Aviation Technical University) 177

[21] Sigorskiy V P and A I Petrenko 1976 *Algorithms for analyzing electronic circuits* (Moscow:

IOP Conf. Series: Materials Science and Engineering **976** (2020) 012011 doi:10.1088/1757-899X/976/1/012011

Soviet radio) 608
[22] Chua L O and Lin Pen-Min 1980 *Machine analysis of electronic circuits. Algorithms and computational methods* (Moscow: Energy) 640
[23] Shary S P *Finite-dimensional interval analysis*. Retrieved from: http://www.nsc.ru/interval

IOP Conf. Series: Materials Science and Engineering **976** (2020) 012012 doi:10.1088/1757-899X/976/1/012012

Static mode of the mathematical model of an electric multipole with memresistive branches in conditions of interval uncertainty

A V Bondarev[1] and V N Efanov[2]

[1]FSBEI HE "Orenburg State University", Kumertau Branch OSU, 3b, 2 Soviet lane, Kumertau, 453300, Russia
[2]FSBEI HE "Ufa State Aviation Technical University", 12, C. Marx str., Ufa, 450000, Russia

E-mail: bondarevav@kfosu.edu.ru

Abstract. This article discusses a method for obtaining a mathematical model of an electric multipole with memresistive branches in finite parameter increments for a large signal mode when calculating a direct current based on a generalized model obtained earlier. A computational algorithm for calculating nanoelectronic circuits based on methods of interval arithmetic is also given.

1. Introduction

Recently, a wide range of nanoelectronic components have emerged that are used for the design of information and computing systems. This relatively new direction in the development of computer technology still has many unsolvable problems of practical use. One of these problems was the use of such elements in conditions other than laboratory ones. In this case, interval uncertainties arise when determining the electrical parameters of these devices, which is superimposed on the uncertainty of the entire system as a whole and complicates the circuit analysis, as well as the entire design process as a whole.

Various authors, of course, have long studied the issues of improving the operational characteristics of computer technology in the design of the basic elements of digital technology. Now there is quite a lot of both new research and academic works of the 20th century on this score. However, a new element base for nanoelectronic technology requires new studies of the robustness of the electromagnetic parameters of the components used.

One of the newest elements of nanoelectronics that appeared not so long ago is the memristor, a passive element in microelectronics capable of changing its resistance depending on the charge flowing through it. For a long time, the memristor was considered a theoretical model [1-7], which could not be realized in practice until the first sample of an element demonstrating the properties of a memristor was created in 2008 by a team of scientists led by R. S. Williams at the Hewlett-Packard research laboratory. The device does not accumulate charge like a capacitor, does not support magnetic flux like an inductor. The change in the properties of the device is provided by chemical reactions in a thin two-layer film of titanium dioxide (5 nm). One layer of the device's film is slightly oxygen-depleted and oxygen vacancies migrate between the layers as the voltage changes. This implementation of the memristor belongs to the

IOP Conf. Series: Materials Science and Engineering **976** (2020) 012012 doi:10.1088/1757-899X/976/1/012012

class of nanoionic devices. The observed phenomenon of hysteresis in the memristor makes it possible to use it, among other things, as a memory cell [8-16].

The well-known properties of memristors make it possible to create computers of a fundamentally new architecture, significantly exceeding semiconductor ones in performance. The projected memristor information-computing systems assume the parallel and independent operation of many modules, and the ability to memorize and operate an unlimited set of values from 0 to 1 means that the executable programs are not limited to binary code [5, 8, 17-20].

In connection with the above, it is necessary to take a fresh look at the mathematical model of an electric multipole using a memristor as one of the elements of thc basic set. The introduction of such an addition to the basic set of elements also determines the appearance of new memresistive branches of the electric multipole.

According to the well-known classification [22], it is customary to distinguish the following typical operating modes of the investigated electronic devices, which is quite true for the elements of nanoelectronics:

- Quasi-linear mode of a small signal when calculating with an alternating current;
- Quasi-linear small signal mode when calculating in the time domain;
- Large signal mode when calculating for direct current (static mode);
- Large signal mode when calculating in the time domain (dynamic mode).

Depending on the selected mode, the initial model of the investigated device will be an electrical multipole of one of the following types: linear resistive, linear reactive, nonlinear resistive and nonlinear reactive. At the same time, the general methodological basis for studying all the listed types of models is their description in the basis of finite deviations of currents and voltages of the branches of the corresponding multipole, since the main goal of our study is to assess the ability of a device to maintain its characteristics within specified limits in the presence of disturbing influences with undefined properties. This article analyzes the third large signal mode in the specified classification when calculating by direct current (static mode) for an electric multipole with memresistive branches in finite increments.

2. Materials and methods

Mathematical models of an electric multipole with memresistive branches in the nominal form and using finite increments were built earlier [1-3] using the method of decomposition of branches of the directed graph of the circuit described in [16; 19-22].

A full-size mathematical model in a full hybrid basis in finite increments of currents and voltages will look like this [19]:

$$\begin{cases} \begin{bmatrix} 0 & \hat{L} \\ \hat{C} & 0 \end{bmatrix} \cdot \frac{d}{dt} \begin{bmatrix} \Delta U_C^P \\ \Delta I_L^X \end{bmatrix} = Q_1 \begin{bmatrix} \Delta U_C^P \\ \Delta I_L^X \end{bmatrix} + Q_2 \begin{bmatrix} \Delta U_H^P \\ \Delta I_H^X \end{bmatrix} + Q_3 \begin{bmatrix} E_N^R \\ J_N^R \end{bmatrix} + Q_4 \begin{bmatrix} \Delta U_M^P \\ \Delta I_M^X \end{bmatrix} - \begin{bmatrix} E_E^L \\ J_E^C \end{bmatrix} \\ \begin{bmatrix} \Delta U_H^P \\ \Delta I_H^X \end{bmatrix} = A_1 \begin{bmatrix} \Delta U_C^P \\ \Delta I_L^X \end{bmatrix} + A_2 \begin{bmatrix} \Delta U_R^P \\ \Delta I_R^X \end{bmatrix} + A_3 \begin{bmatrix} \Delta U_M^P \\ \Delta I_M^X \end{bmatrix} - H \cdot F^I \\ \begin{bmatrix} \Delta U_R^P \\ \Delta I_R^X \end{bmatrix} = G_1 \begin{bmatrix} \Delta U_C^P \\ \Delta I_L^X \end{bmatrix} + G_2 \begin{bmatrix} \Delta U_H^P \\ \Delta I_H^X \end{bmatrix} + G_3 \begin{bmatrix} \Delta U_M^P \\ \Delta I_M^X \end{bmatrix} - W \begin{bmatrix} E_N^R \\ J_N^R \end{bmatrix} \\ \begin{bmatrix} \Delta U_M^P \\ \Delta I_M^X \end{bmatrix} = S_I \begin{bmatrix} \Delta U_C^P \\ \Delta I_L^X \end{bmatrix} + S_2 \begin{bmatrix} \Delta U_H^P \\ \Delta I_H^X \end{bmatrix} + S_3 \begin{bmatrix} \Delta U_R^P \\ \Delta I_R^X \end{bmatrix} - K \begin{bmatrix} E_N^M \\ J_N^M \end{bmatrix} \end{cases} \tag{1}$$

where $\hat{L} = \begin{bmatrix} diag\{L_k + \Delta L_k\} & 0 \\ 0 & diag\{L_i^N + L_i^N(\Delta i)\} \end{bmatrix}$ is the matrix of inductances ($k = \overline{1, n_{\text{Л}L}^X}$, $l = \overline{1, n_{HL}^X}$, $n_{\text{Л}L}^X + n_{HL}^X = dim\Delta U_L^X$); $\hat{C} = \begin{bmatrix} diag\{C_k + \Delta C_k\} & 0 \\ 0 & diag\{C_i^N + C_i^N(\Delta i)\} \end{bmatrix}$ is the capacity matrix

$(k=\overline{1,n_{ЛC}^X},\ l=\overline{1,n_{HC}^X},\ n_{ЛC}^X+n_{HC}^X=dim\Delta I_C^P)$; ΔU_L^X, ΔI_L^X - increments of voltages and currents on inductive chords; ΔU_C^P, ΔI_C^P - voltage and current increments on the capacitive edges; ΔU_R^X, ΔU_R^P, ΔI_R^P, ΔI_R^X - voltage and current increments on resistive edges and chords; ΔU_H^X, ΔU_H^P, ΔI_H^P, ΔI_H^X - voltage and current increments on nonlinear edges and chords; ΔU_M^X, ΔU_M^P, ΔI_M^X, ΔI_M^P - voltage and current increments on memresistive edges and chords; $Q_1=\begin{bmatrix}-B_I & Z_E\\ V_E & -D_I\end{bmatrix}+\begin{bmatrix}-B_{IV} & 0\\ 0 & -D_{IV}\end{bmatrix}\cdot G_1$, $Q_2=\begin{bmatrix}-B_{II} & 0\\ 0 & -D_{II}\end{bmatrix}+\begin{bmatrix}-B_{IV} & 0\\ 0 & -D_{IV}\end{bmatrix}\cdot G_2$, $Q_3=\begin{bmatrix}-B_{II} & 0\\ 0 & -D_{II}\end{bmatrix}\cdot W$ and $Q_4=\begin{bmatrix}-B_{III} & 0\\ 0 & -D_{III}\end{bmatrix}+\begin{bmatrix}-B_{IV} & 0\\ 0 & -D_{IV}\end{bmatrix}$; $Z_E=\left[[0]_{n_L^X\times n_{ЛL}^X}\quad diag\{(Z_E^L+Z_E^L(\Delta i))_l\}\right]$ - matrix of equivalent resistances and $V_E=\left[[0]_{n_C^P\times n_{ЛC}^P}\quad diag\left\{\left(G_N^C+G_N^C(\Delta u)\right)_l\right\}\right]$ - equivalent conductivities of nonlinear inductances and capacities; D_i is the matrix of the main sections of the directed graph of the circuit; B_i - matrix of the main contours of the directed graph of the circuit; n is the number of branches and k is the number of nodes; $E_E^L=\left[[(E_N)_k]_{n_{ЛL}^X\times 1}\quad [(E_N^L+E_Z^L(\Delta i))_k]_{n_{HL}^X\times 1}\right]^T$, $J_E^C=\left[[(J_N)_k]_{n_{ЛC}^P\times 1}\quad \left[\left(J_N^C+J_Z^C(\Delta u)\right)_k\right]_{n_{HC}^P\times 1}\right]^T$ - equivalent vectors of current and voltage sources; where $A_1=\begin{bmatrix}\Delta M^{-1} & Z_X(\Delta I_H^X)\\ G_P(\Delta U_H^P) & \Delta M\end{bmatrix}\cdot\begin{bmatrix}-B_V & 0\\ 0 & -D_V\end{bmatrix}$, $A_2=\begin{bmatrix}\Delta M^{-1} & Z_X(\Delta I_H^X)\\ G_P(\Delta U_H^P) & \Delta M\end{bmatrix}\cdot\begin{bmatrix}-B_{VII} & 0\\ 0 & -D_{VII}\end{bmatrix}$, $A_3=\begin{bmatrix}-B_{VIII} & 0\\ 0 & -D_{VIII}\end{bmatrix}$ and $H=\begin{bmatrix}\Delta M^{-1} & Z_X(\Delta I_H^X)\\ G_P(\Delta U_H^P) & \Delta M\end{bmatrix}$; $F^I=\begin{bmatrix}E & E & 0 & 0 & -E & -E & 0 & 0\\ 0 & 0 & E & E & 0 & 0 & -E & -E\end{bmatrix}\cdot\begin{bmatrix}E_H^H\\ E_X^H\\ J_H^H\\ J_P^H\\ E_H^Z\\ E_H^N\\ J_H^Z\\ J_H^N\end{bmatrix}$; $Z_X(\Delta I_H^X)$ and $G_P(\Delta U_H^P)$ - diagonal matrices of equivalent resistances and conductivities; ΔM^{-1} and ΔM are diagonal matrices of equivalent inverse and forward memresistivities; E is the identity matrix; E_H^H, E_X^H, J_H^H, J_P^H, E_H^Z, E_H^N, J_H^Z and J_H^N - equivalent sources of EMF and current on nonlinear chords and edges; $G_1=\begin{bmatrix}B_{XI} & \hat{R}\\ \hat{Y} & D_{XI}\end{bmatrix}^{-1}\cdot\begin{bmatrix}-B_{IX} & 0\\ 0 & -D_{IX}\end{bmatrix}$, $G_2=\begin{bmatrix}B_{XI} & \hat{R}\\ \hat{Y} & D_{XI}\end{bmatrix}^{-1}\cdot\begin{bmatrix}-B_X & 0\\ 0 & -D_X\end{bmatrix}$, $G_3=\begin{bmatrix}B_{XI} & \hat{R}\\ \hat{Y} & D_{XI}\end{bmatrix}^{-1}\cdot\begin{bmatrix}-B_{XII} & 0\\ 0 & -D_{XII}\end{bmatrix}$ and $W=\begin{bmatrix}B_{XI} & \hat{R}\\ \hat{Y} & D_{XI}\end{bmatrix}^{-1}$; $S_I=\begin{bmatrix}-B_{XVI} & \hat{M}\\ \hat{M}^{-1} & -D_{XVI}\end{bmatrix}^{-1}\cdot\begin{bmatrix}-B_{XIII} & 0\\ 0 & -D_{XIII}\end{bmatrix}$, $S_2=\begin{bmatrix}-B_{XVI} & \hat{M}\\ \hat{M}^{-1} & -D_{XVI}\end{bmatrix}^{-1}\cdot\begin{bmatrix}-B_{XIV} & 0\\ 0 & -D_{XIV}\end{bmatrix}$, $S_3=\begin{bmatrix}-B_{XVI} & \hat{M}\\ \hat{M}^{-1} & -D_{XVI}\end{bmatrix}^{-1}\cdot\begin{bmatrix}-B_{XV} & 0\\ 0 & -D_{XV}\end{bmatrix}$ and $K=\begin{bmatrix}-B_{XVI} & \hat{M}\\ \hat{M}^{-1} & -D_{XVI}\end{bmatrix}^{-1}$; $\hat{M}=M+\Delta M$ is the matrix of the branch memresistivities, $\hat{M}^{-1}=M^{-1}+\Delta M^{-1}$ is the matrix of inverse memresistivities of the branch; E_N^M, J_N^M - equivalent vectors of independent voltage and current sources on memresistive branches.

Mathematical model of a multipole with memresistive branches in a static mode. Multipole calculation in this mode, DC calculation is carried out to determine the parameters of the operating point of nonlinear electronic devices, as well as static corresponding electronic devices, for example, DC transmission ratios, input, and output resistances, etc.

The equivalent circuit of the investigated device in this mode does not contain reactive elements - capacitors or inductors, therefore it is a nonlinear resistive-memresistive multipole. When compiling mathematical models in a reduced hybrid basis, we will assume that the multipole topology satisfies the following two constraints:

- Non-monotonic voltage-controlled resistances do not form a loop;
- Non-monotonic current-controlled memresistivities do not form cross sections.

Non-monotonic voltage-controlled resistances and non-monotonic current-controlled resistances belong to special branches of an electrical multipole, with the first of the mentioned resistances being assigned to the edges, and the second to chords. To ensure the formulated requirement, the numbering of the multipole branches starts with nonmonotonic voltage-controlled resistances and ends with nonmonotonic current-controlled memresistivities.

Earlier in [21-22] it was shown that when composing equivalent circuits in nominal and perturbed operating modes, the circuit topology does not change. Consequently, using the already known expressions of the generalized model of an electric multipole, we will immediately write down expressions for currents and voltages of the branches in finite increments.

Considering the remarks made, the vectors of currents and voltages increments for all branches of the investigated electric multipole are divided into the following characteristic groups: nonlinear edges, nonlinear chords, linear edges, linear chords. Based on the above, the mathematical model of a multipole with memresistive branches (1) in a static mode will take the form:

$$\begin{cases} \begin{bmatrix} \Delta U_H^P \\ \Delta I_H^X \end{bmatrix} = A_2 \begin{bmatrix} \Delta U_R^P \\ \Delta I_R^X \end{bmatrix} + A_3 \begin{bmatrix} \Delta U_M^P \\ \Delta I_M^X \end{bmatrix} - H \cdot F^I \\ \begin{bmatrix} \Delta U_R^P \\ \Delta I_R^X \end{bmatrix} = G_2 \begin{bmatrix} \Delta U_H^P \\ \Delta I_H^X \end{bmatrix} + G_3 \begin{bmatrix} \Delta U_M^P \\ \Delta I_M^X \end{bmatrix} - W \begin{bmatrix} E_N^R \\ J_N^R \end{bmatrix} \\ \begin{bmatrix} \Delta U_M^P \\ \Delta I_M^X \end{bmatrix} = S_2 \begin{bmatrix} \Delta U_H^P \\ \Delta I_H^X \end{bmatrix} + S_3 \begin{bmatrix} \Delta U_R^P \\ \Delta I_R^X \end{bmatrix} - K \begin{bmatrix} E_N^M \\ J_N^M \end{bmatrix} \end{cases} \quad (2)$$

Expression (2) belongs to the class of interval systems of linear algebraic equations, for the solution of which it is permissible to use Kahan's interval arithmetic and the interval methods of Newton and Krawczyk.

The essence of Kahan's interval arithmetic is that operations with intervals containing zero have the same result as in the case of other intervals and allows you to keep the interval expansion of functions unchanged, and also guarantees, under certain conditions, not only the distributiveness of operations, but also monotonicity with respect to inclusion [20-23].

3. Results

Computational algorithm for external estimation of the solution sets of the mathematical model of a multipole with memresistive branches in a static mode. To solve such interval systems of nonlinear algebraic equations, one can use Newton's method [23]. For this, we bring system (2) to the form:

$$f_p(x, y) = 0 \quad (3)$$

Let us denote $IGA\ (A, b)$ - the result of applying the Gaussian method to the interval system $Ax = b$. Then for the first equation of system (2), we get:

$$A = A_2;\ x = \begin{bmatrix} \Delta U_R^P \\ \Delta I_R^X \end{bmatrix};\ b = A_3 \begin{bmatrix} \Delta U_M^P \\ \Delta I_M^X \end{bmatrix} - \begin{bmatrix} \Delta U_H^P \\ \Delta I_H^X \end{bmatrix} - H \cdot F^I \quad (4)$$

$$(\Delta U)_{k+1} = \aleph((\Delta U)_k) \cap (\Delta U)_k \quad (5)$$

where $\aleph(\Delta U)$ is the Newton operator of the following form:

$$\aleph(\Delta U) = mid(\Delta U) - IGA\{F'(\Delta U), f(mid(\Delta U))\} \tag{6}$$

On the direct course of calculations, we find the Jacobi matrix:

$$\hat{A} = \begin{bmatrix} \frac{\partial F_1^{(1)}(\Delta U)}{\partial U_H^P} & \frac{\partial F_1^{(1)}(\Delta U)}{\partial I_H^X} \\ \frac{\partial F_2^{(1)}(\Delta U)}{\partial U_H^X} & \frac{\partial F_2^{(1)}(\Delta U)}{\partial I_H^P} \end{bmatrix} \tag{7}$$

We calculate:

$$r_{ij} \leftarrow \frac{a_{ij}}{a_{jj}} \tag{8}$$

For all $j = 1$ to $n-1$ and for all $i = j + 1$ to n. Here a are the elements of the Jacobi matrix.
On the reverse, using the elements of the matrix A and the vector b according to (4), we find that:

$$y_2 = \frac{b_2^{**}}{a_2^{**}} \tag{9}$$

Then

$$y_1 = b_1^{**} - a_{21}^{**} y_2^{**} \tag{10}$$

Newton's interval method has a drawback, the essence of which is that even if the matrix of the natural interval expansion of the function is non-singular, in the general case it can be calculated $IGA\{F'(\Delta I), f(mid(\Delta I))\}$ only if the width $mid(\Delta I)$ is small enough.

This drawback is eliminated in the method for solving systems of nonlinear interval equations, called the Krawczyk method, by introducing the mapping of the interval function (the Krawczyk operator) [24].

The solution of system (3) by the Krawczyk method in the general case will have the following form:

$$\Delta U_{K+1} = \Re(\Delta U_K) \cap \Delta U_K \tag{11}$$

where $\Re(\Delta I)$ is the Krawczyk operator of the following form:

$$\Re(\Delta U_K) = mid(\Delta U_K) - C \cdot f(mid(\Delta U_K)) + (E - C \cdot f'(\Delta U_K)) \cdot (\Delta U_K - mid(\Delta U_K)) \tag{12}$$

where C is a non-singular real matrix of weight coefficients, the elements of which are selected empirically in such a way as to improve the convergence of the method; f is function (3).

For a fixed matrix C, the Krawczyk method is defined by the expression:

$$\Delta U^{K+1} = \Re(\Delta U^K) \cap \Delta U^K \tag{13}$$

The algorithm for solving the second and third equations of system (2) is implemented in a similar way.

4. Discussion

The static regime considered in this article for an electric multipole with memresistive branches confirms the applicability of the chosen approach to the analysis of nanoelectronic circuits of both memory elements and the architecture of information-measuring systems.

5. Conclusion

The form of presentation of the mathematical model for each mode of operation is quite consistent with the forms required for using the methods of interval analysis. Of course, it is necessary to check the adequacy of the presented model of an electric multipole with memresistive branches both in general form and in dynamic and static modes for real nanoelectronic circuits, which is the subject of further research.

References

[1] Bondarev A and Efanov V 2019 The Principles of Forming of the Mathematical Model of Nanoelectronic Components of Quantum Computer Systems with Memresistance Branches. *Atlantis Highlights in Computer Sciences* **3** 17-22
[2] Bondarev A V 2015 Research problem of a robustness of electronic schemes by methods of interval calculations in the conditions of uncertainty. *Materials of the CSIT'2015 Proceedings of the 17th International Workshop on Computer Science and Information Technologies* 145-9
[3] Bondarev A V, Muravyova E A, Kadyrov R R and Rahman P A 2016 The analysis of opportunities of construction and use of avionic systems based on COTS-modules. *ARPN Journal of Engineering and Applied Sciences* **11 1** 78-92
[4] Borghetti J, Snider G S, Kuekes P J et al. 2010 Memristive switches enable stateful logic operations via material implication. *Nature letters* **464** 873-6
[5] *Bourzac K Memristor Memory Readied for Production.* Retrieved from: www.technologyreview.com/computing/25018/
[6] Chua L O 1971 Memristor – the missing circuit element. *IEEE Trans. Circuit Theory* **18** 507-19
[7] Alefeld G and Mayer G 2000 Interval analysis: theory and applications. *Journal of Computational Applied Mathematics* **121** 421-64
[8] *Memristor: "the missing element".* Retrieved from: http://www.nanonewsnet.ru/articles/2011/memristor-nedostayushchii-element
[9] *Strogy V V Application of memristors in pc architecture.* Retrieved from: https://eduherald.ru/ru/article/view?id=17265
[10] *Johnson R C End of the CPU? HP demos configurable memristor.* Retrieved from: www.eetimes.com/electronicsnews/4088557/End-of-the-CPU-HP-demos-configurablememristor
[11] Kuekes P J, Snider G S and Williams R S 2005 Crossbar nanocomputers. *Scientific American*, **293** 72-8
[12] *Markoff J HP Sees a Revolution in Memory Chip.* Retrieved from: www.nytimes.com/2010/04/08/science/08chips.html?_r=1
[13] *Memristor*. Retrieved from: en.wikipedia.org/wiki/Memristor
[14] *Merritt R HP researcher predicts memory-centric processors*. Retrieved from: www.eetimes.com/electronicsnews/4199856/HP-researcher-predicts-memory-centricprocessors
[15] Strukov D B, Snider G S, Stewart D R and Williams R S 2008 The missing memristor found. *Nature letters* **453** 80-3
[16] Sigorskiy V P and A I Petrenko 1976 *Algorithms for analyzing electronic circuits* (Moscow: Soviet radio) 608
[17] Bondarev A V 2019 A review of quantum computing algorithms. *Prospects for Science* **7(118)** 27-31
[18] Bondarev A V 2019 Review of the element base of quantum computers. *XXI century: results of the past and problems of the present plus* **8 3(47)** 96-100
[19] Bondarev A V 2019 Features of building the architecture of quantum computers. *Modern science: Series Natural and technical sciences* **6** 52-5
[20] Bondarev A and Efanov V 2020 The Principles of Forming of the Mathematical Model of Nanoelectronic Components of Quantum Computer Systems with Memresistance Branches. *Control systems and information technologies* 1(79) 4-10
[21] Bondarev A V 2011 *Decision support system for assessing the robustness of complex airborne*

electronic systems based on COTS products: dissertation for the degree of candidate of technical sciences (Ufa: Ufa State Aviation Technical University) 177

[22] Chua L O and Lin Pen-Min 1980 *Machine analysis of electronic circuits. Algorithms and computational methods* (Moscow: Energy) 640

[23] Alefeld G and Mayer G *Interval analysis: theory and applications* Retrieved from: http://www.nsc.ru/interval/Introduction/ISurveyRus.pdf

Dynamic mode of the mathematical model of an electric multipole with memresistive branches in conditions of interval uncertainty

A V Bondarev[1] and V N Efanov[2]

[1]FSBEI HE "Orenburg State University", Kumertau Branch OSU, 3b, 2 Soviet lane, Kumertau, 453300, Russia
[2]FSBEI HE "Ufa State Aviation Technical University", 12, C. Marx str., Ufa, 450000, Russia

E-mail: bondarevav@kfosu.edu.ru

Abstract. This article discusses a method for obtaining a mathematical model of an electric multipole with memresistive branches in finite parameter increments for a large signal mode in time domain calculation when calculating a direct current based on a generalized model obtained earlier. A computational algorithm for calculating nanoelectronic circuits based on methods of interval arithmetic is also given.

1. Introduction

Information and computing technologies are used in almost all spheres of human activity. The modern development of computer technology has led to the emergence of a whole spectrum of new nanoelectronic components that greatly increase the computing power. However, the tasks of increasing productivity and reducing heat dissipation of modern computing devices without increasing cost remain urgent.

New nanoelectronic components, which have recently been used in the design of information and computing systems, require research aimed at improving the operational characteristics of computing technology.

One of the newest components of this kind is the memristor - a passive element in microelectronics capable of changing its resistance depending on the charge flowing through it. For a long time, the memristor was considered a theoretical model [1-7], which could not be realized in practice until the first sample of an element demonstrating the properties of a memristor was created in 2008 by a team of scientists led by R. S. Williams at the Hewlett-Packard research laboratory. The observed phenomenon of hysteresis in the memristor makes it possible to use it, among other things, as a memory cell [8-16].

Modern computers are built on the basis of the von Neumann architecture: both data and programs are stored in the memory of the machine in binary code, with the computational module separated from the storage devices, and the programs are executed sequentially, one after the other. The already studied properties of memristors allow us to say that on their basis it is possible to create computers of a fundamentally new architecture, significantly exceeding semiconductor ones in performance. The use of memristors in the near future will make, in principle, unnecessary individual hardware components of a computer - processors, video chips, memory and hard drives; the machine will be an architecturally homogeneous device where all data will be simultaneously stored and all operations performed with them. For an upgrade, it will be enough to install additional memristor modules, and for repairs, replace those that have failed [5; 8; 16-20].

All the above makes us look at the electric multipole in a new way, including in its already known mathematical model a memristor, as one of the elements of the basic set. An additional element in the basic set also determines the appearance of new memresistive branches of the electric multipole. In accordance with the adopted classification [21-23], it is customary to distinguish the following typical operating modes of the investigated electronic devices, which is quite true for the elements of nanoelectronics:

- Quasi-linear mode of a small signal when calculating with an alternating current;
- Quasi-linear small signal mode when calculating in the time domain;
- Large signal mode when calculating for direct current (static mode);
- Large signal mode when calculating in the time domain (dynamic mode).

Depending on the selected mode, the initial model of the investigated device will be an electrical multipole of one of the following types: linear resistive, linear reactive, nonlinear resistive and nonlinear reactive. At the same time, the general methodological basis for studying all the listed types of models is their description in the basis of finite deviations of currents and voltages of the branches of the corresponding multipole, since the main goal of our study is to assess the ability of a device to maintain its characteristics within specified limits in the presence of disturbing influences with undefined properties. This article analyzes the large-signal mode in time-domain calculations (dynamic mode) for an electrical multipole with finite-increment memresistive branches.

2. Materials and methods

Mathematical models of an electric multipole with memresistive branches in the nominal form and using finite increments were built earlier [1-3] using the method of decomposition of branches of the directed graph of the circuit described in [19-22].

A full-size mathematical model in a full hybrid basis in finite increments of currents and voltages will look like this [20]:

$$\begin{cases} \begin{bmatrix} 0 & \hat{L} \\ \hat{C} & 0 \end{bmatrix} \cdot \dfrac{d}{dt} \begin{bmatrix} \Delta U_C^P \\ \Delta I_L^X \end{bmatrix} = Q_1 \begin{bmatrix} \Delta U_C^P \\ \Delta I_L^X \end{bmatrix} + Q_2 \begin{bmatrix} \Delta U_H^P \\ \Delta I_H^X \end{bmatrix} + Q_3 \begin{bmatrix} E_N^R \\ J_N^R \end{bmatrix} + Q_4 \begin{bmatrix} \Delta U_M^P \\ \Delta I_M^X \end{bmatrix} - \begin{bmatrix} E_E^L \\ J_E^C \end{bmatrix} \\ \begin{bmatrix} \Delta U_H^P \\ \Delta I_H^X \end{bmatrix} = A_1 \begin{bmatrix} \Delta U_C^P \\ \Delta I_L^X \end{bmatrix} + A_2 \begin{bmatrix} \Delta U_R^P \\ \Delta I_R^X \end{bmatrix} + A_3 \begin{bmatrix} \Delta U_M^P \\ \Delta I_M^X \end{bmatrix} - H \cdot F^I \\ \begin{bmatrix} \Delta U_R^P \\ \Delta I_R^X \end{bmatrix} = G_1 \begin{bmatrix} \Delta U_C^P \\ \Delta I_L^X \end{bmatrix} + G_2 \begin{bmatrix} \Delta U_H^P \\ \Delta I_H^X \end{bmatrix} + G_3 \begin{bmatrix} \Delta U_M^P \\ \Delta I_M^X \end{bmatrix} - W \begin{bmatrix} E_N^R \\ J_N^R \end{bmatrix} \\ \begin{bmatrix} \Delta U_M^P \\ \Delta I_M^X \end{bmatrix} = S_I \begin{bmatrix} \Delta U_C^P \\ \Delta I_L^X \end{bmatrix} + S_2 \begin{bmatrix} \Delta U_H^P \\ \Delta I_H^X \end{bmatrix} + S_3 \begin{bmatrix} \Delta U_R^P \\ \Delta I_R^X \end{bmatrix} - K \begin{bmatrix} E_N^M \\ J_N^M \end{bmatrix} \end{cases} \tag{1}$$

where $\hat{L} = \begin{bmatrix} diag\{L_k + \Delta L_k\} & 0 \\ 0 & diag\{L_i^N + L_i^N(\Delta i)\} \end{bmatrix}$ is the matrix of inductances ($k = \overline{1, n_{\text{Л}L}^X}$, $l = \overline{1, n_{HL}^X}$, $n_{\text{Л}L}^X + n_{HL}^X = dim\Delta U_L^X$); $\hat{C} = \begin{bmatrix} diag\{C_k + \Delta C_k\} & 0 \\ 0 & diag\{C_i^N + C_i^N(\Delta i)\} \end{bmatrix}$ is the capacity matrix ($k = \overline{1, n_{\text{Л}C}^X}$, $l = \overline{1, n_{HC}^X}$, $n_{\text{Л}C}^X + n_{HC}^X = dim\Delta I_C^P$); ΔU_L^X, ΔI_L^X - increments of voltages and currents on inductive chords; ΔU_C^P, ΔI_C^P - voltage and current increments on the capacitive edges; ΔU_R^X, ΔU_R^P, ΔI_R^P, ΔI_R^X - voltage and current increments on resistive edges and chords; ΔU_H^X, ΔU_H^P, ΔI_H^P, ΔI_H^X - voltage and current increments on nonlinear edges and chords; ΔU_M^X, ΔU_M^P, ΔI_M^X, ΔI_M^P - voltage and current increments on memresistive edges and chords; $Q_1 = \begin{bmatrix} -B_I & Z_E \\ V_E & -D_I \end{bmatrix} + \begin{bmatrix} -B_{IV} & 0 \\ 0 & -D_{IV} \end{bmatrix} \cdot G_1$, $Q_2 = \begin{bmatrix} -B_{II} & 0 \\ 0 & -D_{II} \end{bmatrix} + \begin{bmatrix} -B_{IV} & 0 \\ 0 & -D_{IV} \end{bmatrix} \cdot G_2$, $Q_3 = \begin{bmatrix} -B_{II} & 0 \\ 0 & -D_{II} \end{bmatrix} \cdot W$ and $Q_4 = \begin{bmatrix} -B_{III} & 0 \\ 0 & -D_{III} \end{bmatrix} + \begin{bmatrix} -B_{IV} & 0 \\ 0 & -D_{IV} \end{bmatrix}$; $Z_E = \left[[0]_{n_L^X \times n_{ЛL}^X} \quad diag\{(Z_E^L + Z_E^L(\Delta i))_l\} \right]$ - matrix of equivalent resistances and $V_E = \left[[0]_{n_C^P \times n_{ЛC}^P} \quad diag\left\{ \left(G_N^C + G_N^C(\Delta u)\right)_l \right\} \right]$ - equivalent conductivities of nonlinear inductances and capacities; D_i is the matrix of the main sections of the directed graph of the circuit; B_i - matrix of the main contours of the directed graph of the circuit; n is the number of branches and k is the number of nodes; $E_E^L = \left[[(E_N)_k]_{n_{ЛL}^X \times 1} \quad [(E_N^L + E_Z^L(\Delta i))_k]_{n_{HL}^X \times 1} \right]^T$, $J_E^C = \left[[(J_N)_k]_{n_{ЛC}^P \times 1} \quad \left[\left(J_N^C + J_Z^C(\Delta u)\right)_k \right]_{n_{HC}^P \times 1} \right]^T$ - equivalent vectors of current and voltage sources; where $A_1 = \begin{bmatrix} \Delta M^{-1} & Z_X(\Delta I_H^X) \\ G_P(\Delta U_H^P) & \Delta M \end{bmatrix} \cdot \begin{bmatrix} -B_V & 0 \\ 0 & -D_V \end{bmatrix}$, $A_2 = \begin{bmatrix} \Delta M^{-1} & Z_X(\Delta I_H^X) \\ G_P(\Delta U_H^P) & \Delta M \end{bmatrix} \cdot \begin{bmatrix} -B_{VII} & 0 \\ 0 & -D_{VII} \end{bmatrix}$, $A_3 = \begin{bmatrix} -B_{VIII} & 0 \\ 0 & -D_{VIII} \end{bmatrix}$ and $H = \begin{bmatrix} \Delta M^{-1} & Z_X(\Delta I_H^X) \\ G_P(\Delta U_H^P) & \Delta M \end{bmatrix}$; $F^I = \begin{bmatrix} E & E & 0 & 0 & -E & -E & 0 & 0 \\ 0 & 0 & E & E & 0 & 0 & -E & -E \end{bmatrix} \cdot \begin{bmatrix} E_H^H \\ E_X^H \\ J_H^H \\ J_P^H \\ E_H^Z \\ E_H^N \\ J_H^Z \\ J_H^N \end{bmatrix}$; $Z_X(\Delta I_H^X)$ and $G_P(\Delta U_H^P)$ - diagonal matrices of equivalent resistances and conductivities; ΔM^{-1} and ΔM are diagonal matrices of equivalent inverse and forward memresistivities; E is the identity matrix; $E_H^H, E_X^H, J_H^H, J_P^H, E_H^Z, E_H^N, J_H^Z$ and J_H^N - equivalent sources of EMF and current on nonlinear chords and edges; $G_1 = \begin{bmatrix} B_{XI} & \hat{R} \\ \hat{Y} & D_{XI} \end{bmatrix}^{-1} \cdot \begin{bmatrix} -B_{IX} & 0 \\ 0 & -D_{IX} \end{bmatrix}$, $G_2 = \begin{bmatrix} B_{XI} & \hat{R} \\ \hat{Y} & D_{XI} \end{bmatrix}^{-1} \cdot \begin{bmatrix} -B_X & 0 \\ 0 & -D_X \end{bmatrix}$, $G_3 = \begin{bmatrix} B_{XI} & \hat{R} \\ \hat{Y} & D_{XI} \end{bmatrix}^{-1} \cdot \begin{bmatrix} -B_{XII} & 0 \\ 0 & -D_{XII} \end{bmatrix}$ and $W = \begin{bmatrix} B_{XI} & \hat{R} \\ \hat{Y} & D_{XI} \end{bmatrix}^{-1}$; $S_I = \begin{bmatrix} -B_{XVI} & \hat{M} \\ \hat{M}^{-1} & -D_{XVI} \end{bmatrix}^{-1} \cdot \begin{bmatrix} -B_{XIII} & 0 \\ 0 & -D_{XIII} \end{bmatrix}$, $S_2 = \begin{bmatrix} -B_{XVI} & \hat{M} \\ \hat{M}^{-1} & -D_{XVI} \end{bmatrix}^{-1} \cdot \begin{bmatrix} -B_{XIV} & 0 \\ 0 & -D_{XIV} \end{bmatrix}$, $S_3 = \begin{bmatrix} -B_{XVI} & \hat{M} \\ \hat{M}^{-1} & -D_{XVI} \end{bmatrix}^{-1} \cdot \begin{bmatrix} -B_{XV} & 0 \\ 0 & -D_{XV} \end{bmatrix}$ and $K = \begin{bmatrix} -B_{XVI} & \hat{M} \\ \hat{M}^{-1} & -D_{XVI} \end{bmatrix}^{-1}$; $\hat{M} = M + \Delta M$ is the matrix of the branch memresistivities, $\hat{M}^{-1} = M^{-1} + \Delta M^{-1}$ is the matrix of inverse memresistivities of the branch; E_N^M, J_N^M - equivalent vectors of independent voltage and current sources on memresistive branches.

Mathematical model of a multipole with memresistive branches in dynamic mode. Calculation of the dynamic characteristics of electronic circuits in the time domain consists in determining the type of transient processes in their structure, arising under the influence of sources of variable signals and pulse sequences. Because of such a calculation, the time intervals necessary to transfer the circuit from one static mode to another are found, or the time during which currents and voltages reach a given level. The form of the transient is often of interest. The development of electromagnetic processes in time is determined not only by the nature of the change in voltages on capacitors and currents in inductors, but also by the final increments of these parameters.

In cases where the levels of alternating signals in the circuit are significantly less than the levels of direct currents and voltages of the static mode, the analysis of transient processes is carried out in a linearized circuit, where their small-signal models replace all nonlinear elements. In this case, the system of equations of state also becomes linear. To obtain such a model of the investigated circuit, it is necessary to form a hybrid basis containing state variables and exclude all other variables related to the resistive branches of the equivalent circuit, described in [19-20].

Using the conditions established in [20] for the existence of mathematical models of electrical multipoles, we construct such a model for the dynamic mode of an electronic circuit.

The reactive elements of the equivalent circuit - capacitance and inductance - are separated into special branches. As a result, the rest of the equivalent circuit will be a linear multipole with resistive and memresistive branches, which also includes dependent and independent sources. The capacities, the parametric description of which corresponds to the z-branches, must be referred to the edges, which can always be ensured if the assumption of the absence of main capacitive circuits is fulfilled. Likewise, inductances should be assigned to chords. To ensure the formulated requirements, the numbering of the branches should begin with capacitors and end with inductors. Based on the foregoing, the mathematical model of a multipole with memresistive branches (1) in the dynamic mode will take the form:

$$\begin{cases} \begin{bmatrix} 0 & \hat{L} \\ \hat{C} & 0 \end{bmatrix} \cdot \frac{d}{dt} \begin{bmatrix} \Delta U_C^P \\ \Delta I_L^X \end{bmatrix} = Q_1 \begin{bmatrix} \Delta U_C^P \\ \Delta I_L^X \end{bmatrix} + Q_3 \begin{bmatrix} E_{\text{HE3}}^R \\ J_{\text{HE3}}^R \end{bmatrix} + Q_4 \begin{bmatrix} \Delta U_M^P \\ \Delta I_M^X \end{bmatrix} - \begin{bmatrix} E_E^L \\ J_E^C \end{bmatrix} \\ \begin{bmatrix} \Delta U_R^P \\ \Delta I_R^X \end{bmatrix} = G_1 \begin{bmatrix} \Delta U_C^P \\ \Delta I_L^X \end{bmatrix} + G_3 \begin{bmatrix} \Delta U_M^P \\ \Delta I_M^X \end{bmatrix} - W \begin{bmatrix} E_N^R \\ J_N^R \end{bmatrix} \\ \begin{bmatrix} \Delta U_M^P \\ \Delta I_M^X \end{bmatrix} = S_I \begin{bmatrix} \Delta U_C^P \\ \Delta I_L^X \end{bmatrix} + S_3 \begin{bmatrix} \Delta U_R^P \\ \Delta I_R^X \end{bmatrix} - K \begin{bmatrix} E_N^M \\ J_N^M \end{bmatrix} \end{cases} \tag{2}$$

Expression (2) belongs to the class of interval systems of linear algebraic equations, for the solution of which it is permissible to use Kahan's interval arithmetic and the Euler and Hook-Jeeves interval methods.

The essence of Kahan's interval arithmetic is that operations with intervals containing zero have the same result as in the case of other intervals and allows you to keep the interval expansion of functions unchanged, and also guarantees, under certain conditions, not only the distributiveness of operations, but also monotonicity with respect to inclusion [20; 23].

3. Results

Computational algorithm for external estimation of the solution sets of the mathematical model of a multipole with memresistive branches in a dynamic mode. Given the interval nature of the resulting model, this algorithm uses the Euler method for integrating differential equations and the configuration method (Hook-Jeeves method), which is a zero-order method and does not require the calculation of interval derivatives when solving a system of nonlinear algebraic equations [23]:

- We assume $t_0 = 0$ and $l = 0$;
- Calculate: $\Delta U_C^P = \int \Delta C^{-1} \lfloor \Delta I_C^P - J_C^{HE3} - J_C^{3AB}(\Delta U_C^P) - J_C^{HE3}(U_C^P) \rfloor dt$ according to the first equation of the system (2), where $\Delta C = diag\{\Delta C_1, \ \Delta C_2, \Delta C_3\}$ is the matrix of finite increments of capacities;
- Calculate: $\Delta I_C^P = \left[-\frac{1}{\Delta} D_I W \Delta U_C^P - \frac{1}{\Delta} D_I V \Delta U_{BX} \right] \times \Delta$;
- Calculate $\Delta I_R^X = \frac{1}{\Delta} W \Delta U_C^P + \frac{1}{\Delta} V \Delta U_{BX}$ with respect to the unknown vector $\lfloor \Delta I_R^X \rfloor_{t_l+1}$ for the moment $t_{l+1} = t_l + \Delta$ using the Newton-Raphson method for interval variables;
- We believe $l = l + 1$, $t_l = t_l + \Delta$;
- Is the process over? If "yes", go to item next point; otherwise, go to item point 2;
- The end.

4. Discussion

The dynamic mode considered in this article for an electric multipole with memresistive branches confirms the applicability of the chosen approach to the analysis of nanoelectronic circuits of both memory elements and the architecture of information-measuring systems as a whole.

5. Conclusion

The form of presentation of the mathematical model for each mode of operation is quite consistent with the forms required for using the methods of interval analysis. Of course, it is necessary to check the adequacy of the presented model of an electric multipole with memresistive branches both in general form and in dynamic and static modes for real nanoelectronic circuits, which is the subject of further research.

References

[1] Bondarev A and Efanov V 2019 The Principles of Forming of the Mathematical Model of Nanoelectronic Components of Quantum Computer Systems with Memresistance Branches. *Atlantis Highlights in Computer Sciences* **3** 17-22

[2] Bondarev A V 2015 Research problem of a robustness of electronic schemes by methods of interval calculations in the conditions of uncertainty. *Materials of the CSIT'2015 Proceedings of the 17th International Workshop on Computer Science and Information Technologies* 145-9

[3] Bondarev A V, Muravyova E A, Kadyrov R R and Rahman P A 2016 The analysis of opportunities of construction and use of avionic systems based on COTS-modules. *ARPN Journal of Engineering and Applied Sciences* **11 1** 78-92

[4] Borghetti J, Snider G S, Kuekes P J et al. 2010 Memristive switches enable stateful logic operations via material implication. *Nature letters* **464** 873-6

[5] *Bourzac K Memristor Memory Readied for Production.* Retrieved from: www.technologyreview.com/computing/25018/

[6] Chua L O 1971 Memristor – the missing circuit element. *IEEE Trans. Circuit Theory* **18** 507-19

[7] Alefeld G and Mayer G 2000 Interval analysis: theory and applications. *Journal of Computational Applied Mathematics* **121** 421-64

[8] *Memristor: "the missing element".* Retrieved from: http://www.nanonewsnet.ru/articles/2011/memristor-nedostayushchii-element

[9] *Strogy V V Application of memristors in pc architecture.* Retrieved from: https://eduherald.ru/ru/article/view?id=17265

[10] *Johnson R C End of the CPU? HP demos configurable memristor.* Retrieved from: www.eetimes.com/electronicsnews/4088557/End-of-the-CPU-HP-demos-configurablememristor

[11] Kuekes P J, Snider G S and Williams R S 2005 Crossbar nanocomputers. *Scientific American,* **293** 72-8

[12] *Markoff J HP Sees a Revolution in Memory Chip.* Retrieved from: www.nytimes.com/2010/04/08/science/08chips.html?_r=1

[13] *Memristor*. Retrieved from: en.wikipedia.org/wiki/Memristor

[14] *Merritt R HP researcher predicts memory-centric processors.* Retrieved from: www.eetimes.com/electronicsnews/4199856/HP-researcher-predicts-memory-centricprocessors

[15] Strukov D B, Snider G S, Stewart D R and Williams R S 2008 The missing memristor found. *Nature letters* **453** 80-3

[16] Sigorskiy V P and A I Petrenko 1976 *Algorithms for analyzing electronic circuits* (Moscow: Soviet radio) 608

[17] Bondarev A V 2019 A review of quantum computing algorithms. *Prospects for Science* **7(118)** 27-31

[18] Bondarev A V 2019 Review of the element base of quantum computers. *XXI century: results of the past and problems of the present plus* **8 3(47)** 96-100
[19] Bondarev A V 2019 Features of building the architecture of quantum computers. *Modern science: Series Natural and technical sciences* **6** 52-5
[20] Bondarev A and Efanov V 2020 The Principles of Forming of the Mathematical Model of Nanoelectronic Components of Quantum Computer Systems with Memresistance Branches. *Control systems and information technologies* 1(79) 4-10
[21] Bondarev A V 2011 *Decision support system for assessing the robustness of complex airborne electronic systems based on COTS products: dissertation for the degree of candidate of technical sciences* (Ufa: Ufa State Aviation Technical University) 177
[22] Chua L O and Lin Pen-Min 1980 *Machine analysis of electronic circuits. Algorithms and computational methods* (Moscow: Energy) 640
[23] Shary S P *Finite-dimensional interval analysis*. Retrieved from: http://www.nsc.ru/interval

Formation of a commodity fuel portfolio of Kuzbass on the basis of coal preparation waste

O P Chernikova, S A Strekalova, N G Zhdanova and O V Grinkevich

Siberian state industrial University, 42, Kirov Str., Novokuznetsk, 654007, Russia

E-mail: chernikovaop@yandex.ru

Abstract. The Kemerovo Region development strategy until 2035, updated in September 2020, is a forecast of the global technological leadership of Kuzbass, diversification of transport communications with other regions and states, adaptation to the conditions of the 2019 coal industry crisis. To solve these strategic tasks, coal mining and processing enterprises must balance their fuel portfolio and look for new ways to market their products, including the possibility of using production waste. Within the framework of the research carried out by the team of authors: the ecological, technological and economic advantages of using a coal-water suspension as a fuel component have been systematized; identified potential areas of its application; an analysis of the world experience of transporting coal through pipelines was carried out; a proposal was formulated for a project for the construction of the Kuzbass-Ural pipeline for the supply of coal-water fuel to Western Siberia and the Eastern Urals, and the advantages and disadvantages of coal hydrotransportation in comparison with its railroad delivery were identified.

1. Introduction

Russia is one of the leading coal mining and exporting countries, with 57 operating coal mines, 130 open-pit mines and 64 processing plants at the beginning of 2020. The total volume of processed coal products in 2019 amounted to 207.6 million tons (1.9% more than in 2018), including 205.9 million tons at the concentration plants (2.8% more than in 2018). Kuzbass is the most powerful coal supplier of coal mining regions, producing 57% of the total coal volume in the country.

One of the problems of industrial use of coal is the formation of waste from its processing. Process waters of mining and processing enterprises with small particles of coal are discharged into sedimentation tanks, hydraulic dumps, mined workings or into the nearest ravines and water bodies. In some cases, the sludge is dehydrated and placed on the territory of residence. At the same time, coal sludge is a highly toxic product, which contains flocculants and coagulants polluting the environment used at the processing plants [1]. The total volume of coal sludge produced by the Kuzbass concentrators is more than 10500 thousand tons per year.

The Kuzbass development strategy until 2035 sets ambitious goals for the growth of industrial production in general and coal production in particular, but a prerequisite for these trends is the strictest compliance with environmental requirements in the region. President of the Russian Federation V.V. Putin has set tasks for the regional authorities to ensure environmental safety. At the same time, the requirements for the leadership of the Kemerovo region separately indicate the need to solve the problems of the city of Novokuznetsk. Since 2019, Novokuznetsk has been implementing a comprehensive program to improve the environmental situation, the activities of which are aimed at reducing environmental pollution by industrial enterprises, heat power engineering, transport, as well as modernizing the state observation network for air pollution and strengthening environmental supervision. By 2021 in Novokuznetsk, it is planned to reduce the volume of harmful emissions by 7%, and by 2024 - by 20%.

Modern ways of using coal products in the energy sector should not be archaic. New strategic approaches are needed to address these challenges, given the global industry crisis that began in 2019. New technologies are needed that would make it possible to use the advantages of coal and minimize the negative consequences of its use. One of the solutions to the above problems is the active use of coal-water fuel (a dispersed system consisting of finely ground coal, water and a plasticizer reagent) within the region and ensuring the possibility of its transportation beyond its borders.

2. Materials and methods

The empirical basis of the scientific research was the analysis of Russian and international experience in the use of products obtained from coal preparation waste-coal-water fuel. A comprehensive study of this issue was carried out based on a systematic approach. The methodological basis of the study was the scientific and educational literature given in the list of sources.

When setting the goal and defining the tasks of scientific research at the initial stage, the team of authors used an abstract-logical research method to study the possibilities and directions of using and transporting Russian coal through pipelines. The methods of comparative analysis and synthesis made it possible to identify, generalize and systematize the advantages of coal-water suspension in comparison with other types of fuel, as well as the advantages of coal-pipeline transport over railway.

3. Results

The analysis carried out in the work made it possible to systematize the environmental, technological and economic advantages of using a coal-water suspension as a fuel component (table 1), as well as to identify its potential applications (table 2).

Table 1. Advantages of using coal-water slurry.

Type of benefits	Content
1. Environmental:	Safety for the environment at all stages of the operating cycle of use, reducing harmful emissions by 1.5-3.5 times.
2. Technological:	Does not require modernization of structures of consumer enterprises; the operational processes of receiving, supplying and burning fuel are subject to mechanization and automation; resource utilization rate over 97%.
3. Economic:	Saves the cost of fuel resources by 1.3-5 times; reduces storage, transportation and incineration costs by 15-30%; the investment payback period is 1-2.5 years; when incinerated, it produces material (fly ash) for use in construction.

The widespread introduction of coal-water fuel can serve as the basis for the effective replacement of natural gas and oil at many thermal power plants and state district power plants with minimal capital costs and ensuring the required level of harmful emissions into the atmosphere.

In addition, due to the need to solve the problems of railroad congestion and the high cost of transportation, as well as the successful practical application of coal pipelines in the world economy, the team of authors considers it expedient to consider the construction project of the Kuzbass-Ural pipeline for the supply of coal-water fuel to Western Siberia and the Eastern Urals and conducting a factor analysis of its implementation. Such a coal pipeline can be laid through the regional centers of Western Siberia - Novosibirsk, Omsk, Tyumen, Yekaterinburg with branches to the north and south to ensure the supply of coal-water fuel to power plants and industrial enterprises in the south of Western Siberia and the east of the Urals. Further, construction can be continued to the western border of Russia to ensure the transport of liquid coal to Europe.

Table 2. Potential areas of application of coal-water fuel.

Potential application of coal-water fuel	Benefits of using
Fuel resource of boiler houses of housing and communal services.	Costs are optimized by 30-60% in comparison with the use of fuel oil and by 20-40% in comparison with coal combustion. Emissions of nitrogen oxides are reduced by 20-30%. Minimum investment for the reconstruction of boiler houses.
Heat and steam production in industrial plants.	Costs for steam generation of sugar, textile and other enterprises are reduced by 20-40%. Fast transformation of steam boilers to the use of VUT.
Additional marketable product for mine and open pit owners.	The use of own coal waste as liquid coal for production and technical needs allows to reduce the consumption of marketable coal products and get additional profit.
Fuels for asphalt plants.	Use in asphalt heating furnaces optimizes fuel costs.

Coal pipelines, in comparison with railroad transport, have a number of advantages: uniformity and continuity of fuel flow; high reliability; high throughput of the system; averaging the quality characteristics of products; independence of transportation from weather conditions; no loss of minerals during transportation; reducing the negative impact on the environment; a small number of service personnel; high potential for automation of loading and unloading and transport operations; the possibility of using the earth's surface above them.

The main disadvantage of using coal pipelines is high capital construction costs, but if we compare these investments with investments in the construction of a railway, they are clearly less.

4. Discussion

Potential consumers of coal-water fuel are industrial organizations of various fields of activity and housing and communal services enterprises. Coal-water fuel can be the main or reserve fuel resource of boiler houses, as well as thermal power plants of various sizes. This fuel may be of particular interest for residents of new micro-districts, cottage settlements, and remote from energy resources.

Studies on the use of ash from thermal power plants as components of building materials and its application to obtain a hardening fill for filling mine workings and leveling the earth's surface [2-5] have shown high efficiency indicators, which can also reduce the volume of waste entering the sludge storage.

Currently, 99% of Kuznetsk coal is transported by rail; while the main drawback of transportation is the geographical location of the region (the average distance to seaports within Russia is 5075 km., to border crossings - 4093 km.).

In connection with the large volumes of transported coal, the issue of the throughput of railways becomes urgent, which is far from always sufficient to ensure rhythmic supplies of coal products. Since the Trans-Siberian Railway is overloaded, traffic jams are constantly formed at the exits from the Kemerovo Region.

In addition, since the early 1990s, there is an outstripping growth of transport tariffs in comparison with the prices for transported products: for the period, 1991-2009 wholesale prices for industrial products increased 34 thousand times, while railway freight rates increased 53 thousand times [6]. Today, transport costs account for up to 60% of the price of Kuznetsk coal, which, when delivered to ports, more than doubles the cost. The growth of railway tariffs negates the competitive advantages of the Kuznetsk Basin (quality characteristics of products, volumes of mineral reserves in the subsoil, the level of demand from consumers).

In recent years, cooperation between mining companies and the railway has been complicated by the constant settlement of issues in the areas of tariff determination, infrastructure development, and coordination of production and transport processes, which in modern conditions has led to the fact that the coal top management sees the main risks of business and industry development in the transport component generally.

The solution to the accumulated production and transport problems can be the implementation of large investment projects by the railway to increase the throughput of its highways, accompanied by a decrease in freight tariffs, or the search for alternative ways of transporting coal.

Kuzbass coking coals and coke are in demand westward from the Kemerovo region due to the location of consumers. All metallurgical plants in the country have a demand for the products of the region, while the enterprises of the Eastern Urals consider our producers as the only suppliers of coal.

Thermal coal is also transported eastward, but most of it is also exported westward on the domestic Russian market. The main consumer organizations in the western regions include the Novosibirsk thermal power plants, the Tomsk power plant, the Barnaul thermal power plants, the Biysk thermal power plant, as well as local boiler houses. In addition, Kuznetsk coal is used for energy purposes in the regions of Central Russia, the Volga region, the Central Black Earth Region, and the Urals.

In the eastern direction, the consumers of Kuznetsk thermal coals are enterprises of the Far East. Kuznetsk coal is not used in the energy sector of Eastern Siberia due to the presence of large hydroelectric power plants, Kansk-Achinsk and other coal basins on its territory.

On the world stage, among the countries - importers of Kuznetsk coal and coke, there are more than 50. The volume of export of thermal coal exceeds that of coking coal. Regarding the location of the Kemerovo region, ¾ export volumes are carried out in the western direction. The main importers of energy grades are Great Britain, Finland, Romania, Poland, coking coal – Germany, Spain, Belgium, Slovakia, Hungary, Kazakhstan, Turkey, India, and the United Arab Emirates.

A quarter of Kuznetsk coal and coke is exported to the east, relative to the location of the region, to Japan, Korea, China and the Republic of China (Taiwan). East Asian countries acquire both energy and coking coals, and their share in exports is gradually increasing.

Long-term scientific research of domestic and foreign scientists substantiated the scientific, technical and economic potential of the main pipeline transportation of coal products [7-13]. In the context of an increasing rate of changes in the energy sector, a detailed analysis that allows identifying the position of a coal-mining region in the industry segment, adjusting the production and sales strategy, and thereby increasing the economic efficiency of its activities is in great demand [14-15].

There are practical examples of the implementation of projects for the hydrotransportation of coal over long distances in the world: in the USA, Canada, France, Italy, Poland, Austria, China, India, etc. [16]. In Russia, there is also an example of the practical implementation of long-distance delivery of coal through pipes: in the early 90s, the Belovo-Novosibirsk coal pipeline operated in the Kemerovo and Novosibirsk regions, which transported coal in the form of coal-water fuel [13]. The project successfully proved the possibility of transporting coal slurry through pipes over long distances and burning it in power boilers of a CHP plant, but in the conditions of a transitional market economy, the project was closed.

In 2020, it became known that for the first time in many years, the volume of production in Kuzbass decreased at the end of 2019 (251 million tons compared to 255 million tons in 2018). In the winter of 2020, Poland, which was one of the largest importers of Kuzbass coal among the EU countries, abandoned Russian coal. Coal mining enterprises of Kuzbass and the regional authorities have made great efforts to get priority in the transportation of coal along the Baikal-Amur Mainline and Transsib in the eastern direction. At the same time, the carrying capacity of the railways in the eastern direction is not enough for everyone, and the work to eliminate the problems is going far behind schedule, or is completely frozen.

Against the background of the intensification of the shipment of products to the east by the enterprises of the Far Eastern region, the Kuzbass coal companies require a radical reduction in the cost of coal transportation. At the same time, Russian Railways, citing an almost 20% rise in quotations in the ports of the North-West, devaluation of the ruble and non-fulfillment of obligations by coal miners in terms of traffic volumes, will cancel the 12.8% discount on the delivery of coal products established in March 2020 from October 2020 energy marks for a distance of over three thousand kilometers, which was supposed to operate until the end of the year.

It makes sense to construct a long coal pipeline only in the direction west of the Kemerovo region, since the Irkutsk region and the Krasnoyarsk Territory are rich in their own coal reserves and large hydroelectric power plants operate on their territories.

In addition, one of the main strategic directions for the development of world energy provides for a significant reduction in the anthropogenic load on the environment with its preservation in a state favorable for human life, through the introduction of "green" energy practices [17-20]. Construction and the further need to ensure constant transportation of coal-water fuel through the pipeline would help to solve environmental problems in Kuzbass.

5. Conclusion

Within the framework of the research carried out by the team of authors, on the basis of the analysis of world and Russian experience in industrial production, the ecological, technological and economic advantages of using coal-water fuel were identified and systematized; potential areas of its application are indicated; the analysis of the world practice of coal transportation through pipelines was carried out; a proposal was formulated for a project for the construction of the Kuzbass-Ural pipeline for the supply of coal-water fuel to Western Siberia and the Eastern Urals, and the advantages and disadvantages of coal hydrotransportation in comparison with its railroad delivery were identified.

References

[1] Yuxing Zhang, Zhiqiang Xu, Yanan Tu, Jinyu Wang and Jie Li 2020 Study on properties of coal-sludge-slurry prepared by sludge from coal chemical industry. *Powder Technology* **366** 552-9

[2] Ertug Aydin 2016 Novel coal bottom ash waste composites for sustainable construction. *Construction and Building Materials* **124** 582-8

[3] Sanchit Gupta and Sandeep Chaudhary 2020 Use of fly ash for the development of sustainable construction materials. *New Materials in Civil Engineering* **21** 677-89

[4] Gang Xu and Xianming Shi 2018 Characteristics and applications of fly ash as a sustainable construction material: A state-of-the-art review. *Resources, Conservation and Recycling* **136** 95-109

[5] Mohammad Shariful Islam, Tausif E Elahi, Azmayeen Rafat Shahriar and Nashid Mumtaz 2020 Effectiveness of fly ash and cement for compressed stabilized earth block construction. *Construction and Building Materials* **255** 119392

[6] Sherin E A 2017 Modernization of the industrial complex from the perspective of the concept of production cycle (a case study of kuznetsk coal use). *Geography and Natural Resources* **3** 281-7

[7] Liangyong Chen, Yufeng Duan, Meng Liu and Changsui Zhao 2010 Slip flow of coal water slurries in pipelines. *FuelMay* **5** 1119-26

[8] Petukhov V N, Svechnikova N Yu, Yudina S V, Gorokhov A V, Lavrinenko A A and Kharchenko V F 2016 Use of coal flotation wastes for energy purposes in conditions of Belovskaya Concentrating mill JSC. *Coke and chemistry* **5** 38-41

[9] Gunung Oh, Ho Won Ra, Sung Min Yoon, Tae Young Mun and Sang Jun Yoon 2019 Syngas production through gasification of coal water mixture and power generation on dual-fuel diesel engine. *Journal of the Energy Institute* **92** 265-74

[10] Ruikun Wang, Jianzhong Liu, Yukun Lv and Xuemin Ye 2016 Sewage sludge disruption through sonication to improve the co-preparation of coal–sludge slurry fuel: The effects of sonic frequency Applied. *Thermal Engineering* **9925** 645-51

[11] Han Hao, Zongwei Liu, Fuquan Zhao, Jiuyu Du and Yisong Chen 2017 Coal-derived alternative fuels for vehicle use in China: A review. *Journal of Cleaner Production* **14110** 774-90

[12] Kurgankina M A, Vershinina, K Y, Ozerova I P and Medvedev V V 2018 On the issue of

transition of thermal power plants from traditional fuels to organic coal fuel compositions. Bulletin of the Tomsk Polytechnic University. *Geo Assets Engineering* **9** 72-82

[13] Malyshev D Yu and Syrodoy S V 2020 Justification of resource efficiency of technologies for burning coal-water fuels with biomass additives. *Bulletin of the Tomsk Polytechnic University. Geo Assets Engineering* **6** 77-85

[14] Glushakova O V, Chernikova O P and Strekalova S A 2020 Integral Assessment of the Corporate Strategy Effectiveness in the Iron and Steel Industry. Steel in translation **5** 309-16

[15] Kononenko E A and Dyachuk O V 2000 Coal mining by dredgers for pipeline transportation. *Ugol'* **11** 60-2

[16] Khodakov G S 1994 The statistical analysis of the pilot commercial operational results of the power generation and delivery Complex Belovo-Novosibirsk. *Conf. on Coal Utilization and Fuel Systems USA Florida* 863-6

[17] Chernikova O, Zlatitskaya Y and Klimashina Y 2019 Benchmarking sustainable energy technologies in crossborder regions: issues of economic efficiency. *IOP Conference Series: Earth and Environmental Science* **395** 012102

[18] Sarah Knuth 2018 «Breakthroughs» for a green economy? Financialization and clean energy transition. *Energy Research & Social Science* **41** 220-229

[19] Su Shuai and Zhang Fan 2020 Modeling the role of environmental regulations in regional green economy efficiency of China: Empirical evidence from super efficiency DEA-Tobit model. *Journal of Environmental Management* **261** 110227

[20] Albert Merino-Saum, Marta Giulia Baldi, Ian Gunderson and Bruno Oberle 2018 Articulating natural resources and sustainable development goals through green economy indicators: A systematic analysis. *Resources, Conservation and Recycling* **139** 90-103

IOP Conf. Series: Materials Science and Engineering **976** (2020) 012015 doi:10.1088/1757-899X/976/1/012015

About control of industrial stability of dams of the fuel and energy complex facilities in Kuzbass on the basis of application of shallow-depth geophysical technologies

A N Solovitskiy, N Yu Nikulin, A V Nastavko, N A Smirnov and K A Makarov

Department of Geology and Geography, Kemerovo State University, 6, Krasnaya Street, Kemerovo, 650000, Russia

E-mail: san.mdig@mail.ru

Abstract. It has been established that the problem of using applied low-depth geophysical technologies to ensure the sustainability of industrial and civil construction in Kuzbass is as follows: on the one hand, there is a significant increase in measurement accuracy and efficiency, and on the other hand, there is the absence of new models that allow interpretation of not only base soils and geological objects, but also their condition. The research provides a generalization of practical experience in creating physical and geological models. Based on the studies, it was found that applied shallow geophysical technologies are more informative and automated than traditional ones. The practical use of this theory is to ensure the sustainability of industrial and civil construction, and, consequently, the development of the Kuzbass territory.

1. Introduction

The analysis performed by the authors indicates that the possibilities of geophysical technologies are much wider than prospecting or survey work [1-9]. Their use implements modern trends in the development of technologies such as improving the accuracy, efficiency and automation of processes in measurements and processing of received information. Therefore, they are in demand for the study of natural and technogenic geological processes in the development of the territory of the Kemerovo region to ensure safety (including objects of the fuel and energy complex). Region is characterized by high urbanization and concentration of production, which has important scientific and practical value for it. This is especially true for fuel and energy complex facilities that are disrupted by mining, which dramatically changes the natural characteristics of rock massifs and reduces the stability of the bases [10]. Information about these processes is essential to ensure the design of fuel and energy complex facilities, as well as their safety during operation [5; 7; 10]. The theory of digital geophysical support for the sustainability of industrial and civil construction of dams of fuel and energy complex facilities has not been developed yet [9]. Traditional technologies of geophysical surveys in areas of territorial development are currently focused on mapping and studying the signs of occurrence of mineral deposits. The development of digital geophysical support for the sustainability of dams of fuel and energy complex facilities is constrained by the lack of new mathematical models of deformation processes that ensure the adequacy of the condition of the base soils, as well as geological structures [1-9]. The main condition for such models is zoning of soils, identification of geological structures and identification of their state. The purpose of this theory is the collection, storage, processing and analysis of information on zoning of soils, the allocation of geological structures and identification of their condition in areas of fuel and energy complex facilities. The results obtained in this way will provide an information basis for obtaining new knowledge about the state of soils and geological structures, which will not only solve the problem of sustainability of dams of fuel and energy complex facilities, but also a number of other national economic problems. The above allows us to consider the development of the theory and methods of the theory of usage of applied shallow geophysical technologies to ensure the industrial sustainability of dams of fuel and energy complex facilities in Kuzbass relevant. The practical use of its results is to ensure the industrial sustainability of dams of fuel and energy complex facilities for the development of territories. The main aspects of multi-purpose application of applied low-depth geophysical technologies to ensure the industrial stability of dams of fuel and energy complex facilities in Kuzbass will be considered on the example of taking into account the water content and seismicity of the territory of Kuzbass.

2. Materials and methods

Currently, the number of dams at fuel and energy complex facilities in Kuzbass is more than two hundred. The model of their industrial stability includes a subsystems of structure and water cut. Therefore, first of all, we will consider the role of information about the water content of dams of fuel and energy complex facilities obtained through the use of applied low-depth geophysical technologies. It is known that the regime of groundwater depends on many factors. The high level of underground water occurrence hinders the development of land development and construction growth, as well as increases the level of its seismicity. Waterlogged rock mass reduces its stability and requires preventive measures. The violation of a significant territory of Kuzbass by mining operations dramatically changes the natural hydrogeological regime [10]. Traditional technologies for searching for underground water are based on drilling wells, which is characterized by significantly higher time and financial costs, as well as the duration of research. These are discrete (spot) measurements with a rare network of exploration lines that make it difficult to build digital models of groundwater occurrence, which does not provide reliability and accuracy in determining their quantitative parameters. Traditional geophysics for the search for groundwater in the difficult geological conditions of mountainous areas and in urban areas is not always effective. Electrical exploration of the method of vertical electrical sounding (VES) in the search for groundwater of crack type in mountainous areas is laborious. The progressive method of pulsed electrical exploration in settlements often does not work due to strong electromagnetic interference.

3. Results

Consider the results obtained on the basis of these subsystems of the model in more detail on a number of examples. The analysis of electrical resistance during shallow applied geophysical studies is of great importance, since it is an indicator of the water content of soils and geological structures. The electrical resistance of most soils is determined mainly by porosity, water saturation and clay content. An increase in soil moisture and clay content leads to a decrease in specific electrical resistivity (SER). Moreover, an increase in the mineralization of the fluid in the pores also leads to a decrease in resistivity. The method of electrotomography allows you to explore the soil massif almost continuously. This makes it possible to fully explore any structure and determine the location of even small-irrigated areas. The method of electrotomography is the development of VES. In the method of electrotomography, first all electrodes are grounded, and then they are connected to current and measuring circuits using a switch. In other words, the same electrodes grounded on the profile are repeatedly used as the supply and measuring electrodes. The number of electrodes in the used equipment is 64, while the electrodes are structurally connected by a multicore cable, thereby forming an electrical reconnaissance spit.

A geoelectric section along profile 4 at one of the objects in the Kemerovo region is shown in figure 1.

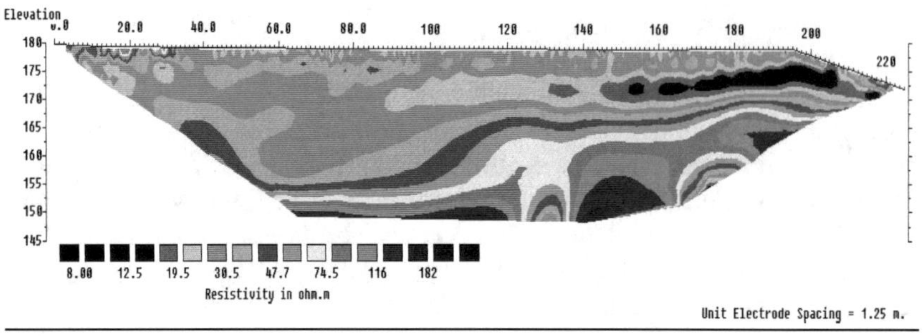

Figure 1. Geoelectric section obtained as a result of interpretation of measurements along profile 4.

Unlike conventional VES, where the interpretation is carried out in the framework of horizontally layered models of the geoelectric section, the data of electrotomography are interpreted in the framework of two-dimensional and three-dimensional models.

At a distance of 25–200 m from the beginning of the profile to a depth of 9–10 m a zone with resistivity is predominantly 20–35 Ohm·m, corresponding to clay soils. In the lower part of the section a zone with resistivity of 25–80 Ohm·m was identified, which corresponds to clay and sandy soils of the base of the object. At a distance of 150-200 m from the beginning of the profile, a region of reduced resistivity (less than 20 Ohm·m) corresponding to the zone with increased soil moisture was revealed.

4. Discussion

Based on the foregoing, we can conclude that traditional technologies for ensuring the industrial sustainability of dams of fuel and energy complex facilities in Kuzbass were based on the principle of minimum costs C, in other words, whose objective function was unambiguous and had the following form:

$$C = \sum a_i x_i = \min \tag{1}$$

where a_i is the factor; x_i is the cost of the i-th technological process.

According to the authors, the mathematical basis of the proposed approach to ensure the stability of construction objects in Kuzbass is the following criterion that describes the zoning of territories:

$$J_i \neq J_0 \tag{2}$$

where J_i - investigated trait; J_0 - its normative value.

5. Conclusion

Based on the research, the following conclusions are made:

- It is established that the model of industrial sustainability of dams of fuel and energy complex facilities in Kuzbass includes subsystems of water cut and construction of territories.
- It has been established that the subsystem of water cut in the territories, implemented on the basis of the method of electrotomography, allows us to study the soil massif almost continuously, which makes it possible to monitor their condition and control the industrial safety of dams.

References

[1] Morela C A, Lapola M M and Carrara A 2016 Comparative analyzes among electrical resistivity tomography arrays in the characterization of flow structure in free aquifer. *Geofísica Internacional* **55(2)** 119-29

[2] Zamri Chik and S M Taohidul Islam 2012 Finding Soil Particle Size through Electrical Resistivity in Soil Site Investigations. *Electronic Journal of Geotechnical Engineering* **17.M** 1867-76

[3] Muhammad Syukri, Rosli Saad, M M Nordiana and I N Azwin 2014 Preliminary Study of Sumatera Fault Using 2-D Resistivity Imaging Method. *Electronic Journal of Geotechnical Engineering* **19/D** 971-9

[4] Amini A and Ramaz H 2016 Application of Electrical Resistivity Imaging for Engineering Site Investigation. A Case Study on Prospective Hospital Site, Varamin, Iran. *Acta Geophysica* **64.6** 2200-13

[5] Bortnikova S, Yurkevich N, Bessonova E, Karin Y and Saeva O 2013 The combination of Geoelectrical Measurements and Hydro-Geochemical Studies for the Evaluation of Groundwater Pollution in Mining Tailings Areas. *The Handbook of Environmental Chemistry (*Berlin: SpringerBerlin Heidelberg) 18

[6] Olenchenko V V, Kucher D O, Bortnikova S B, Gaskova O L, Edelev A V and Gora M P 2016 Vertical and lateral spreading of highly mineralized acid drainage solutions (Ur dump, Salair): electrical resistivity tomography and hydrogeochemical data. *Russian Geology and Geophysics* **57(4)** 617-28

[7] Mahato P P 2018 Detection of Cavity Using Electrical Resistivity Tomography (ERT) at Patherdih, Jharia Coal Field, Dhanbad, India. *Universal Journal of Geoscience* **6.4** 114-7

[8] Hassan A A 2017 Numerical Modelling of Subsurface Cavities Using 2D Electrical Resistivity Tomography Technique. *Diyala Journal for Pure Science* **13.2** 197-216

[9] Dahlin T and Zhou B 2004 A numerical Comparison of 2D Resistivity Imaging with Ten Electrode Arrays . *Geophysical Prospecting* **52** 379-98

[10] Solovitskiy A N, Brel O A, Nikulin N YU, Nastavko E V and Meser T A 2017 Land Resource Management as the Ground for Mining Area Sustainable Development. *The second international innovative mining symposium "Devoted to Russian Federation year of environment" Kemerovo* 2102012

IOP Conf. Series: Materials Science and Engineering **976** (2020) 012016 doi:10.1088/1757-899X/976/1/012016

Modelling of mass transfer in separation of hydrocarbon mixtures and modernization of industrial columns

A G Laptev[1], E A Lapteva[1] and A D Gilyazov[2]

[1]Kazan State Power Engineering University, 51, Krasnoselskaya Str., Kazan, 420066, Russia
[2]LLC EPC "Inzhekhim", 14/83, Shalyapin Str., Kazan, 420049, Russia

E-mail: grivka100@mail.ru

Abstract. A system of differential equations of mass transfer with an interfacial mass source for multicomponent distillation in a column with random and structured packings in the film mode is presented. The interfacial source is associated with the matrix of volumetric mass transfer coefficients as well as a driving force for mass transfer. The diffusion fluxes of mass of components are written in the form of Fick's law with elements of the matrix of coefficients of molecular diffusion in the gas (vapor) phase. To simplify the mathematical model, one-dimensional equations for the diffusion model of the structure of flows in the gas and liquid phases with the interfacial sources and coefficients of back (longitudinal) mixing of gas and liquid flows are used. The Dankwerts boundary conditions are applied to the system of equations.To calculate the back mixing coefficients, experimental data obtained for the selected types of packings are used.When modelling mass transfer in hydrocarbon mixtures consisting of several dozen components, the system of equations for the diffusion models is solved for the key components and fractions of the oil and gas condensate (OGC) mixture stabilization column at an industrial enterprise.A brief description of the technological scheme is given.Scientific and technical solutions are developed to replace obsolete sieve trays in the stabilizer column with a new structured corrugated roll packing having a rough surface.It is found that a packing height of eight meters is sufficient for separating OGC in a given composition range.As a result of implementation of the new packing, the content of isopentane in the stable condensate decreases and its concentration in the broad fraction of light hydrocarbons increases from 9 wt.% to 11.7 wt.%.The implementation of packing in the stabilizer column meets the requirements of the technical specifications.

1. Introduction

The problem of modelling the separation of multicomponent mixtures (MCM) into whole components is important and relevant for both fundamental and applied science in various industries [1]. To solve this problem, various approaches are used in chemical engineering [2-4] with the introduction of various types of contact devices [5-7], mainly bubble trays, structured and random packings [5; 8-10].

When modelling processes in multicomponent mixtures, it is necessary to take into account the influence of the main and cross effects, as well as, in some cases, heat transfer between phases. In this regard, the mathematical description of mass transfer processes is a complex problem, in the solution of which it is necessary to use matrix equations.

2. Mathematical model

The purpose of the present work is mathematical modelling of the separation of a multicomponent mixture (MCM) in a column for stabilizing oil and gas condensate.

In the numerical modelling of transfer of momentum, mass of components and heat in film devices with a random packing, a system of partial differential equations for the continuous phase, i.e. gas (vapor), is written and the interaction with a falling liquid film on the surface of random elements is taken into account using volumetric interfacial sources. The sources depend on characteristics of the boundary layer at the interface and the driving forces of the process [7-8]. For example, with a known profile of gas velocity in the cross-section of the packed column, the equations of mass transfer in the MCM in a cylindrical coordinate system have the form:

$$w_g(r)\frac{\partial c_{gi}}{\partial z}+\mathrm{div}j_i=\sum_{i=1}^{n-1})(K_{vg\,ij}\,a)\,(c_{gi}^{*}-c_{gi}) \tag{1}$$

where c_{gi} is concentration of the ith component in the gas phase; $w_g(r)$ is gas velocity as a function of the radial coordinate, m; z is vertical coordinate, m; $(K_{vg\,ij}\,a)$ are elements of the matrix of volumetric mass transfer coefficients, s^{-1}; c_{gi}^{*} is equilibrium concentration; n is number of components in the mixture; j is mass flux density.

Equation (1) is supplemented by the expression for the mass balance of components through the interface and the conditions of thermodynamic equilibrium.

Diffusion fluxes of components are written in the form of Fick's law, taking into account the molecular turbulent transfer:

$$j_{ir}=-\rho_g\sum_{i=1}^{n-1}(D_{ij}+\delta_{ij}D_T)\frac{\partial c_i}{\partial r} \tag{2}$$

$$j_{iz}=-\rho_g\sum_{i=1}^{n-1}(D_{ij}+\delta_{ij}D_T)\frac{\partial c_i}{\partial z} \tag{3}$$

where D_{ij} is element of the square matrix of molecular diffusion coefficients [D] with dimension (n-1)(n-1); D_T is turbulent diffusion coefficient, m^2/s; δ_{ij} is Kronecker's delta; ρ_g is gas density, kg/m^3; n is number of components.

To simplify the presented mathematical model, the flow structure models can be used [7-8].

The most widely used for calculating the concentration profiles of components in packed columns is the diffusion model of the flow structure, which takes into account the back mixing of gas and liquid flows. In the equations considered below, it is assumed that the interphase transfer is determined by equimolar mass transfer, which leads to the constancy of mass fluxes of vapor and liquid along the column height during distillation.

The levels of one-dimensional steady-state diffusion models written for the gas (vapor) and liquid phases have the form molecular turbulent transfer:

$$\frac{dc_{gi}}{\partial z}=\frac{S}{G}\sum_{i=1}^{n-1}(K_{vgij}\,a)(c_{gj}^{*}-c_{gj})+\frac{d_{eq}}{Pe_g}\frac{d^2c_{gi}}{dz^2}, \tag{4}$$

$$\frac{dc_{li}}{\partial z} = \frac{S}{L}\sum_{i=1}^{n-1}(K_{vg\,ij}a)(c^*_{gj} - c_{gj}) - \frac{d_{eq}}{Pe_l}\frac{d^2c_{li}}{dz^2}, \tag{5}$$

where L, G are mass flowrates of liquid and gas, respectively, kg/s; S is cross-sectional area of the column, m^2. d_{eq} is equivalent diameter of the packing, m; Pe_g, Pe_l are Peclet numbers (criteria) for back mixing in the gas and liquid phases, respectively; $Pe_g = w_g d_{eq} / D_{mg}; Pe_l = u_l d_{eq} / D_{ml}$; w_g, u_l are average gas and liquid velocities, respectively, m/s; D_{mg}, D_{ml} are coefficients of back mixing of flows, m^2/s.

The Dankwerts boundary conditions are applied to equations (4), (5):

- at z = 0 (gas inlet and liquid outlet):

$$c_{gi} = c_{gsi} - \frac{d_{eq}}{Pe_g}\frac{dc_{gi}}{dz}; \quad \frac{dc_{li}}{dz} = 0, \tag{6}$$

- at z = H (gas outlet and liquid inlet):

$$c_{li} = c_{lsi} + \frac{d_{eq}}{Pe_l}\frac{dc_{li}}{dz}; \quad \frac{dc_{gi}}{dz} = 0, \tag{7}$$

where H is height of the packed bed, m; indices: s are initial (starting) values of concentration.

To calculate Peclet numbers, generalizations of experimental data for each type of packing are used. In the gas phase, expressions derived from the Taylor model [8] can be applied.

An expression for determining the matrix of mass transfer coefficients follows from the additivity equation of phase resistances:

$$[K_{vg}] = [[m][\beta_l]^{-1} + [\beta_g]^{-1}]^{-1}, \tag{8}$$

where [m] is matrix of distribution (equilibrium) coefficients; $[\beta_l]$, $[\beta_g]$ are matrices of mass transfer coefficients in the liquid and gas phases, respectively, which are determined in accordance with the mathematical model of the boundary layer [8].

3. Distillation column modernization

An example of modernization of an oil and gas condensate (OGC) mixture stabilization column at an oil refinery is considered.

The stabilizer column has a variable cross-section (the upper part is smaller than the lower one) with 38 multi-pass sieve trays. In the upper part there are 19 four-pass trays with a diameter of 2.6 m, and in the lower part there are 19 six-pass trays with a diameter of 3.2 m. The load interval for the oil and gas condensate mixture is from 160 to 250 m^3/h. The distillation process in the column takes place at a pressure of 0.6-1.4 MPa and temperatures: column bottom - no more than 250 °C; column top - no more than 150 °C.

The heated OGC mixture is fed to the central part of the column. In the upper part of the column, vapors of a broad fraction of light hydrocarbons (BFLH) are removed, where they are further condensed in a refrigerator-reflux condenser and enter the reflux tank. Part of the BFLH from the tank is fed for further processing or storage, while the other part goes back to the column in the form of cold reflux. Stable condensate (up to 250 °C) is removed from the stabilizer bottom through a heat exchanger-recuperator for heating the initial mixture. Part of the bottom mixture is heated in the furnace and returned to the column in the form of a "hot stream".

Based on the use of a mathematical model, calculations are performed and an analysis of the stabilizer column operation is made. It is found that the change in the composition of C_6 and higher in the initial mixture from 65 wt. interest to 49 wt. interest leads to a decrease in the maximum productivity from 550 to 410 m^3/h due to insufficient throughput of the trays in the upper part of the column. Replacement of sieve trays in the upper part of the column with structured roll packing "Inzhekhim" having a corrugated surface and roughness is considered. The packing is made of metal strips and, depending on the corrugation height, can have a specific surface area of 200 to 350 m^2/m^3. To operate at increased loads, calculations are made for the packing with a_v = 200 m^2/m^3 [8]. The task of the calculation is to determine the height of the packing for the given operating and design characteristics and the quality of mixture separation. The input concentrations of the components and the flow rate of the mixture into the packed section are determined from the tray-by-tray calculation by the method of theoretical trays and the results are compared versus the data of the industrial operation of the column.

However, modelling of mass transfer in a mixture consisting of several dozen components is laborious; therefore, the calculation is performed for the following key components: propane, isobutane, n-butane, isopentane and a fraction of 40-50 °C.

As a result of calculations, it is found that the height of the packed bed H = 8.0 m is sufficient for a given quality of the mixture separation when operating with an increased load up to 550 m^3/h, and for the feedstock load of the OGC mixture - up to 250 m^3/h. As a result of the introduction of structured packing, the content of isopentane in the stable condensate is reduced. The data of industrial operation of the column after the introduction of packing show that the content of isopentane in BFLH increases from 9-10 wt. interest up to 11.7 wt. interest and the content of the $\sum C_6$ components decreases to 0.36 %.

4. Conclusion

This article shows an example of mathematical modelling of a multicomponent hydrocarbon mixture based on the use of flow structure models and the introduction of a new roll packing in a stabilizer column of an oil and gas condensate mixture. The calculation is carried out for the key components and the required packed bed height is selected to provide the specified amount of mixture separation.

The developed mathematical model and the considered roll packing can be used to solve scientific and technical problems of designing new columns or modernizing existing ones at oil and gas refining enterprises.

Acknowledgments

The study was carried out within the framework of the scientific project of the Russian Science Foundation No. 18-79-101-36.

References

[1] Kulov N N 2017 International Scientific and Technical Symposium. *The Second International Kosygin Readings Timed to the 100th Anniversary of the Russian State University Named After A.N. Kosygin* 30-9

[2] Komissarov Yu A and Dam K Sh 2019 *Chemical engineering: Multicomponent distillation* (Moscow: Urait) 255

[3] Komissarov Yu A, Gordeev L S and Vent D P 2019 *Chemical engineering: Scientific foundations of distillation processes. Part 1* (Moscow: Urait) 416

[4] Sister V G and Lagutkin M G 2018 *Processes and apparatus of chemical engineering. In two parts. The main course and promising processes. Calculation of equipment strength and residual life. Reactors and chemical processes* (Moscow: Moscow Polytech Press) 369

[5] Vitkovskaya R F, Pushnov A S and Shinkunas S 2019 *Aerohydrodynamics and heat & mass transfer of packed apparatus* (Moscow: Lan-Press) 288

[6] Korovin P I, Lagutkin M G, Korovina N A and Pushnov A S 2019 Proceedings of the IV.

International Scientific and Practical Conference 88-92
[7] Razinov A I, Klinov A V and Dyakonov G S 2017 *Processes and apparatus of chemical engineering* (Kazan: Kazan National Research Technological University Press) 860
[8] Laptev A G, Basharov M M, Lapteva E A and Farakhov T M 2020 *Models of interphase transport and calculation of process efficiency. Part 2. Heat and mass transfer processes* (Kazan: Center for innovative technology) 565
[9] Ivanov I V, Lotkhov V A, Kulov N N and Moiseeva K A 2016 Mass transfer in a packed extractive distillation column. *Theoretical Foundations of Chemical Engineering* **50.5** 667-77
[10] Kagan A M, Pushnov A S and Ryabushenko A S 2007 Attachment contact devices. *Khimicheskaya Tekhnologiy* **8** **5** 232-40

Distillation of vacuum gas oil in a column with new packings

M I Farakhov[1], A G Laptev[2], T M Farakhov[2] and A A Akhmitshin[1]

[1] LLC EPC "Inzhekhim", 14/83, Shalyapin Str., Kazan, 420049, Russia
[2] Kazan State Power Engineering University, 51, Krasnoselskaya Str., Kazan, 420066, Russia

E-mail: Tvt_kgeu@mail.ru

Abstract. A description of the technological scheme of the unit for separating the fraction of vacuum gas oil (FVG) is given, where the process of separation of heavy gas oil from FVG takes place in the column. The main focus of the work is on the design of a distillation column using the new structured roll packing. The fractional composition of the feedstock, by-product, light gas oil, and the target product, which is heavy gas oil, is presented. Requirements for the design of a structured packing in terms of pressure drop and separation efficiency are formulated. A structured roll packing with corrugations and rough surface having specific surface area of 300 m^2/m^3 is chosen. The diagram and photograph of the packing are presented. Comparative values of specific pressure drop for several structured packings are given. It is shown that the selected packing provides the lowest hydraulic resistance. For mathematical modeling and calculation of the multicomponent distillation process, a system of mass transfer equations with interfacial bulk sources of transfer components is used. In this case, for the conditional component, a fraction at the specified boiling points is used. As a result of solving the system of equations with boundary conditions, the profiles of the fractions along the height of the packed bed are found and the structural characteristics of the structured packing are selected that meet the requirements for the top and bottom of the column. Calculations of a vacuum distillation column with this packing are carried out and it is found that the bed height of 6 meters is sufficient to meet the requirements for the fractional composition of the upper and lower products, while meeting the pressure drop requirements. The developed scientific and technical solutions have been implemented at a refinery in Malaysia and correspond to the technical specifications for the design of the unit.

1. Introduction

Mathematical modeling and selection of instrumentation for the separation of oil mixtures by distillation are relevant for both fundamental science and applied research for oil and gas refining [1-3]. It is known that distillation is one of the most energy-intensive processes, and mass transfer columns have a complex design and increased metal consumption [4-5]. Depending on the conditions of the process of separating mixtures in the columns, bubbling, jet, packed or combined gas-liquid contact devices are used [6-8]. If the column operates at reduced pressure, the columns are called vacuum columns and are used mainly for the distillation of heavy hydrocarbon mixtures [9-12].

2. Materials and methods

The unit for separating the fraction of vacuum gas oil (FVG) is intended to separate heavy gas oil from the broad FVG in order to provide the extract normalization unit with raw materials. The feed capacity of the unit is 65 thousand tons per year.

The FVG separation unit consists of the following modules (figure 1).

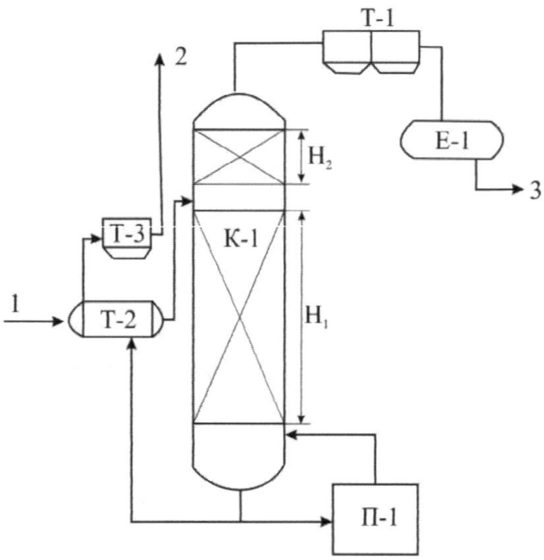

Figure 1. Schematic technological scheme of the vacuum distillation unit for the fraction of vacuum gas oil: 1 - input of the feed (FVG); 2 - output of the target bottom product (380–540 °C); 3 - output of the top by-product (300–420 °C); K-1 - distillation packed column; T-1, T-3 - air heat exchangers; T-2 - heat exchanger-recuperator; П-1 - furnace

- Module for separation of FVG by vacuum distillation in a packed column K-1;
- Module for distillate condensation and cooling in air heat exchanger T-1 with distillate collection tank E-1;
- Module for cooling the finished product in the heat exchanger-recuperator T-2 and in the air heat exchanger T-3;
- Module for heating the bottom liquid in the П-1 furnace;
- Vacuum creating system.

The purpose of the present work is to design the K-1 vacuum distillation column with a scientifically grounded choice of a high-efficiency packing of a structured type, as well as to use a plate heat exchanger with a high specific surface area developed by the authors as a heat exchanger-recuperator T-2.

According to the design, the vacuum distillation unit with the K-1 column has a feed capacity of 194.3 tons per day, i.e. 8.1 tons per hour. Column diameter is 1200 mm; column height is 16336 mm.

The commissioning of this unit will ensure the release of light fraction of FVG and heavy fraction of FVG (fractional composition of the mixtures is given in table 1):

- Target bottom product is a heavy FVG with a boiling point of 380–540 °C, which is a feed for a unit for the oil extracts normalization;
- Top by-product is a light FVG with a boiling point of 300–420 °C, which is used as a feed for petrochemical processes.

The initial raw material with an initial temperature of 60 °C is pumped into the heat exchanger-recuperator T-2, where it is heated by the heat of the finished product to the required temperature of 220-280 °C, i.e. close to the initial boiling point of the initial mixture under vacuum. Then the mixture is fed through the throttle to the upper part of the K-1 vacuum column through the flow distributor. The temperature regime of the feed supply at the inlet to the column is maintained by controlling the flowrate of the hot heat carrier (heavy FVG).

At the entrance to the column K-1, partial evaporation of the lightest fractions of the feedstock occurs due to heat when the pressure is reduced to vacuum. The unevaporated liquid phase of the feedstock enters the section with packing and flows in the form of a film over its surface into the column bottom, interacting with the vapor flow that rises from the column bottom after the П-1 furnace. The process of vacuum distillation takes place on the structured packing of height H1. To separate the droplet liquid that can be carried away by the rising vapor, a separating section of the packing of height H2 is provided in the upper part of the column.

Table 1. Fractional composition of gas oil.

Fractional composition, %	Feedstock, °C	By-product - light gas oil, °C	Target product - heavy gas oil, °C
5	351.76	302.86	384.74
10	368.43	319.90	398.17
15	380.92	331.44	407.64
20	391.36	339.95	415.16
25	399.87	346.82	421.59
30	407.28	352.44	427.59
35	414.30	357.56	433.61
40	421.26	362.17	439.80
45	427.99	366.69	445.97
50	434.73	371.14	452.21
55	441.63	375.40	458.63
60	448.87	379.85	465.38
65	456.48	384.38	471.81
70	464.66	389.09	478.51
75	473.22	394.33	486.21
80	482.42	399.92	494.97
85	492.99	406.31	504.98
90	505.84	414.19	516.92
95	522.44	424.88	533.39

The vapor flow from the top of the column K-1 enters the air cooler-condenser T-1, where it is condensed and cooled to 70 °C. The fraction of waste is 11.68 wt.%. The cooled condensate is collected in tank E-1 and then pumped to the storage vessel.

The vapor flow at the bottom of the column is created by intensively heating part of the bottom liquid in the П-1 furnace. Evaporation in the column occurs only due to the heat of heating the flow (up to 310 °C) when the pressure is reduced to vacuum.

Part of the bottom liquid from the П-1 furnace enters the heat exchanger-recuperator T-2, in which it gives off heat to the feedstock, after which it is additionally cooled in the air cooler T-3 and then fed to the storage vessel.

The vacuum in the column is created using a vacuum creating system based on a vacuum pump. The vacuum line is connected to the E-1 tank (the line is not shown in the scheme).

3. Selecting the packing design

For a vacuum distillation column, the packing, in addition to its high separation capacity, must provide a low hydraulic resistance due to restrictions on the pressure drop in the upper and lower parts of the apparatus. Therefore, the choice of packing type is limited to structured roll or plate type designs. During distillation, the resistance to mass transfer is concentrated both in the liquid and in the vapor phases (somewhat more in the vapor phase), and then the packing should provide a turbulent regime of vapor movement and wave flow of the film. This mode is ensured by corrugated metal sheets, with the corrugations at an angle to the horizon (30-45°), and an artificially rough surface. The roughness ensures the wave flow of the film even at low Reynolds numbers.

These requirements are met by a structured roll corrugated packing with an applied roughness on the surface (protrusions with a height of about 1 mm and with a pitch of 3 mm). A structured packing is shown in figure 2, and a photograph of the manufactured packing for an industrial column is shown in figure 3 [11].

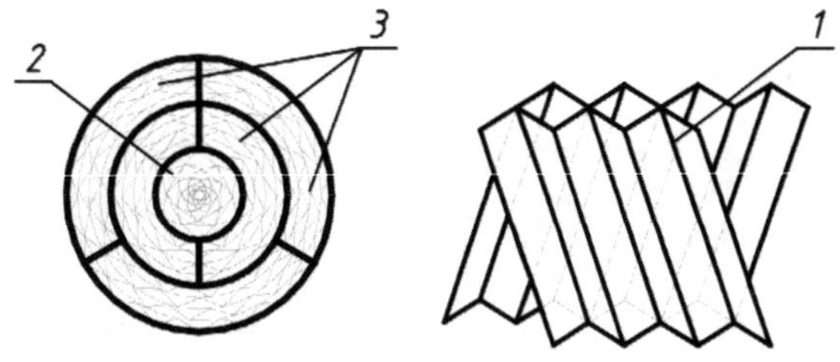

Figure 2. Type of structured packing: 1 - corrugations of adjacent sheets; 2 - central pack; 3 - other packs.

Figure 3. Packing "Inzhekhim" for industrial columns.

Depending on the height of the corrugation, the packing can be manufactured with a specific surface area of 200 to 350 m2/m3 and has a free volume of 0.95 m^3/m^3 [11].

$$\frac{\Delta p_{\text{irrigated}}}{H}, \frac{\text{Pa}}{\text{m}} \quad (1)$$

Figure 4. Hydraulic resistance of irrigated packings depending on gas velocity and irrigation density:
1, 2 - roll packing, irrigation density;
3 - structured packing INTALOX 3T at full reflux (isooctane-toluene system);
4 - roll packing with a rough surface.

Figure 4 shows comparative values of the specific pressure drop for structured packings for the air-water system under normal conditions. As follows from the data presented, the selected design of the structured packing has the lowest hydraulic resistance.

4. Conclusion

To calculate the process of separating vacuum gas oil into specified fractions, a system of differential equations of the diffusion model of the flow structure is used [11]. The one-dimensional equations of the diffusion model are written for the liquid and vapor phases, where mass transfer is considered due to volumetric interfacial sources, together with the conditions of thermodynamic equilibrium. As a result of calculations, it is found that six meters of the packing height is sufficient for a given quality of separation of the FVG (table 1). The considered design of the packing can be used in the modernization or design of industrial distillation apparatus.

Acknowledgments
The study was carried out within the framework of the scientific project of the Russian Science Foundation No. 18-79-101-36.

References

[1] Golubeva I A, Khudyakov D S and Rodina E V 2020 *Chemistry and Technology of Fuels and Oils* **56.3** 395-404
[2] Ledenev S M, Shibitova N V and Zhirnov V V 2020 *Chemistry and Technology of Fuels and Oils* **56.1** 9-11
[3] Kulov N N , Maksimov V V, Maljusov V A and Zhavoronkov N M 1979 *The Chemical Engineering Journal* **18.3** 183-8
[4] Ivanov I V, Lotkhov V A, Kulov N N and Moiseeva K A 2016 *Theoretical Foundations of Chemical Engineering* **50.5** 667-77
[5] Kagan A M, Pushnov A S and Ryabushenko A S 2007 *Khimicheskaya Tekhnologiya* **8.5** 232-40
[6] Gorodilov A A, Pushnov A S and Berengarten M G 2014 *Chemical and Petroleum Engineering* **50.1-2** 84-90
[7] Dmitrieva G B, Berengarten M G, Kagan A M, Pushnov A S and Klimov A G 2007 *Chemical and Petroleum Engineering* **1** 11-4
[8] Komissarov Y A, Ravichev L V and Kiselev M S 2019 *Theoretical Foundations of Chemical Engineering* **53.5** 747-59
[9] Telyakov E S, Ponikarov A S and Osipova L E 2018 *Theoretical Foundations of Chemical Engineering* **52.1** 11-23
[10] Telyakov E S, Osipova L E and Ponikarov A S 2018 *Teoreticheskie Osnovy Khimicheskoi Tekhnologii* **52.1** 13-25
[11] Laptev A G, Basharov M M, Lapteva E A and Farakhov T M 2020 *Models of interphase transport and calculation of process efficiency. Part 2. Heat and mass transfer processes* (Kazan: Center for innovative technology) 565
[12] Babiev V A, Korovin P I, Lagutkin M G, Generalov M B and Sidelnikov I I 2018 *Chemical Industry Today* **5** 26-9

Features of the formation of environmental protection mechanisms during the operation of objects of the electric power complex

E V Provodina[1], O Yu Krasovskaya[1], T A Gerasimova[1], Yu V Predeus[2] and D P Tserpento[2]

[1] Saratov State Law Academy, 1, Volskaya street, Saratov, 410056, Russia
[2] Socio-economic Institute of Yuri Gagarin State Technical University of Saratov, 89, Radishcheva street, Saratov, 410003, Russia

E-mail: lysenko-elena12@yandex.ru

Abstract. Based on the analysis of the methods of greenhouse gas stations used for environmental pollution, their systematization into a single mechanism for ensuring environmental safety at various levels is presented: from the state level to the technological system. The content of the constructed mechanism for substantiating the need for the interdependent functioning of its elements. The peculiarity of the proposed mechanism is that it contains elements that ensure the safety of the environment, guaranteeing the means and methods of implementing environmental protection issues.

1. Introduction

The topic of environmental safety and environmental protection in connection with the operation of energy facilities is in demand and is widely discussed, since these facilities, already by their very existence, are a source of increased danger to the environment. This is primarily due to the problem of greenhouse gas emissions, "… which is most relevant for companies in the fuel and energy complex" [1].

The problem of environmental pollution in connection with the functioning of energy facilities in the long term may become decisive for the existence of future generations. Therefore, the increased concern about it at the level of an independent direction of state policy is quite understandable and natural. "... Significant restraint of the growth of greenhouse gas emissions and the reduction by organizations of the fuel and energy complex of harmful emissions into the environment" is considered an important consequence of the energy conservation policy in the Energy Strategy of the Russian Federation for the period up to 2035 [2]. In science, it is proposed to solve the existing problem, in particular, by rejecting the use of outdated equipment, development and implementation of innovative technologies in the field of energy production [3].

The attention of scientists and politicians, in the light of its solution, is riveted to the digital transformation and intellectualization of the fuel and energy complex, which is also provided for in the Energy Strategy of the Russian Federation for the period up to 2035 as significant characteristics of the chosen course of development.

"The more the public sees the environment as a priority policy issue, the more likely it is that corresponding policies will be implemented" [4].

Studies have shown that "... if individuals perceive a policy instrument as fair, they are more likely to support that policy instrument" [5]. It is no coincidence that environmental disasters are called a potential engine of public opinion related to the environment [4].

"At the same time, in order to switch to new conditions for the functioning of the industry, it is necessary to form new subsystems to ensure the implementation of digital and intelligent networks in the electric power complex, which, in the future, should provide economic, environmental, social and technological effect for the national economy" [6].

In this connection, the transition to new conditions for the functioning of the entire energy system (extensive modernization) is complicated by the need to achieve a balance of economic benefits and environmental safety.

It is proposed to solve the problem of reducing harmful emissions into the environment proceeding from the fact that "... the electric power complex is currently not only a branch of the national economy, but also a system forming element of the national economy" [3]. Therefore, economically profitable energy facilities continue to remain in demand, despite their potential danger to the environment "... the historically established system for the production of electricity at coal-fired thermal power plants is economically beneficial; however, such power plants have a negative impact on the environment" [3].

In this direction, the researchers identified problems "... among which are the high level of moral and physical deterioration of equipment, a decrease in efficiency and resource saving in the industry, a high proportion the use of obsolete technologies, a low share of the introduction of innovative and advanced technologies in the industry, a high share of the environmental load on the environment and other problems" [7].

2. Materials and methods

The main purpose of the study is to systematize the elements and form a mechanism for environmental protection in connection with the functioning of the electric power system, to identify areas that require enhanced modernization.

The following tasks were formulated:

- To define the elements of the mechanism ensuring environmental safety during the operation of energy facilities;
- To present the elements of the mechanism that guarantee compliance with safety requirements during the operation of power system facilities.

The study of the stated problem was based on an assessment of existing solutions at various levels of organization in the field of energy (political, economic, legal) using modern scientific approaches, as well as the method of comparative, logical, systemic analysis and other methods.

3. Results

In light of the stated problem, the proposals of scientists, developed in the spirit of preventing environmental pollution about "... switching from coal to gas and renewable energy sources, electrification of transport" [1], are quite natural, which, along with "sequestration technologies" [1], are considered in quality of modern emission reduction technologies. However, the solution to this problem is not limited to technological aspects and necessitates a more detailed approach to the prevention of environmental pollution in connection with the activities of energy facilities. Therefore, the method of state stimulation of the reduction of greenhouse gas emissions - carbon pricing, as well as the legal mechanism within the framework of determining the conditions of anthropogenic confrontation to the problem, acquire its own significance. Such an aggregate positioning of means and methods of counteracting further escalation of the problem gives reason to consider them as independent elements of a preventive mechanism, the object of protection of which is environmental safety. New technologies, such as switching from coal to gas and renewable energy sources, allow for an overall reduction in global emissions, which is one of the main requirements of the Paris Agreement.

However, the understanding of the safety of functioning of energy facilities from the point of view of ecology should not be limited to the transition to new technologies. It should not be forgotten that energy facilities are a priori a source of increased danger, and their operation requires special control over the observance of special safety rules "... the number of accidents has been decreasing in recent years, but their negative consequences are increasing" [8]. At the same time, the reliability of power supply includes "... maintainability, uninterrupted operation, durability and maintainability of the system ..." [8].

In addition, the causes of accidents, in addition to the use of worn out equipment and weather conditions, are "A lot of the reasons of system accidents are coming due to organizational work. The situation is complicated by insufficient qualify" [9]. Therefore, the decrease in the accident rate observed in recent years [10], due to a number of factors, should be considered as one of the significant independent elements gaining positive momentum in the direction of environmental protection.

The mechanism responsible for the safe operation of these facilities must include warranty elements that ensure the safe operation of the relevant facilities. Such protective guaranteeing mechanisms exist and are presented at the state level in the form of imperatives and are of a pronounced preventive nature. Constructions containing responsibility allow the implementation of the stimulating properties of the entire mechanism that ensures the safety of the environment in the area under consideration, a serious attitude in the operation of energy facilities and equipment. Consequently, the aspect of safe operation and the guarantee of its observance should be considered as one of the equally important and relevant elements of the proposed mechanism, along with, for example, the pricing mechanism, since it presupposes not only the protection of the environment as a human habitat, but also directly provides guarantees of the highest values established by the Basic Law of the state - the right to life and health in connection with the operation of energy facilities that serve as a source of increased danger.

The safe functioning of the electric power complex is not only the most important condition for positioning this industry as a component of the state economy, but also provides a favorable environment for the existence and development of future generations.

Safety as a necessary specific property of the power system is realized through a whole complex of interdependent elements, only the combination of which is able to guarantee the preservation of the environment.

The environmental protection mechanism in relation to energy facilities includes the following main components:

- Innovative equipment, which allows, first of all, to reduce harmful emissions into the environment and to implement the provisions contained in the legal mechanisms and regulating the goals to reduce the content of carbon dioxide in the air;
- Safe operation of the relevant equipment at energy facilities;
- Guaranteeing means of compliance with the requirements for safe operation of the energy complex (imperative mechanisms of responsibility).

Thus, the safety of the electric power system, including the implementation of a set of interdependent elements that ensure it at various levels, creating the possibility of stable, continuous development of the electric power economy in the state, is a set of preventive measures in order to prevent negative impact on the environment with high quality and uninterrupted power supply consumers.

4. Discussion

Our proposed model of the safe existence and functioning of the electric power system takes into account both the technological safety of the equipment used at the energy complexes from the point of view of maintaining the ecology, and the factor of accidents, considered as an independent element of the threat to environmental safety.

In a broad sense, the mechanism for the safe operation of the relevant facilities cannot be limited only to harm to the environment, but must include the safe existence of people, without endangering their highest values of the right to life and health, guaranteed by the fundamental law of the state. This is directly related to compliance with safety requirements and rules at energy facilities. In the light of the above, special attention should be paid to the mechanisms that guarantee the exact implementation of the requirements of the legal and regulatory framework.

The procedure for the operation of such facilities is under special state protection, due to the nature and level of public danger of possible consequences both for the environment and for the population, therefore, this factor must be taken into account when developing a mechanism for the safe functioning of the entire energy system.

At the same time, the elements of this mechanism should not abstractly exist independently of each other, but constantly dynamically interact, ensuring the management of factors that influence the state of the environment in the implementation of preventive measures.

5. Conclusion

As a result of the study, dedicated to the peculiarities of the formation of the mechanism of environmental protection during the operation of energy facilities, it was found that the mechanism for ensuring environmental safety, aimed at reducing greenhouse gas emissions, should include the following elements:

- Provision of the energy system with modern technologies (switching from coal to gas), which create the technical possibility of reducing emissions, which is one of the main conditions for the implementation of the environmental protection mechanism;
- Widespread use of sequestration technology aimed at absorbing and reducing carbon content in the atmosphere;
- Public policy, the elements of which are a carbon pricing mechanism as the main method of government incentives for emission reductions and a legal mechanism that regulates targets for reducing carbon dioxide in the air;
- A set of measures aimed at reducing accidents as an independent factor of environmental pollution in connection with the operation of energy facilities, including ensuring safety requirements and rules at energy facilities by means of guaranteeing mechanisms, that is, legal structures that stimulate compliance with the stated requirements.

The presented mechanism is complex and is based on the interdependent functioning of the constituent elements, including specific guaranteeing features. Its application is not limited to the territorial boundaries of one state, in connection with which it can be equally promisingly implemented in order to reduce the level of greenhouse gas emissions and preserve the environment.

References

[1] Tsopanova Z et al 2020 *IOP Conf. Ser.: Mater. Sci. Eng.* **837** 012002
[2] *Energy Strategy of Russia until 2035* Retrieved from: https://minenergo.gov.ru/node/1920
[3] Gibadullin A A et al 2019 *IOP Conf. Ser.: Mater. Sci. Eng.* **537** 042065
[4] Tobias Böhmelt 2020 Environmental disasters and public-opinion formation: A natural experiment. *Environmental Research Communications* **2** 081002
[5] Robert A Huber, Michael L Wicki and Thomas Bernauer 2020 Public support for environmental policy depends on beliefs concerning effectiveness, intrusiveness, and fairness.

Environmental Politics **29:4** 649-73
[6] Zimnukhova D I et al 2019 Management and development of digital technologies in the electric power industry of Russia. *Journal of Physics*: *Conf. Ser.* **1399** 033097
[7] Hojiev J N et al 2020 *IOP Conf. Ser.: Mater. Sci. Eng.* **837** 012008
[8] Gibadullin A A 2020 *IOP Conf. Ser.: Mater. Sci. Eng.* **837** 012007
[9] Udintsev D N et al 2020 *IOP Conf. Ser.: Mater. Sci. Eng.* **837** 012004
[10] *Ministry of Energy of the Russian Federation* Retrieved from: https://minenergo.gov.ru

IOP Conf. Series: Materials Science and Engineering **976** (2020) 012019 doi:10.1088/1757-899X/976/1/012019

Ontology of experiment planning for obtaining a probabilistic model of single-phase electricity consumers

V A Mokhov, D V Shaikhutdinov, T N Kruglova, R G Oganyan and A N Tkachev

Platov South-Russian State Polytechnic University (NPI), 132, Prosveshcheniya, Novocherkassk, 346428, Russia

E-mail: mokhov_v@mail.ru

Abstract. The paper presents the rationale and the problem of forming a probabilistic model of single-phase electricity consumers. To solve the problem, the corresponding mathematical apparatus is stated and the ontology of experiment planning is developed. The applicability of the prepared ontology was demonstrated on a specific example of an open dataset from the Intelligent Systems Subcommittee (ISS) IEEE. At the same time, the authors identified the characteristic modes of energy consumption for a particular residential building and data sets for direct construction of a probabilistic model of single-phase electricity consumers with total data loss caused by filtering less than 10%.

1. Introduction

Often, when solving practical problems, the effectiveness of computer modeling is determined by the degree of similarity of the characteristics of the model of the investigated object with its real physical characteristics [1-2]. This is especially important in instances when a system of automatic control of a physical object is built based on the results of modeling. To solve the problem of controlling interphase switching of single-phase electricity consumers [3], a model of the consumer behavior is required. For this purpose, this work presents the developed mathematical apparatus and ontology for planning an experiment to obtain a version of such a model with specific examples of calculations.

2. Materials and methods

Previously, in the work of the authors [3], the problem of controlling interphase switching of single-phase electricity consumers was solved, the essence of which is formulated as follows:

- n non-industrial single-phase consumers (SC) of electricity are connected to the distribution network powered by the distribution transformer substation (TS);
- each of the SC is provided with dynamic switching tools in relation to the phases L1, L2, L3 within the TS;
- the implementation of the solution to the task of switching the SC is performed dynamically at regular intervals t.

The input data for finding a solution is:

- original number combination of SC to TS connection phases:

$$L^0 = \{l_1^0, l_2^0, \dots, l_n^0\},\ l_r^0 \in \{1,2,3\}, \qquad r = \overline{1,n} \tag{1}$$

- a set of digitized load current oscillograms corresponding to each combination of connections (1) for all SCs

$$I=\{\overline{i_1},\overline{i_2},...,\overline{i_n}\} \tag{2}$$

(in the form of arrays of k instantaneous values of currents $\overline{i_s}=\{i_{s1},i_{s2},...,i_{sk}\}$, obtained for a specified integer number of periods).

The solution to the problem is to find a new combination of numbers of SC connections to TS phases for the time t' $(t'<t)$ in real time:

$$L^*=\{l_1^*,l_2^*,...,l_n^*\},\ l_r^*\in\{1,2,3\},\ r=\overline{1,n}, \tag{3}$$

in the implementation of which, in comparison with the original combination (1), the smallest values are provided: the number of switching of the SC in phases, the values of the indicators characterizing the asymmetry of currents and their total harmonic distortions with the same set of oscillograms I(2). To find the combination (3), the presented problem is reduced to a combinatorial optimization problem, and the search for its solution is performed using binary algorithms of agent-based metaheuristics. The objective function for the problem is calculated based on (2) and is presented in [3]. Obviously, to perform the calculations in the stated problem, a system for generating sets (2) is required.

The most widespread spectral decomposition of periodic functions is their representation in the form of a truncated segment of the Fourier series [4, 5]:

$$\tilde{a}(t)=a_0+\sum_{k=1}^{n}a_k\sin k\omega t+\sum_{k=1}^{n}a_{n+k}\cos k\omega t \tag{4}$$

where $x(t)$ is a periodic function with main period T; $\omega=2\pi f$; $f=\frac{1}{T}$; $a_0={}^{1}/_{T}\int_0^T x(t)dt$; $a_k={}^{2}/_{T}\int_0^T x(t)\sin k\omega t\,dt$; $a_{n+k}={}^{2}/_{T}\int_0^T x(t)\cos k\omega t\,dt$; $k=\overline{1,n}$.

In accordance with (4), the elements of the set (2) can be uniquely determined by the set of values of the Fourier coefficients $\bar{a}=(a_0,a_1,...\,a_n,a_{n+1},...\,a_{2n})$.

We assume $a_0,a_1,...\,a_n,a_{n+1},...\,a_{2n}$ to be random variables. In this case, each SC or element in (2) represents a random process that can be described by the corresponding random vector $\bar{a}=(a_0,a_1,...\,a_n,a_{n+1},...\,a_{2n})$.

Then, to build the model (2), it becomes necessary to preliminarily estimate the statistical parameters of the distribution $a_0,a_1,...\,a_n,a_{n+1},...\,a_{2n}$. The solution to the assessment problem assumes the presence of γ corresponding statistical observations, for which, by means of (4), a statistical sample of the form must be formed:

$$\begin{matrix}\bar{a}_1\leftrightarrow\{a_{0_1},a_{1_1},...\,a_{n_1},a_{n+1_1},...\,a_{2n_1}\}\\ ...\\ \bar{a}_\gamma\leftrightarrow\{a_{0_\gamma},a_{1_\gamma},...\,a_{n_\gamma},a_{n+1_\gamma},...\,a_{2n_\gamma}\}\end{matrix}. \tag{5}$$

The statistical estimates obtained on the basis of (5) will make it possible to determine the parameters of the distribution functions

$$\eta_k=\mathcal{F}_k^x(\zeta_k),\ \ \zeta_k\in[0;1] \tag{6}$$

for each of $2n$ values as well as to provide the opportunity to generate random vectors

$$\overline{A}\leftrightarrow\{a_0,a_1,...\,a_n,a_{n+1},...\,a_{2n}\}, \tag{7}$$

corresponding to the statistical sample (5) and determining the characteristics of the load currents (2) for a given number of p individual consumers in solving the problem of search (3).

As a statistical sample for the formation (5), we will use the information on the amplitude spectrum of the current consumption from the open dataset Private Home 1 [6], provided by the Intelligent Systems Subcommittee (ISS) IEEE at [7].

3. Results

Generally, a means of constructing mathematical models of various processes in order to increase the efficiency of experimental research is the planning of an experiment [8]. It allows to increase the reliability of the research results [9] and to reduce the time and funds for the experiment. The main stages of the planned experiment [10] can be described as follows.

1) The goal of planning an experiment in this case is to obtain parameters for distribution functions (6) at a given level of confidence.
2) With regard to the conditions of the experiment, information is available that the tests were carried out on the basis of installation single-phase 5.75 kVA in the period from 3rd to 18th June 2011 for a single family housing, with a sampling period of 5 minutes by means of the analyzer quality of electrical energy Chauvin Arnoux 8335 (France). Single family housing has the following characteristics:

    a) number of people per house: 3 adult people;
    b) number of rooms in the house: 3 bedrooms, 3 bathrooms, 1 living room, 1 kitchen, 1 laundry room, 1 hallway and corridor;
    c) main electrical equipment used: 1 washer / dryer, 1 dishwasher, 1 electric stove, 1 electric oven, 1 refrigerator, 1 microwave, 1 range hood, 1 vacuum cleaner, 1 hair dryer, 3 TVs, 1 LCD display, 2 laptops , 1 router, 2 electric heaters.

3) Input data is time counts with a period of 5 minutes (registered parameter) and corresponding random combinations of equipment active/switched on during this period (non-registered parameter). The output data is the data of the amplitude spectrum of the current consumption.
4) The output data is presented as a set of 4320 measurements for each harmonic, their values are measured in amperes with an accuracy of representation to the fifth decimal place.
5) Obviously, when using the data from the open ISS IEEE sets indicated in the previous paragraph, the experiment can be considered passive, based on the registration of input and output parameters characterizing the object of research, without interfering with the experiment in the course of its conduct. In this case, experimental data is processed after the end of the experiment.
6) Statistical processing of the experimental results to construct a mathematical model of the behavior of the characteristics under study was performed as follows.

    a) In the general dataset, all measurements related to the inactive period of residents were excluded: from 00:00 hours to 09:00 hours. This period of time is characterized by the fact that during its course the elements of electrical equipment with low energy consumption operate mainly in automatic mode, which does not determine the main trend of daily energy consumption. As a result, the amount of data for research was reduced to 2700 measurements for each harmonic.
    b) Based on RMS statistics for current consumption I_{rms}, calculated through the effective values of harmonics of various orders I_h according to the formula (8)

$$I_{rms} = \sqrt{\sum_{h=1}^{\infty} I_h^2}, \tag{8}$$

    all measurements were ranked.
    c) For the sequence of data obtained in 6.b, subsamples were identified that define the typical energy consumption patterns for a residential building. For this:

        i. using the method of a sliding window of variable length [11, 12], estimates of the significance level were calculated at which the set of measurements within the bounds obeyed one of the known distribution laws;
        ii. for each of the laws involved, possible bound configurations were recorded (numbers of boundary elements and estimates of the level of significance);

iii. from the set of all possible configurations, those were selected that, with their minimum number and a given level of significance α ensured maximum coverage of calculated values I_{rms} with minimal intersection;

iv. for uncovered bounds (with a size of 100 or more elements and an insufficient level of significance), based on binary optimization algorithms, their elements were filtered to obtain a configuration corresponding to one of the known distribution laws with a significance level α while maintaining the maximum possible number of elements.

The set of configurations obtained in this way determined the size, boundaries, data sets and distribution laws that determine the characteristic modes of energy consumption.

d) In order to obtain configurations for the effective values of harmonics (within the limits of the power consumption modes defined in clause 6.c), the measurements were also filtered for compliance with one of the known distribution laws with a given level of significance α using the maximum number of elements (similar to n. 6.c.iv).
The resulting set of configurations determined the numerical composition, boundaries, data sets and distribution laws that determine the characteristic modes of energy consumption at the level of effective harmonic values.

7) An explanation of the results obtained and the formulation of recommendations for their use assumes that, based on the results of the implementation of the stages 6.a÷6.d obtained data sets are associated with statistical samples (5), on the basis of which estimates are calculated to determine the parameters of the distribution functions (6). In this case, the generation of random vectors (7) within each characteristic mode of energy consumption for a residential building with level of significance α becomes trivial.

Graph illustrating the curve of change of 2700 ranked values of the calculated value I_{rms}, is shown in the figure 1.

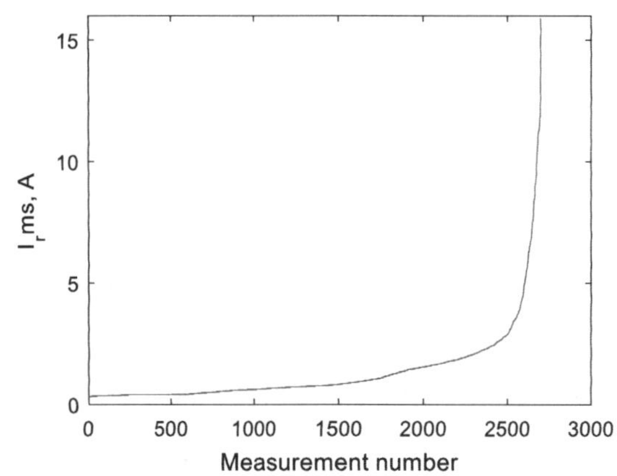

Figure 1. Ranged Values Curve I_{rms}.

Figure 2 is a graph showing an initial set of window configurations selected for subsequent filtering and extraction of typical power consumption modes.

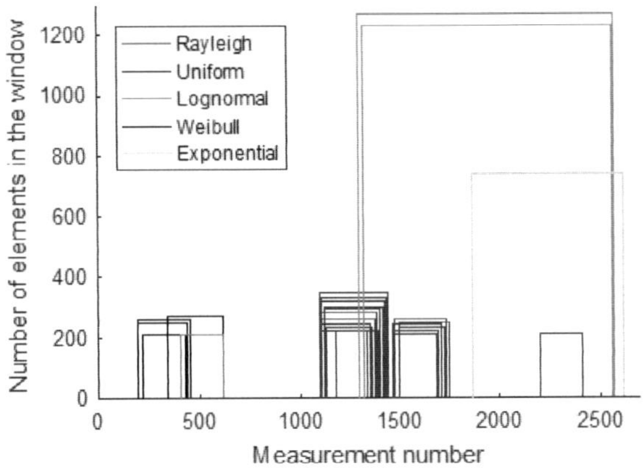

Figure 2. Initial set of bound configurations.

When constructing the graph (figure 2), bounds with the following configuration were displayed: window width - not less than 205 elements, confidence level - not less than 0.7. The distribution laws that sub-samples within bounds obey are indicated by colors: magenta for Rayleigh, blue for Uniform, red for Lognormal, black for Weibull, green for Exponential.

Subsequent I_{rms} data processing based on the graph (figure 2) made it possible to identify 6 typical power consumption modes, the configurations of which are presented in table 1.

Table 1. Configurations of typical power consumption modes of the object under study.

Mode	Boundary indices of elements	Number of elements, units	Range of values I_{rms}, A	Share in the total range of values I_{rms}, %	Law of distribution	Confidence probability α	Filtering loss, units	Filtering loss, %
1	1-190	190	0.323-0.3788	0.36%	Weibull	0.91343	65	34.21%
2	191-449	259	0.3791-0.4030	0.15%	Weibull	0.90267	10	3.86%
3	450-610	161	0.4032-0.4218	0.12%	Lognormal	0.90312	5	3.11%
4	611-1210	600	0.4224-0.7009	1.79%	Uniform	0.90252	77	12.83%
5	1211-2560	1350	0.7012-3.7011	19.27%	Rayleigh	0.90722	79	5.85%
6	2561-2700	140	3.7064-15.8924	78.27%	Exponential	0.90157	3	2.14%
Overall	1-2700	2700	0.323-15.8924	100%	--	--	239	8.85%

The first mode corresponds to the first (from left to right in figure 2) uncovered bound, which, according to the filtering results, best matches the Weibull distribution law. The second mode is also defined on the basis of the Weibull distribution law (in figure 2 - the first covered, black bound from left to right. The third mode is determined on the basis of the Lognormal distribution law (in figure 2 - the first red box on the left). The fourth mode is determined on the basis of the Uniform distribution law (figure 2 shows a frame between the red and magenta bounds, partially covered with many blue bounds). The fifth mode is defined on the basis of the Rayleigh distribution law (in figure 2, the larger magenta frame). The sixth mode is determined on the basis of the Exponential distribution law (in figure 2 - the rightmost one, starting from the left border of the magenta frame and partially covered with a green frame). The complete final configurations of power consumption modes are presented in the corresponding rows of table 1.

Total data filtering loss for a collection of 2700 values I_{rms} was less than 9%. In this case, the highest percentage of relative losses was obtained for the first mode, the share of value range of which in the total value range is only 0.36%.

The set of configurations for the effective values of harmonics (within the bounds determined by the configurations of the modes from Table 1) was obtained in a similar way by filtering the corresponding data sets based on modifications of the binary algorithms of agent-based metaheuristics [13-15] with total data losses within 10%.

4. Discussion

The applicability of the described experiment planning ontology for obtaining a probabilistic model of single-phase electricity consumers is demonstrated with the example of one of the open ISS IEEE sets. Its applicability in relation to other datasets, datasets of a harmonic nature from other subject areas, as well as the study of the possibility of constructing appropriate probabilistic models with high specified values of significance levels is the subject of further research.

5. Conclusion

The obvious results of the work presented by the authors are:

1) ontology of experiment planning to obtain a probabilistic model of single-phase electricity consumers;
2) a demonstration of its application in the example of one of the open ISS IEEE sets;
3) a list of typical energy consumption modes for a particular residential building, obtained on the basis of processing statistics of its energy consumption;
4) datasets for direct construction of a probabilistic model of single-phase electricity consumers, obtained on the basis of filtering measurement results by means of modifications of binary algorithms of agent-based metaheuristics with total losses less than 10% of the total volume of measurements.

Acknowledgments

The reported study was funded by RFBR according to the research project № 18-38-20188 «Development and research of tools for intellectual modeling and synthesis of electric power net-works technological processes trajectories» using equipment of shared facility «Diagnosis and energy-efficient electrical equipment» (NPI).

References

[1] Weisberg M 2012 *Simulation and Similarity: Using Models to Understand the World* (Oxford: Oxford University Press) 288
[2] Diedrichsen J and Kriegeskorte N 2017 Representational models: A common framework for understanding encoding, pattern-component, and representational-similarity analysis. *PLoS computational biology* **13(4)** e1005508
[3] Mokhov V, Tkachev A and Shaykhutdinov D 2019 Optimal Control of Phase-to-Phase Switching of Single-Phase Power Consumers Based on Binary Algorithms of Agent Metaheuristics. *2019 International Russian Automation Conference (RusAutoCon) Sochi: IEEE* 1-6
[4] Elbi M D and Kizilkaya A 2020 Multicomponent signal analysis: Interwoven Fourier decomposition method. *Digital Signal Processing* **104** 102771
[5] Singh P, Joshi S D, Patney R K and Saha K 2017. The Fourier Decomposition Method for nonlinear and non-stationary time series analysis. *Proceedings of the Royal Society A: Mathematical, Physical and Engineering Sciences* **473(2199)** 20160871
[6] Canizes B et al 2015 Resource Scheduling in Residential Microgrids Considering Energy Selling to External Players. *Clemson University Power Systems Conference 2015 (PSC 2015 Clemson: USA)* 10-13
[7] *Open Data Sets* Retrieved from: http://www.gecad.isep.ipp.pt/Ies2/Consumption/PrivateHome_1/2011_June_03_to_17.xls

[8] Wu C J and Hamada M S 2011 *Experiments: planning, analysis, and optimization* (New York: John Wiley & Sons) 552

[9] Baird D C 1962 *Experimentation: an introduction to measurement theory and experiment design* (USA: Prentice Hall) 212

[10] Antony J 2014 Design of experiments for engineers and scientists (Elsevier) 672

[11] Sedyakin I, Ulyanov M V 2018 Time-efficient algorithm for the generalized entropy calculating of two-dimensional words by the method sliding window. *Modern information technologies and IT education* **14(3)** 560-566

[12] Du L, Song Q and Jia X 2014 Detecting concept drift: an information entropy based method using an adaptive sliding window. *Intelligent Data Analysis* **18(3)** 337-364

[13] Mirjalili S, Mirjalili S M and Yang X S 2014 Binary bat algorithm. *Neural Computing and Applications* **25(3-4)** 663-681

[14] Mirjalili S and Lewis A 2013 S-shaped versus V-shaped transfer functions for binary particle swarm optimization. *Swarm and Evolutionary Computation* **9** 1-14

[15] Mafarja M, Aljarah I, Heidari A A, Faris H, Fournier-Viger P, Li X and Mirjalili S 2018 Binary dragonfly optimization for feature selection using time-varying transfer functions. *Knowledge-Based Systems* **161** 185-204

IOP Conf. Series: Materials Science and Engineering **976** (2020) 012020 doi:10.1088/1757-899X/976/1/012020

Testing the software and hardware complex for managing the power supply system of stationary consumers of railways

E A Tretyakov

Omsk State Transport University, 35, Marksa, Omsk, 644046, Russia

E-mail: eugentr@mail.ru

Abstract. A prototype of an intelligent electronic device in the form of a software and hardware complex has been developed and tested, which provides identification of modes, automatic maintenance of optimal (quasi-optimal) voltage levels in the distribution network of the controlled area, automatic coordinated control of voltage regulation means and reactive power compensation means, FACTS devices, storage devices, generation, active consumers based on the involved components of the concepts of "virtual power plant" and "Internet of energy", automatic reconfiguration of the electrical network in normal mode. The results of numerical experiments on simulation models and physical measurements at operating facilities of the electric power industry confirm the reliability of the scientific and practical provisions presented in the article.

1. Introduction

Digital modeling of electric power systems in real time with physical connection of relay protection devices to the model is the most effective method used in the development of equipment, the study of transient and emergency modes. A power system simulator for real-time simulation (Real Time Digital Simulator, RTDS) allows you to combine real devices with a virtual model of their operating environment and conduct complex tests of relay protection and automation devices, including those operating under the IEC-61850 protocol with full feedback and response to change their state [1-3].

The study of the stability of the power system when changing modes of generating and consuming electrical equipment requires testing using digital simulators in a closed loop mode [4-6]. The software of the test devices includes specialized programs that provide selection, playback and viewing of pre-prepared oscillograms of emergency processes in the COMTRADE format [7-10].

To test the performance of relay protection devices without taking into account the instantaneous response of the electric power system, tests can be carried out in an open loop mode using hardware and software test complexes [11-12].

2. Materials and methods

The developed prototype of the software and hardware complex of the intelligent system for automatic control of transport and distribution of electrical energy in the power supply system of stationary consumers of railways [13] is a superstructure (addition) to the existing (prospective) automated control system of the power grid infrastructure, providing the following main functions:

- Automatic maintenance of optimal (quasi-optimal) voltage levels in the controlled area of the distribution network;

- Automatic coordinated intelligent control of distributed voltage regulation and reactive power compensation devices, FACTS devices, storage devices, generation, demand of active consumers based on the involved components of the concepts of "Virtual Power Plant" and "Internet of Energy" for voltage stabilization, elimination of overload (increased throughput) power facilities of the electrical network, reducing electricity losses and costs;
- Automatic reconfiguration of the electrical network.

The information support of the PTK includes:
- Description of input analog and discrete signals;
- Description of output discrete signals;
- Description of digital input and output signals;
- Description of the organization of information exchange with separate substation systems.

The IED of the substation level is designed to implement the algorithms under consideration as a control controller and can be performed on the basis of an industrial computer due to the absence of strict requirements for the development time of control actions and rather complex optimization calculations and data processing methods that require significant computing resources.

The IED of the bay level has freely programmable logic; it can be implemented on the basis of the controller and is designed to implement the automation algorithms for the elements of the power supply system, i.e. acts as a local controller of executive devices (power facilities).

The communication architecture of the software and hardware complex (PTC) for the supply substation and the controlled section of the electrical distribution network includes an automated place (AWP) with servers, switches, communication and control controllers, network switches, intelligent electronic devices (substation level), intelligent electronic devices (connection level), control objects, interface devices with measuring devices (field level).

The enlarged structural and functional diagram of the PTC is shown in figure 1.

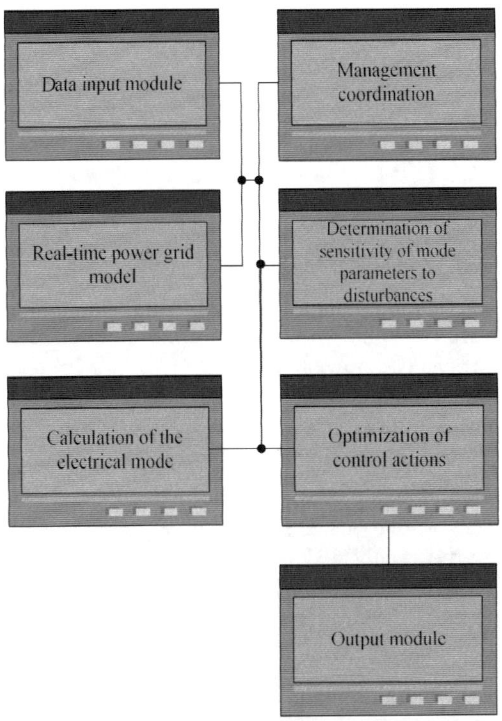

Figure 1. Simplified structural and functional diagram of the PTC.

The developed software (SW) of the PTC prototype for the substation-level IED is designed to determine the optimal coordinated control actions of generation sources, storage devices, reactive power compensation devices, FACTS devices, energy routers for voltage stabilization, reduction of overload in the electric power system based on a multi-agent approach.

Testing the PTC to check the developed algorithms is performed on the basis of a mathematical (digital) model of the power supply system (parameters of lines, transformers and other elements, their mathematical models, a system of equations of states of the electric power system for calculating the steady state, etc.) [14].

The mathematical algorithm of the program based on the assessment of the state of the electric power system (topology, mode parameters, including the results of measurements), taking into account the ontology (knowledge base), which contains process models, criteria and constraints, determines, according to the voltage stabilization condition, the value of control actions for implementation by executive devices based on a multi-agent approach.

The program for the electric network model provides the following functions: input and receipt of initial data from external measuring devices; output of solution results on the monitor screen and data transfer to external devices, including executive ones.

The structure of the hardware and software complex for testing the PTC is shown in figure 2.

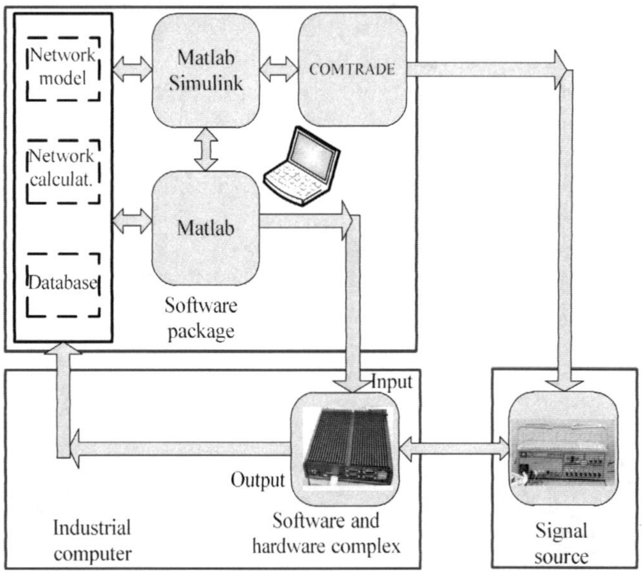

Figure 2. The structure of the hardware and software complex for testing the software and hardware complex.

The program can be used in software and hardware complexes for controlling intelligent electronic devices, connection controllers, and energy routers of electric power systems.

The main stages of simulation based on the developed algorithms [14] can be conventionally represented in the form of the following enlarged steps:

- Determination of control actions for voltage stabilization, optimization of power flows based on a multi-agent approach by performing sensor analysis of the electrical network (in normal and post-emergency modes) by determining the sensitivity of the electrical network nodes to assess the boundaries of the influence of linear and centralized regulators on the voltage mode and determine the actual and permissible power flows through lines limiting the throughput;
- Description of local agents and coordinating agents, knowledge base in the power supply system;
- Simulation of electrical network modes with the calculated settings of the regulators, the calculated topology of the electrical network for the purpose of voltage regulation and

overload limitation in the power supply system of stationary railway consumers in Matlab Simulink.

For research and approbation of the developed scientific and technical solutions and methods of intelligent control of transport and distribution of electricity in the power supply system of stationary consumers of railways, a prototype of a software and hardware complex of an intelligent electronic device based on an industrial computer tested using physical signals of currents and voltages was created.

The PTC test procedure included the following stages:

- Development of a mathematical model of a given section of power supply systems in normal mode from 12 buses with a full description;
- Setting the module for calculating the steady state of the electrical network based on nonlinear equations of state to simulate the response of the mode parameters to external disturbances;
- Setting the database module required for the operation of the software package;
- Connection to a software package that simulates the external environment via Ethernet of an industrial computer with an installed PTC;
- Connection to an industrial computer to a parallel port through a verified measuring transducer of three analog signals from the Retom 41M software and hardware complex;
- Adaptation of the software and hardware complex for the mathematical model of the electrical network;
- Software simulation with a step of 1-5 seconds of changing the parameters of the power supply system by loads and in the form of calculated voltages, power flows, etc. based on the mathematical model of the network;
- PTC reading of mode parameters, network topology state and determination of control actions (voltage in three nodes of the electrical network is simulated by analog signals with Retom 41M) according to the condition of voltage stabilization;
- Transfer of control actions from the PTC to a software package with a mathematical model of the network and obtaining new mode parameters.

3. Results and Discussion

In figure 3-5 shows the test results of a prototype of a PTC based on an industrial computer in the form of the obtained increments of the active and reactive power of regulators, voltages without control and with control based on a given network circuit for voltage stabilization (figure 4).

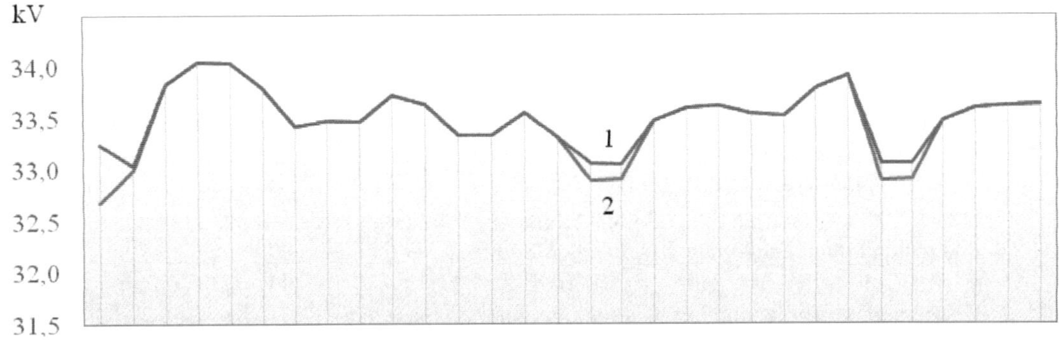

Figure 3. Voltage 1 ssh TP 16 without (2) and with control (1) voltage.

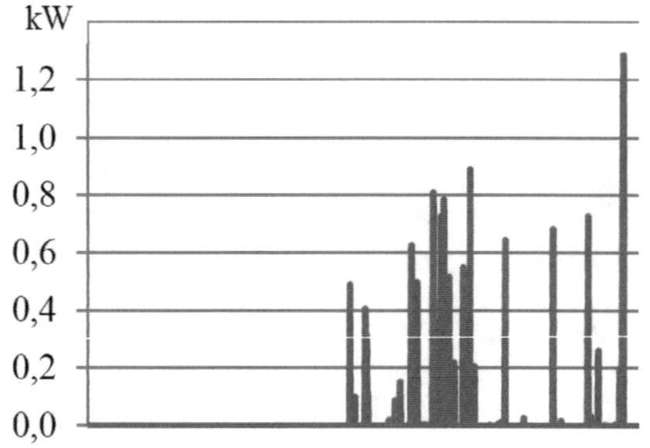

Figure 4. Increment of active generation of TS 1.

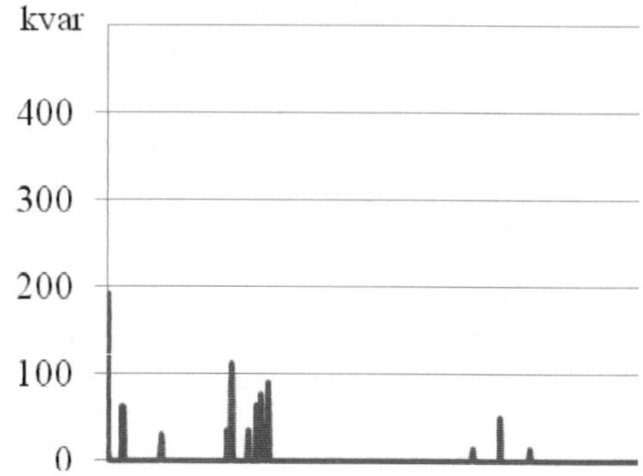

Figure 5. Reactive power 1 ssh TS 16.

4. Conclusion

The research results have shown the practical feasibility and effectiveness of the presented algorithms for coordinated voltage control in the power supply system of stationary railroad consumers.

Integration in one software product of the components of system dynamics and agent behavior makes it possible to holistically simulate the behavior of a multi-agent voltage control system in electric distribution networks of railways.

The results of numerical experiments on simulation models and physical measurements at operating facilities of the electric power industry confirm the reliability of the scientific and practical provisions presented.

The presented studies create a theoretical and practical foundation for the development of a software platform with an open architecture for intelligent control of the railway power supply system as a whole, which is the topic of further research by the author.

References

[1] Novash I V 2017 Modeling of power systems and testing of relay protection devices in real and model time. *Energetika. Proceedings of higher educational institutions and energy associations of the CIS* **3** 198-210

[2] Mochalov D O, Zakonshek Ya V and Shamis M A 2013 Complexes of modeling in real time for modern power systems. *Relay protection and automation* **1** 10

[3] Common Format for Transient Data Exchange (COMTRADE) for Power Sys-tems 1999 *IEEE Std* **37** 56

[4] Monti A, Stevic M, Vogel S, De Doncker R W, Bompard E, Estebsari A, Profumo F, Hovsapian R, Mohanpurkar M, Flicker J.D, et al. 2018 A global real-time superlab: enabling high penetration of power electronics in the electric grid. *IEEE Power Electr Mag.* **5(3)** 35-44

[5] Stevic M, Monti A and Benigni A 2015 Development of a simulator-to-simulator interface for geographically distributed simulation of power systems in real time. *IECON 2015-41st annual conference of the IEEE* 005020-5

[6] Casella F, Leva A and Bartolini A 2017 Simulation of large grids in openmodelica: reflections and perspectives. *Linköping University Electronic* **132** 227-33

[7] Mirz M, Estebsari A, Arrigo F, Bompard E and Monti A 2017 Dynamic phasors to enable distributed real-time simulation. *6th international conference on Clean electrical power* 139-44

[8] Schutt-Ainé J E 2001 Latency insertion method (LIM) for the fast transient simulation of large networks. *IEEE Trans Circuits Syst* **48(1)** 81-9

[9] Celeita David et al. 2016 Implementation of an educational real-time platform for relaying automation on smart grids. *Electric Power Systems Research* **130** 156-66

[10] Milton M, Benigni A and Monti A 2019 Real-Time Multi-FPGA Simulation of Energy Conversion Systems. *IEEE Transactions on Energy Conversion* **99** 1-1
[11] Demiray T, Andersson G and Busarello L 2008 Evaluation study for the simulation of power system transients using dynamic phasor models. *Transmission and distribution conference and exposition* 1-6
[12] Shen Z and Dinavahi V 2016 Real-time device-level transient electrothermal model for modular multilevel converter on FPGA. *IEEE Trans. Power Electron* **31(9)** 6155-68
[13] Tretyakov E A and Cheremsin V T 2020 Increase of passage capacity of distributive electric networks with renewable energy sources through their reconfiguration. *Bulletin of the Tomsk Polytechnic University, Geo Assets Engineering* **3(331)** 112-22
[14] Tretyakov E and Golovnev G 2019 Voltage control in distribution electric rail networks using reactive power sources based on agent-based approach. *Proceedings - 2019 International Ural Conference on Electrical Power Engineering* 8877620

IOP Conf. Series: Materials Science and Engineering **976** (2020) 012021 doi:10.1088/1757-899X/976/1/012021

Identification of the root causes of accidents at hazardous production facilities of the fuel and energy complex

A V Alekseeva, A T Volokhina and E V Glebova

Department of Industrial Safety and Environmental Protection, Russian State University of Oil and Gas (NRU) named after I.M. Gubkina, 65, Leninsky Prospect, Moscow, 119991, Russia

E-mail: sanechkaaleks@mail.ru

Abstract. The article presents the results of the analysis of statistical data on accidents at the facilities of the main pipeline gas transportation for the period from 2008 to 2018. It has been established that the dynamics of accident rates has a cyclical wave-like character. In the course of analyzing information on accidents that occurred during the study period, groups of the main causes of accidents in the main pipeline gas transportation were identified. In addition, in the course of the study, the nature of the occurrence of stress corrosion cracking was studied in detail, and the root factors that led to the occurrence of accidents for this reason were identified. With the help of regression analysis, a functional relationship was established between the number of accidents due to corrosion destruction and hidden defects, which included defects of the pipe manufacturer and defects in construction and installation works.

1. Introduction

The gas transmission system is a complex technical system strategically important for the development of industry and energy supply in Russia. As the technical and economic conditions for the development of the oil and gas complex change, it is necessary to solve more and more complex problems to ensure the uninterrupted and safe operation of gas trunk lines. One of the priority areas of companies engaged in gas transportation activities is to prevent accidents and reduce the amount of potential material damage from accidents.

Thus, the operating organizations are required to carry out large-scale work to ensure the industrial safety of gas transmission facilities at all stages of its life cycle. To determine the goals, priority areas and main tasks, as well as to develop a strategy to increase the level of industrial safety of facilities, a detailed analysis of all possible factors and conditions affecting the accident rate of gas transmission network facilities is required.

2. Materials and methods

To establish the reasons that have a negative impact on the level of industrial safety of gas transmission system facilities, an analysis of the statistical data on accidents in this industry was carried out on the basis of information provided in the annual reports of the Federal Service for Environmental, Technological and Nuclear Supervision for the period 2008 - 2018 [1]. In addition, the data on accidents presented on the official website of Rostekhnadzor in the section "Lessons learned from accidents" [2], as well as information bulletins [3-8] were considered.

Since the absolute number of accidents does not allow obtaining an objective picture of the accident rate at gas transportation facilities, since the length of the main gas pipelines is not constant, this study uses the specific accident rate, which is the number of accidents per thousand kilometers of a gas pipeline.

To identify the functional relationship between the number of accidents that occurred as a result of corrosion destruction and the number of hidden defects (factory defects and defects in construction and installation works), a regression analysis was carried out using the MS Excel analysis package, as a result of which a mathematical model was built and its statistical reliability was determined.

3. Results

During the analysis of the statistical data on accidents, it was found that the dynamics of the specific frequency of accidents on the main gas pipelines for the period under study has a wave-like character (figure 1).

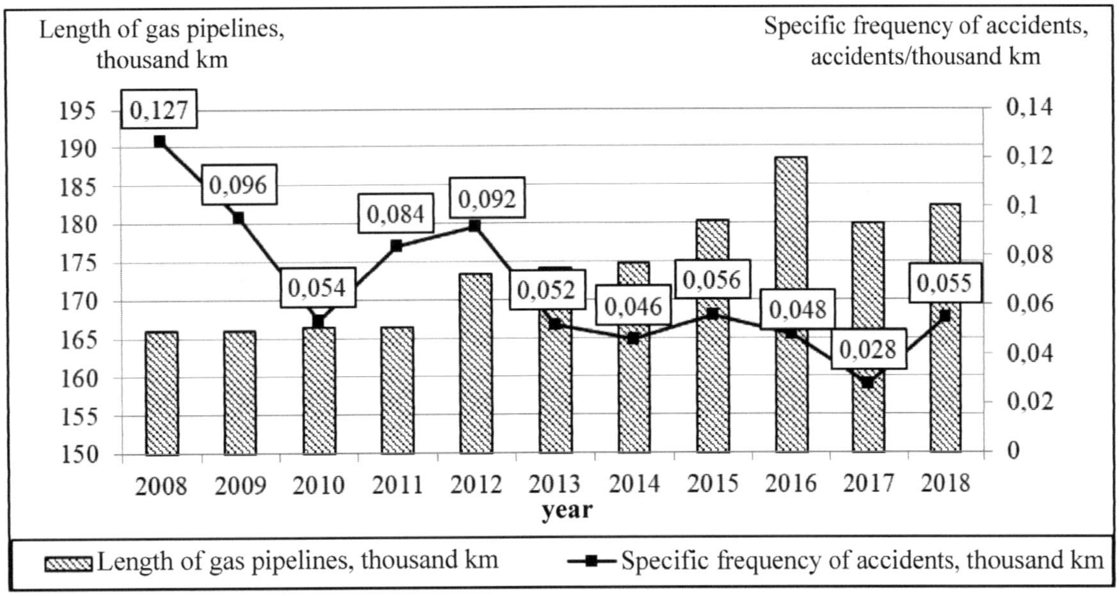

Figure 1. Dynamics of the specific indicator of accidents on main gas pipelines for the period 2008-2018.

In total, 127 accidents were recorded during the study period. From the diagram shown in figure 1, it can be seen that the highest specific accident rate is observed in 2008 (0.127 accidents / thousand km). The decrease in the specific indicator of accidents on main gas pipelines from 2008 to 2010 (0.054 accidents / thousand km.) Is associated with the strengthening of measures for technical diagnostics using the in-line inspection method. In-line diagnostics is the most effective way to monitor the technical condition of main gas pipelines to detect defects in the pipe body, welded joints and eliminate them in the course of repair and maintenance work. Because of overhaul and current repairs using in-line diagnostics, a significant decrease in the number of accidents occurred.

In 2011 and 2012 an increase in the investigated accident rate is observed, which is associated with the discrepancy of some of the pipelines with the structural and technical requirements, which, in turn, makes it difficult to carry out in-line diagnostics (the presence of unequal pipes and shut-off valves, the inconsistency of the radii of the pipe turning angles, the presence of backing rings previously used in pipe welding, lack of chambers for launching and receiving in-line diagnostics devices).

After 2010, there has been an increase in the rate of wear of the fixed assets of the gas transmission system. The same reason is due to a significant increase in the accident rate at gas transmission facilities in 2018 - if in 2011 the share of gas pipelines with a service life of 50 years or more was 3%, then in 2018 the share of these pipelines was 18%.

The analysis of the accident rate for 2008-2018 shows that the main reason for the violation of the integrity of the main gas pipeline is the intensive development of corrosion processes, mostly stress corrosion cracking (SCC) or the so-called stress corrosion. The share of accidents that occurred in the period 2008 -2018 for this reason, it is 52.4%.

The dynamics of accidents due to corrosion destruction (figure 2) has a cyclical wave-like character. Since 2015, there has been a downward trend in the number of accidents for this reason, which can be associated with the active overhaul of insulating coatings since the early 2000s, as well as current repairs based on the results of in-line diagnostics.

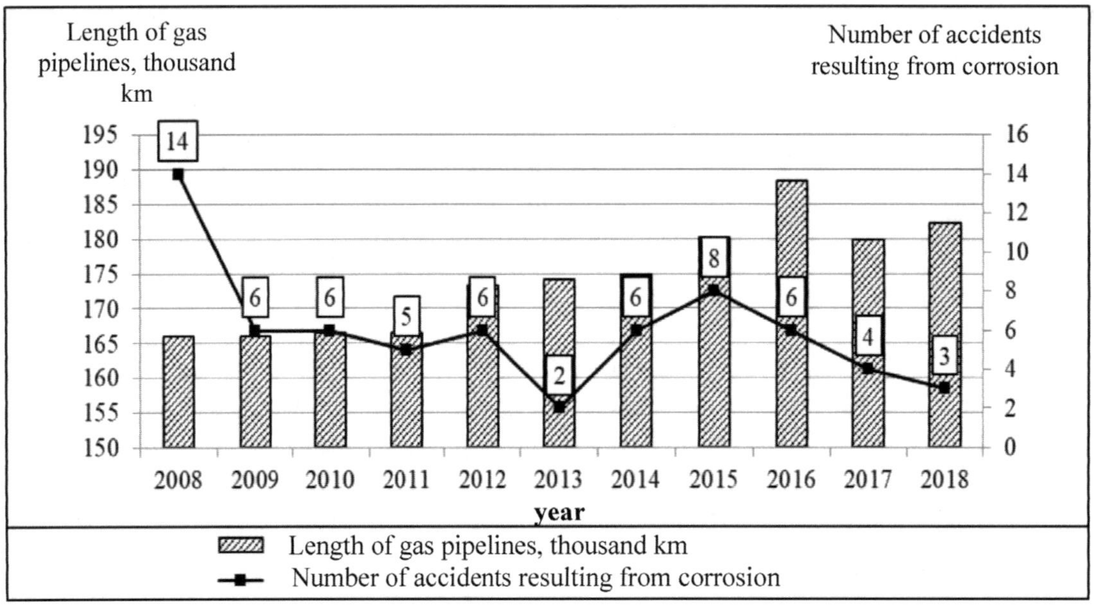

Figure 2. The number of accidents caused by corrosion damage on main gas pipelines, 2008-2018.

Insufficient protection of main gas pipelines from corrosion processes is mainly associated with the loss of the quality of the film insulating coating on gas pipelines commissioned over 25 years ago. So, in the period from 1991 to 1996, the share of accidents due to corrosion destruction in the total accident rate balance of PJSC "Gazprom" was about a quarter, from 1998 to 2003 - a third, and from 2008 to 2018 - more than half of the total number of accidents.

The main factors that determine the conditions for the occurrence of stress corrosion are: the presence of high tensile stresses in the pipeline wall; unsuitable condition and poor quality of insulating materials; contact of the pipe with the environment (soil, water) [9].

Analysis of the statistical data on accidents on domestic and foreign gas pipelines shows that cracking of the metal of pipelines most often occurs near compressor stations (at a distance of 10-30 km along the gas flow), and in some cases in places where pipelines turn, that is, where the temperature gas, respectively - hot walls of the gas pipeline and the highest working pressure, which cause stresses in the wall of the pipeline [10].

The acting loads play a decisive role in the emergence and development of SCC. That is, to activate the process of corrosion cracking of the metal, the presence of surface or internal tensile stresses is necessary. An increased stress-strain state in the pipe metal can be caused both by the effect of the working pressure and gas temperature, and by construction and installation works (CMP) with deviations from the project, as a result of which the position of the pipe profile does not correspond to the design solution, which leads to additional loads. In addition, in practice, there are fractures caused by the presence of residual stresses arising during the production and processing of metal.

Thus, stress corrosion cracking is possible under the combined effect of external and internal factors, such as the low quality of the pipe metal, characterized by the peculiarities of the chemical and phase composition that contribute to the formation of micro cracks, the contact of the metal surface with an aggressive corrosive medium, and the critical level of external and internal loads. It should be noted that with the simultaneous effect of static stresses and a corrosive environment, a stronger deterioration in the mechanical properties of the metal is observed than would occur during the separate action of these factors.

The initiation of cracks during stress corrosion can be caused by the presence of a poor-quality welded joint in the presence of lack of fusion and lack of penetration in the root of the weld, which cause stress concentration; external mechanical damage to the pipe metal that does not go beyond the tolerances, the stressed state of the gas pipeline section, created as a result of deviation from the project during construction (deviation of the pipe axes).

4. Discussion

When analyzing the causes of accidents in main gas pipelines, it is necessary to take into account that, as a rule, each accident is the result of a combination of a number of undesirable events. The multifactor nature of the causes of stress corrosion on gas pipelines makes it difficult to find out the mechanism and patterns of pipeline corrosion.

In order to identify the root causes of accidents caused by corrosion destruction, the results of investigations of emergency failures of the main gas pipeline published on the Rostekhnadzor website in the section "Lessons learned from accidents in the oil and gas complex" were considered and analyzed [2]. The pie chart in figure 3 shows the distribution of accidents due to corrosion damage, depending on the root cause of their occurrence.

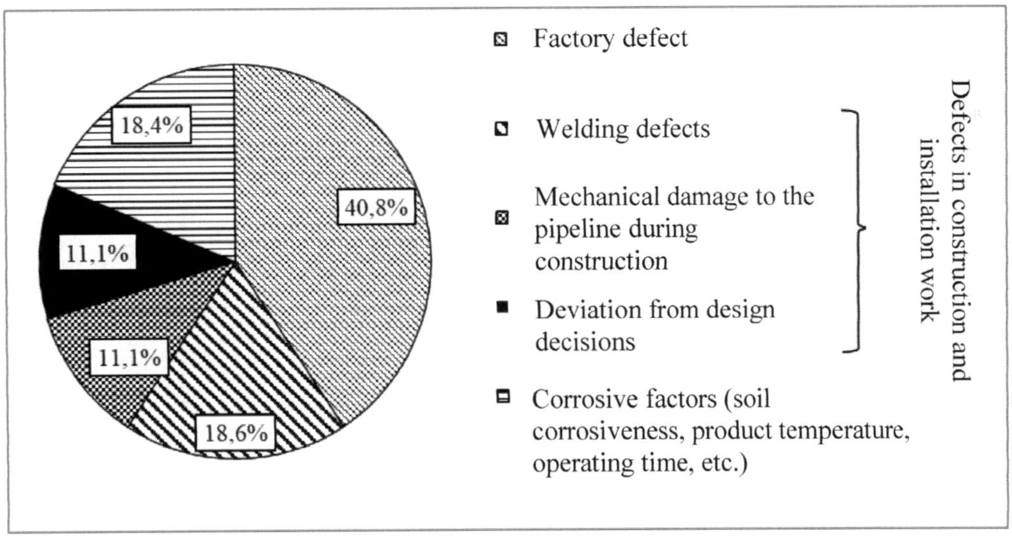

Figure 3. Factors leading to the destruction of main gas pipelines due to corrosion.

Of all accidents that occurred in the period from 2014 to 2018, in 81.6% of cases, corrosion destruction occurred due to internal hidden defects, among which, in turn, the following stand out: 40.8% of cases - a factory defect (as a rule, this is a defect of a longitudinal factory seam), 18.5% - welding defects (non-compliance with welding technology during construction), 11.1% - mechanical damage to the pipeline during construction (dents, scuffs, scratches), 11.1% - retreat from design decisions. The last three groups can be attributed to defects in construction and installation work (caused by poor quality work during construction and subsequent repairs).

The functional relationship between the number of accidents that occurred on the main gas pipeline as a result of corrosion destruction and the number of hidden defects (the number of factory defects and the number of violations during construction and installation works) can be described by the following equation obtained as a result of regression data analysis:

$$Y = 2{,}7368+0{,}4737\times X_1+0{,}7368\times X_2, \qquad (1)$$

where Y is the number of accidents that occurred on the main gas pipeline as a result of corrosion damage; X_1 is the number of accidents, the root of which are factory defects; X_2 is the number of accidents, the root cause of which is defects in construction and installation works.

The statistical significance of equation (1) was checked using the coefficient of determination R_2, F-Fisher's test and the confidence interval for the regression coefficient. It was found that the calculated parameters of the model obtained in the course of the regression analysis explain the relationship between the above parameters by 98%. During the assessment of the reliability of the results of the regression analysis, it was found that the values of the lower and upper boundaries of the confidence interval of the predictors X_1 and X_2 do not overlap 0, and the Fisher's calculated test is higher than the critical one ($F_{fact}> F_{cr}$), therefore, the model is statistically significant.

It should be noted that in order to prevent the occurrence of the above factors, in which there is a risk of accidents due to corrosive destruction, careful construction control and supervision are required during the installation of gas pipelines, the obligatory conduct of regular in-line diagnostic examinations using modern methods and means, as well as the proper quality of inspection and technical control during the supply of materials and equipment.

5. Conclusion

At the facilities of the main pipeline gas transportation for the period from 2008 to 2018 in 127 accidents were recorded, the dynamics of a wave-like nature was noted: several periods with peaks of accidents in 2008, 2012, 2015 and 2018 were identified. The accidents are classified into four groups depending on the main causes of accidents at the facilities of the main pipeline gas transportation: corrosion destruction, incl. stress corrosion cracking, construction defects (defects in construction and installation works, including welding defects), a generalized group of mechanical damage, including accidental damage to the pipeline during excavation in the buffer zone and mechanical damage resulting from the effects of natural phenomena), manufacturing defects. More than half of the registered accidents (52.4%) are caused by metal corrosion; due to construction defects (marriage of construction and installation works, non-compliance with welding technology) - 26.2% of accidents, due to external hazards associated with mechanical damage to the gas pipeline - 12% of accidents and 6.4% of accidents occurred as a result of factory defects.

In the framework of this study, a regression analysis of data on the number of accidents that occurred on the main gas pipeline as a result of corrosion destruction and the number of hidden internal defects was carried out. As a result of assessing the statistical reliability of the obtained model, it was found that the equation is statistically significant and explains by 98% the relationship between the number of accidents that occurred on the main gas pipeline as a result of corrosion destruction, the number of factory defects and the number of violations during construction and installation works.

References

[1] *Annual reports on the activities of the Federal Service for Environmental, Technological and Nuclear Supervision* Retrieved from: http://www.gosnadzor.ru/public/annual_reports/
[2] *Lessons learned from accidents* Retrieved from: http://www.gosnadzor.ru/industrial/oil/lessons/index.php?sphrase_id=1830672
[3] *Newsletter of the Federal Service for Environmental, Technological and Nuclear Supervision 2008* (Moscow: Federal Service for Environmental, Technological and Nuclear Supervision)

77
[4] *Newsletter of the Federal Service for Environmental, Technological and Nuclear Supervision 2009* (Moscow: Federal Service for Environmental, Technological and Nuclear Supervision) 27
[5] *Newsletter of the Federal Service for Environmental, Technological and Nuclear Supervision 2010* (Moscow: Federal Service for Environmental, Technological and Nuclear Supervision) 49
[6] *Newsletter of the Federal Service for Environmental, Technological and Nuclear Supervision 2011* (Moscow: Federal Service for Environmental, Technological and Nuclear Supervision) 26
[7] *Newsletter of the Federal Service for Environmental, Technological and Nuclear Supervision 2017* (Moscow: Federal Service for Environmental, Technological and Nuclear Supervision) 37
[8] *Newsletter of the Federal Service for Environmental, Technological and Nuclear Supervision 2018* (Moscow: Federal Service for Environmental, Technological and Nuclear Supervision) 20
[9] Nazarova M N et al 2017 *IOP Conf. Ser.: Earth Environ. Sci.* **87** 062011
[10] Moskovkina V N and Zagidulin R V 2020 *IOP Conf. Ser.: Earth Environ. Sci.* **459** 042009

IOP Conf. Series: Materials Science and Engineering **976** (2020) 012022 doi:10.1088/1757-899X/976/1/012022

Features and problems of environmental pollution from decentralized heating systems (on the example of Chita)

Yu O Rikker and M V Kobylkin

Transbaikal State University, 30, Aleksandro-Zavodskaya street, Chita, 672039, Russia

E-mail: mkchita@gmail.com

Abstract. This article is devoted to the study of the environmental load of decentralized energy facilities using the example of furnace heating in Chita. Existing residential and non-residential buildings heated with household stoves and low-power boilers increase the environmental burden on the environment. There are tendencies to switch to long-burning boilers. An analytical assessment of the change in the amount of emissions of the main pollutants when changing the operating modes of boilers is presented. Mechanisms for solving problems are proposed.

1. Introduction

Ambient air quality has a direct impact on human health. In 2018, Chita entered the list of cities with the dirtiest air. In 2019, a significant excess of the MPC of harmful emissions was also recorded, including for phenol, dust and suspended solids, benzo(a)pyrene, the vast majority of emissions of which are accounted for by numerous energy enterprises - Chita CHP-1 and Chita CHP-2, and also a number of boiler houses, also fueled by coal [1]. Despite the proven harm of carcinogenic benzo(a)pyrene, other emissions in the air also cause significant damage to human health, including: carbon monoxide, nitrogen dioxide, sulfur oxide, heavy metals, formaldehyde, dioxin, fine particles, ash, phenol, the emissions of which, in turn, are incorrectly attributed only to energy facilities. In addition to the obvious contribution of CHP and boiler houses, as well as numerous road transport, there is a problem of stove heating that is rarely considered in scientific research.

In the winter of 2019-2020, cases of high and extremely high air pollution were recorded in Chita and the Zabaikalsk, an excess of MPCs for phenol, dust and suspended solids was noted, dense smog with a characteristic odor, and reduced visibility were observed [2]. The geographical location of the city in a lowland (foundation pit) and calm weather contributes to the stagnation of smoke and combustion products and the formation of dense smog.

Despite the existing program for the closure and modernization of boiler houses in the city of Chita and the ambitious plans of the KarboFEC company to equip the Zabaikalsk with thermo-robot boilers in particular [3], and the Clean Air project within the framework of the Ecology national project in general [4], at present, in the Zabaikalsk, a rather alarming situation has developed in terms of the incidence of respiratory organs.

In 2018, an increase in the incidence of oncological diseases was noted in the Zabaikalsk. In the Aginsky district, an increase was noted by 1.76 times, in Chita - by 1.21, in Petrovsk-Zabaikalsk by 1.24, in Ulyotovsky, Shelopuginsky and Shilkinsky - by 1.1. At the same time, according to the analytics of the regional Rospotrebnadzor, in the list of oncology incidence in the period from 2011 to 2015, the first place was taken by cancer of the trachea, bronchi and lung [5].

2. Materials and methods

The problem of furnace heating emissions from the private sector is currently insufficiently considered in the scientific literature.

Wood burning stoves and low power coal boilers are widespread in the private sector of the city. Burning wood releases the following chemical compounds into the air: carbon monoxide, nitrogen dioxide, sulfur oxide, heavy metals, formaldehyde, ash, phenol, and fine particles. At the same time, in addition to wood, coal fuel is often used in heating stoves, however, unlike energy facilities, chimneys of private residential buildings are not equipped with ash collectors, and emissions from domestic heating stoves are not standardized and are not measured.

The problem of incineration of household waste (including plastic, highly toxic during combustion) in household heating furnaces is a separate issue. So, residents of Chita have repeatedly reported about the incineration of household waste at the bases on Lazo Street, but no significant measures were taken on this problem.

It is difficult to estimate the volume of housing in Chita heated with stove heating, and it can be called sufficient to be one of the causes of air pollution in Chita [6], while the option voiced by the head of the city E. Yarilov to minimize furnace heating emissions is gasification - does not seem very likely in the current economic conditions.

At the same time, the federal government deprived the region of 860 million rubles. previously assumed for the implementation of the "Clean Air" program within the framework of the national project "Ecology", which in fact led to the suspension of all work under the program [7].

In addition, at present, there is a growing demand for boiler plants that ensure the duration of combustion. Their feature is work with a reduced excess air ratio. Combustion with a low excess air ratio is longer in time, however, it increases the amount of harmful emissions.

It is possible to assess the danger of a transition in the private sector to long-burning boilers analytically, using the methodology for determining emissions of pollutants into the atmosphere when burning fuel in low-power boilers [8]. The methodology is intended to determine emissions of pollutants into the atmospheric air with flue gases from boiler units with a steam capacity of up to 30 t/h and hot water boilers with a capacity of up to 35 MW (30 giga calories/h) based on the data of periodic measurements of their concentrations in flue gases or by calculation when burning solid, liquid and gaseous fuels.

Nitrogen oxides (NO_x), carbon monoxide (CO) and benzo(a)pyrene ($C_{20}H_{12}$) are of the greatest interest in estimating emissions, since for them the intensity of the formation processes correlates with a change in the fuel combustion mode. The emission of sulfur oxide (SO_2) is mainly due to the chemical composition of the fuel and the processes of binding of oxide to fly ash, and therefore, within the framework of the method, it is impossible to assess the effect of combustion modes in the boiler on SO_2 emissions without additional experimental studies.

The excess air coefficient in the furnace is taken as a parameter characterizing the combustion mode.

The calculation of NO_x in terms of NO_2 emitted into the atmosphere with flue gases from boilers with solid fuel layer combustion is carried out according to the formula (1):

$$M_{NOx} = B_p \cdot Q_i^r \cdot 0.35 \cdot 10^{-3} \cdot \alpha_T \cdot (1 + 5.46 \cdot \frac{100 - R_6}{100}) \cdot \sqrt[4]{Q_i^r \cdot q_r \cdot \beta_r \cdot k} \tag{1}$$

where: B_p - estimated fuel consumption, kg/s; Q_i^r - lowest heat of combustion of fuel, MJ/kg; α_T - coefficient of excess air in the furnace; R_6 - characteristic of coal particle size distribution, %; q_r - thermal stress of the combustion mirror, MW/m^2; β_r - dimensionless coefficient taking into account the recirculation of flue gases; k - is the conversion factor.

IOP Conf. Series: Materials Science and Engineering **976** (2020) 012022 doi:10.1088/1757-899X/976/1/012022

According to the formula, NO_x emissions are directly proportional to the combustion air excess ratio.

In the absence of instrumental measurement data, CO emissions can be estimated by the ratio (2):

$$M_{co} = 10^{-3} \cdot B_p \cdot q_3 \cdot R_1 \cdot Q_i^r, \tag{2}$$

where: q_3 - heat loss due to chemical incompleteness of fuel combustion,%; R_1 - coefficient that takes into account the share of heat loss due to chemical incompleteness of fuel combustion due to the presence of carbon monoxide in the products of incomplete combustion.

A feature of the CO calculation is an implicit dependence on the excess air ratio in the furnace. With standard estimated calculations, CO emission will not depend on α_T since losses q_3 practically do not change within the normal combustion mode with excess air greater than 1.1. However, long-burning boilers operate in modes of oxygen deficiency, in such conditions, an exponential increase in losses $q_3$occurs [9]. Accounting for underburning will indirectly assess the impact of excess air on CO emissions.

The $C_{20}H_{12}$ emission calculation is estimated by the ratio (3):

$$M_{C20H12} = 10^{-3}\left(\frac{A \cdot Q_i^r}{\exp(2.5 \cdot \alpha_T)} + \frac{R_2}{t_n}\right) \cdot \left(\frac{D_n}{D_f}\right)^{1.2} \cdot (1 - \eta_z) \cdot \frac{\alpha_T}{1.4} \cdot V_{cr} \cdot B_\mathrm{p} \cdot k, \tag{3}$$

where: A - coefficient characterizing the type of grate and the type of fuel; R_2 - coefficient characterizing the temperature level of the screens; t_n - saturation temperature of water at the outlet from the boiler, °C; D_n / D_f - ratio of nominal to actual boiler load; η_z - degree of gas purification in the ash collector for ash, %; V_{cr} - the volume of dry flue gases formed during complete combustion of 1 kg of fuel, with an excess air ratio of 1.4, m^3/kg.

3. Results

Using the above dependences, the calculation of a low-power boiler at 90% of the load with a thermal stress of the combustion mirror of ≈0.500 MW/m^2, burning 0.62 kg/s of fuel with a net calorific value of 13.44 MJ / kg, was carried out. The calculation was carried out in a wide range of changes in the excess air ratio. The calculation results are shown in figure 1.

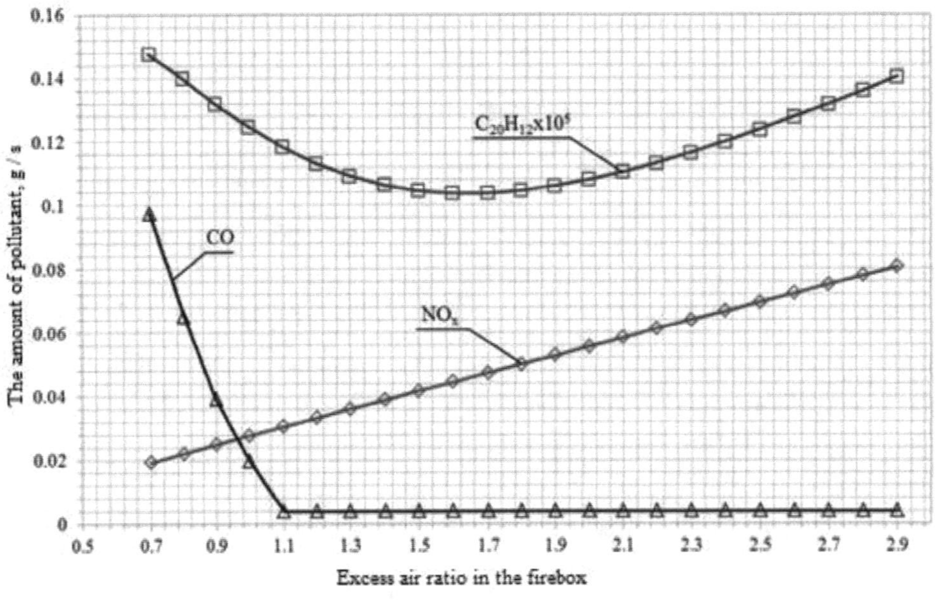

Figure 1. Change in the amount of pollutants in the flue gases from the boiler at different values of the excess air ratio in the furnace.

For correct visualization of the dynamics of changes in various substances, the M_{C20H12} emission in Figure 1 is presented with a factor of 10^5.

Analyzing the emissions of harmful substances at different air excess ratios and based on the calculations made, the following can be noted:

If the excess air ratio of 1.2 is taken as the normal operating mode, then when the excess air ratio decreases to 1, nitrogen oxide emissions decrease by 16%, CO emissions increase by 400%, benzo(a)pyrene emissions increase by 10%.

When the excess air ratio drops to 0.9, nitrogen oxide emissions decrease by 25%, CO emissions increase by 900%, and benzo(a)pyrene emissions increase by 16%.

Considering the hazard classes of substances, it can be concluded that reducing the excess air in the boiler furnace increases the environmental load on the atmosphere, even despite the reduction in NO_x emissions.

It is worth noting that in long-burning boilers from responsible manufacturers, a system is organized for the afterburning of fuel pyrolysis products, which solves the problem of CO emissions. Working with such boilers requires increased attention to the combustion and afterburning processes, which is not always done in domestic conditions.

The problem associated with an increase in the emission of benzo(a)pyrene is currently difficult to solve and requires additional study [10-12]. For small boilers, there is no effective solution to suppress benzo(a)pyrene.

In practice, there are modernizations of standard boilers for long-burning boilers without post-combustion, and the operation of standard boilers with a low excess of air. The main goal pursued by people with such modernizations is to reduce the number of fuel fillings. The environmental factor is ignored.

4. Discussion

To solve such a large-scale problem, several solutions can be proposed.

First, the use of cleaner fuels. For example, in the Krasnoyarsk Territory, there is experience in the manufacture of smokeless coal (smokeless briquettes) by SUEK. This type of fuel is positioned by the manufacturer as environmentally friendly; fuel is pressed under high pressure and high temperature, which allows removing harmful substances immediately during production. Smokeless briquettes burn without waste, due to which coal dust is not poured through the grate into the ash pan and does not fly out into the chimney [13]. At the same time, the caloric content of smokeless briquettes reaches 6000 - 7000 giga calories/kg.

Secondly, the introduction of administrative responsibility for the incineration of highly toxic waste in household stoves and low-power boilers, intensive work of the supervisory authorities to identify the facts of such emissions; development of a program to support and subsidize the purchase of smokeless fuel for low-income groups of the population.

Third, the development of new models of low-power energy-saving boilers (for example, with heat storage) and the introduction of energy-saving measures.

It is possible to combine the duration of the boiler operation on one fuel tab and prevent an increase in harmful emissions by quickly burning fuel under optimal conditions with the accumulation of heat in accumulators with its further use. In this connection, the technology of boilers with heat storage seems to be the most promising for use for these purposes.

5. Conclusion

The problem of furnace heating emissions from the private sector currently requires attention and detailed study.

There is a tendency to switch to boilers that burn in an environment with a low oxygen content, which negatively affects the environmental situation in the residential area.

Analytical calculations show an increase in the emission of the most dangerous substance - benzo(a)pyrene when switching to combustion with a small excess of air in the furnace.

Of particular interest is the direction of development of new models of low-power energy-saving boilers with heat accumulation, which can combine the advantages of long-burning boilers and boilers operating with an optimal excess of air.

Unfortunately, these measures are quite costly and are hardly feasible without centralized funding.

This topic presupposes a great research potential not only for power engineers and ecologists, but also for economists, because, as Academician M.A. Styrikovich said, "Energy is physics plus economics".

References

[1] *Five Russian cities with the dirtiest air* Retrieved from: https://news.rambler.ru/ecology/37554801-5-gorodov-rossii-s-samym-gryaznym-vozduhom/
[2] *Chita suffocates in smog* Retrieved from: https://www.eastrussia.ru/material/chita-zadykhaetsya-v-smoge/
[3] *Zabaikalka intends to equip the region with thermal robots to combat dirty air* Retrieved from: https://www.chita.ru/news/150192/
[4] *Harmful emissions from ZabZhD boilers in Chita decreased by 12%* Retrieved from: https://www.chita.ru/news/146528/
[5] *Six districts of the region are recognized as unfavorable in oncology* Retrieved from: https://www.chita.ru/news/104000/
[6] *The head of Chita announced a project to purify air from car exhaust* Retrieved from: https://www.chita.ru/news/146178/
[7] *Why Chita will continue to choke* Retrieved from: https://www.chita.ru/blogs/editorial/150615/
[8] *Methodology for determining emissions of pollutants into the atmosphere when burning fuel in boilers with a capacity of less than 30 tons of steam per hour or less than 20 Gcal per hour* Retrieved from: https://meganorm.ru/Data2/1/4294849/4294849657.pdf
[9] Belousov V N, Smorodin S N and Smirnova O S 2011 *Fuel and combustion theory. Part II. Combustion theory: textbook* (St. Petersburg: SPbGTURP) 142
[10] Rikker Yu O and Kobylkin M V 2019 Review of technologies for suppressing emissions of benzo(a)pyrene at energy enterprises. *IOP Conference Series: Materials Science and Engineering* **791** 012044
[11] Rikker Y O, Kobylkin M V and Batukhtin A G 2019 Improving the efficiency of flue gas desulphurization of TPS. *IOP Conference Series: Materials Science and Engineering* **552** 012024
[12] Shelepova V S and Zvyagintseva A V 2017 Benzapirene - chemical and biological problems of our time ($C_{20}H_{12}$). *Fire safety: problems and prospects* **8** 477-80
[13] *Coal briquette* Retrieved from: https://ugol-briket.ru/

The use of the basic parameters of water-physical properties of oil-contaminated soils in the technology of the biological remediation stage (Western Siberia)

M V Nosova[1,2], V P Seredina[2] and A S Rybn[3]

[1] Joint Stock Company Tomsk Research and Design Institute of Oil and Gas, JSC TomskNIPIneft, 72, Mira ave, Tomsk, 634027, Russian Federation
[2] National Research Tomsk Polytechnic University, TSU, Biological Institute, Department of Soil Science and Soil Ecology, 36, Lenin ave, Tomsk, 634050, Russian Federation
[3] National Research Tomsk Polytechnic University, TPU, 30, Lenin ave, Tomsk, 634050, Russian Federation

E-mail: nsmvsh@mail.ru

Abstract. This report presents the results of field and experimental studies of water-physical properties of soils of oil-contaminated floodplain ecosystems of the mid-taiga subzone of Western Siberia. Features and basic patterns of their change in various contamination zones (epicenter - impact zone) are revealed. The state of water-physical properties of technologically contaminated soils is compared with background counterparts. Recommendations on the use of basic soil-hydrological constants (range of active moisture, wilting moisture, and field moisture capacity) in the technology of phytomelioration are suggested. Study objective: to analyze characteristics of the impact of oil pollution on water-physical properties of alluvial soils and the possibility of using obtained data for the purposes of remediation works.

1. Introduction

To solve the problem of technogenic soil formation and assess its manifestation, reliable quantitative characteristics are needed that reflect the impact of human economic activity on the diagnostic signs of root-inhabited soil horizons that are of top interest in the course of remediation works. Therefore, special attention should be paid to the study of water-physical properties of oil-contaminated soils, since these parameters affect a number of soil-biochemical processes and have a large impact on the main integral property of the soil - its fertility [1-3].

The hazard of oil pollution is associated with a high sensitivity of higher plants used for the final stage of remediation to it - phyto-reclamating sowing of oil-resistant herbs. It was determined that soil water-physical properties, which are the most important parameters for the favorable growth and development of plants, radically change under the impact of oil.

2. Materials and methods

Objects of this study are full-profile soil sections laid in the territory of crude oil spill in various pollution zones (epicenter, impact zone). The key area is located in the mid-taiga subzone of Western Siberia on the left bank of Ob River in the central part of the floodplain.

IOP Conf. Series: Materials Science and Engineering **976** (2020) 012023 doi:10.1088/1757-899X/976/1/012023

The methodology included a cross-spectrum comparative analysis of background soils and soils subject to oil pollution. The background was an unpolluted alluvial sod standard gleyey mid-fine heavy loamy soil. Due to the damage of soil horizons and the impossibility of their diagnosis, all specimens of technogenic soils were taken by layers and were identified in accordance with the genetic horizons of their background counterpart. Objects of research are 2 full-profile soil sections and 26 specimens of soil pits. The selected specimens were used to determine the following parameters: field moisture capacity - using the Kachinsky method; maximum hygroscopic content - according to the Nikolaeva method, when saturating soil with potassium sulfate; plant wilting moisture - by the calculation method using a coefficient of 1.5 from the maximum hygroscopic content; active moisture range - by the calculation method using the difference between the field moisture capacity and the plant wilting moisture; maximum water capacity and structural coefficient (Cstr) - according to generally accepted methods [4]. Petroleum products in the soil were determined by the fluorimetric method using Fluorat-02 fluid analyzer. The obtained results were processed by the method of mathematical statistics using STATISTICA 6.0 software.

3. Results

The methodology included a cross-spectrum comparative analysis of background soils and soils subject to oil pollution. The background was an unpolluted alluvial sod standard gleyey mid-fine heavy loamy soil. Due to the damage of soil horizons and the impossibility of their diagnosis, all specimens of technogenic soils were taken by layers and were identified in accordance with the genetic horizons of their background counterpart. Objects of research are 2 full-profile soil sections and 26 specimens of soil pits. The selected specimens were used to determine the following parameters: field moisture capacity - using the Kachinsky method; maximum hygroscopic content - according to the Nikolaeva method, when saturating soil with potassium sulfate; plant wilting moisture - by the calculation method using a coefficient of 1.5 from the maximum hygroscopic content; active moisture range - by the calculation method using the difference between the field moisture capacity and the plant wilting moisture; maximum water capacity and structural coefficient (Cstr) - according to generally accepted methods [4]. Petroleum products in the soil were determined by the fluorimetric method using Fluorat-02 fluid analyzer. The obtained results were processed by the method of mathematical statistics using STATISTICA 6.0 software (figure 1).

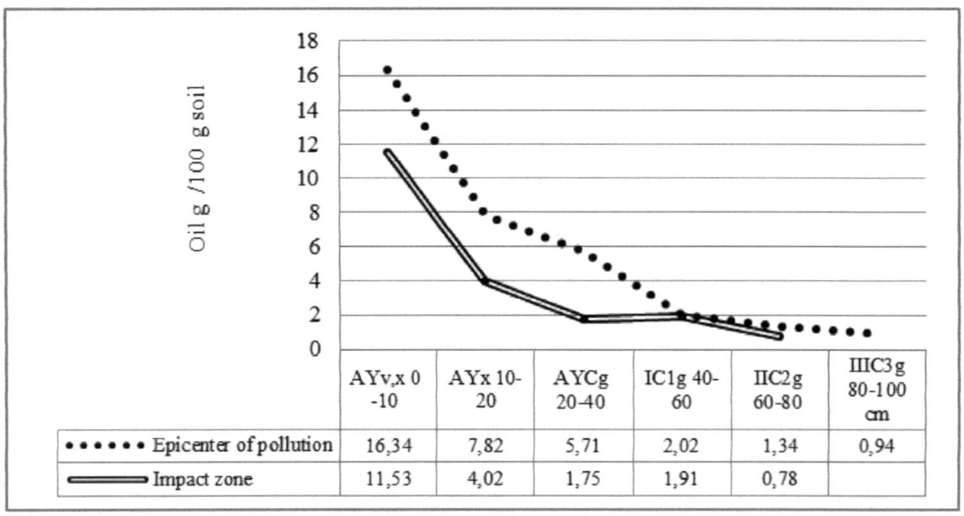

Figure 1. Oil content in oil-contaminated soils.

Therefore, during pollution, water-physical parameters acquire properties unfavorable for plant growth and development, and the biogenic-accumulative horizon of soils is subject to the greatest transformations (figure 2).

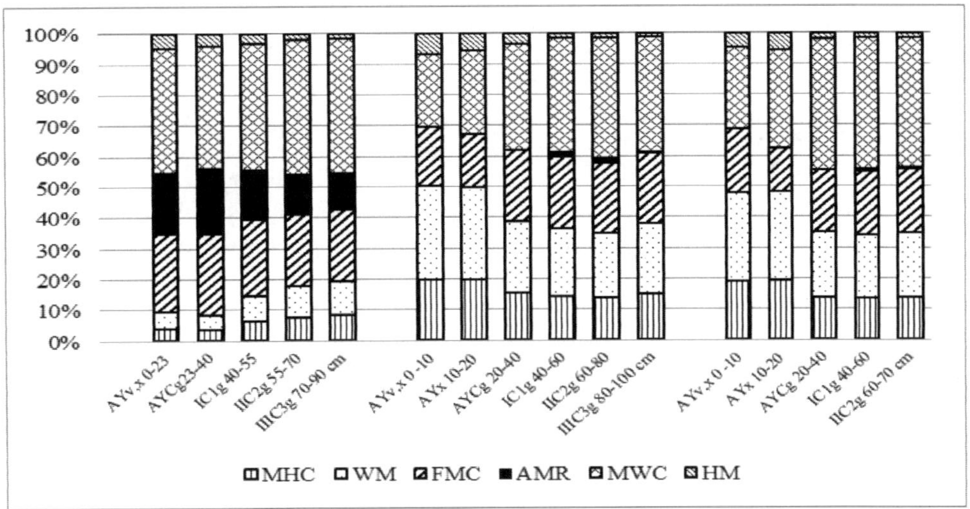

Figure 2. Oil content in oil-contaminated soils. Water-physical properties of alluvial soils: A - background, B - epicenter of pollution, C – impact zone of pollution. Note: HM - hygroscopic moisture, MWC - maximum water capacity, AMR – active moisture range, FMC- field moisture capacity, WM- wilting moisture, MHC - maximum hygroscopic content.

Under the influence of oil pollutants significant changes in the parameters of hygroscopic moisture and maximum water capacity of oil-contaminated soils occur. The values of hygroscopic moisture and maximum water capacity are lower than background values, which is associated with the absence of organic pollutants in the pore space of initial soils. As far as the depth decreases, the values of these parameters also decrease, and it is associated with the compaction of soil horizons and an increase in their granulometric composition. Depending on the maximum water capacity of soils, the water application rate is established, however, at the final stage of remediation works - phytomeliorative sowing of oil-resistant herbs - these rates are not regulated by land reclamation plans [6].

Excessive moisture contributes to the development of anaerobic processes that reduce soil fertility and plant layer density - a visual parameter for the assessment of the quality of remediation works. The maximum hygroscopic content is a parameter characterizing the water resistance of the soil structure, and therefore, its anti-erosion resistance. In chemically-transformed soils this parameter is higher (14.99%) as compared to background soils (7.70%), which is associated with oily films covering soil aggregates. However, these changes cannot be called positive due to the fact that they are associated with the anaerobic effect of oil, the weakening of intermolecular forces between soil and water particles, and the acquisition of hydrophobic properties by the soil. In the background specimens the wilting moisture is 2 times lower (10.30%) than in a newly contaminated soil at the epicenter of the oil slick (22.48%) and at the periphery of the oil spill (20.98%). High values of this parameter show a significant amount of moisture in the soil, which is physiologically inaccessible to plants [7-8]. Plant wolting moisture monitoring is especially relevant in the vegetative phase of growth. Due to the fact that the seedlings are extremely sensitive to the critical state of soil moisture. In the process of contamination there is a complete absence of the active moisture range in the upper horizons and minimal values of this parameter in the lower part of the soil profile (1.9% -2.5%) in the center of the oil slick. The picture is different with the soil correlated with the impact zone of the newly contaminated area. Probably, in the process of lateral migration of oil from the epicenter of pollution to the impact zone, hydrocarbons lost most of their heavy (resinous-asphaltenic materials) fractions. Therefore, as you move away from the source of pollution, the range of active moisture begins to recover (up to 6.5%), and, accordingly, the productivity of the soil increases. This regularity is confirmed by the results of statistical analysis - there is a negative correlation between the concentration of oil and the range of active moisture ($r = -0.59$ at $p = 0.289$ at a depth of 0-10 cm), which shows a reliable feedback between these signs.

4. Discussion

Assessment of the fertility of the upper layers of alluvial soils made it possible to establish that their humus-accumulative horizon is the most productive one and is of the greatest interest for the remediation of disturbed lands, however, this particular horizon is subject to the greatest transformation [9-11]. To restore its initial state, significant operational investments are required, therefore, the use of roll geotextile can become a promising method for restoring water-physical properties of soils. This material will restore the projective top soil within one summer season, eliminate the risks associated with inadequate agrotechnical treatment of soils, and also provide natural detoxification of residual oil pollutants in the soil. The structure of geotextile makes it possible to fix fertilizers and plant mixtures in it, which will not be washed out by sediments, as well as by waters during periodic flooding of the floodplain. The possibility of creating a projective top soil stable in terms of the area and density directly depends on the remediation mixture, which should be selected in accordance with climatic characteristics, the soil cover and hydrological characteristics of the area of geotextile application. An additional measure that provides the greatest effect from the use of roll geotextile can be the introduction of oil-resistant grasses, crops that improve the soil structure, into the biological stage of remediation - *(Lupinus polyphyllus)*, rye *(Secale cereale)*, yellow sweet clover *(Melilotus officinalis)*, meadow clover *(Trifolium pretense).*

5. Conclusion

Therefore, in humid areas the formation of solid bituminous crusts on the surface, as well as in the soil profile, under conditions of oil pollution, dramatically changes the air and water regime, which negatively affects all soil characteristics. Changes in the main soil-hydrological constants (active moisture range, wilting moisture capacity, maximum water capacity) show the acquisition of hydrophobic properties by the soil, the emergence of moisture in humus horizons inaccessible to plants, and show the need for additional reclamation works with the possibility of further monitoring the dynamics of changes in water-physical properties, which are not studied to the fulest extent in typical land remediation plans

Acknowledgments

We are grateful to JSC «TomskNIPIneft» for the opportunity to publish this article to the sponsors, as well as the staff of the Department of Soil Science and Soil Ecology of the National Research Tomsk State University.

References

[1] Pikovskiy Y I, Gennadiev A N, Kovach R G et al 2017 Hydrocarbon status of alluvial soils in the istra morphostructural node (Moscow oblast). *Eurasian Soil Science* **50(12)** 1363-74
[2] Seredina V P and Sadykov M E 2011 The soils of West Siberia middle taiga oil deposits and a predictive estimate of contamination hazard with organic pollutants. *Contemporary Problems of Ecology* **5** 457-61
[3] Pikovskii Y I, Smirnova M A, Gennadiev A N et al 2019 Parameters of the native hydrocarbon status of soils in different bioclimatic zones. *Eurasian Soil Science* **52(11)** 1333-46
[4] Shein E V and Goncharov V M 2006 *Agrophysics* (Rostov-na-Donu: Fenix publ.) 400
[5] Gennadiyev A N 2016 Oil and environment. *Bulletin of the Moscow state University. Seriya 5. Geography* **6** 30-9

[6] Seredina V P, Kolesnikova E V, Kondykov V A, Nepotrebny A I and Ognev S A 2017 Peculiarities of oil pollution influence on soils of middle taiga in Western Siberia. *Oil-industry* **5** 108-12
[7] Shein E V 2009 The particle-size distribution in soils: problems of the methods of study, interpretation of the results, and classification. *Eurasian Soil Science* **3** 284-291
[8] Vodyanitskii Y N, Savichev A T, Avetov N A and Shishkonakova E A 2018 Soil pollution in sites of oil extraction in Western Siberia, Russia. *Oil Pollution. Issues, Impacts, Outcomes: Nova Science publishers Inc New York* **1** 32
[9] Eckmeier E and Wiesenberg G L B 2009 Short-chain n-alkanes (C16-20) in ancient soil are useful molecular markers for prehistoric biomass burning. *J. Archaeological Science* **36** 1590-6
[10] Pinedo J, Ibáñez R, Lijzen J P A and Irabien A 2014 Human risk assessment of contaminated soils by oil products: total TPH content versus fraction approach. *Hum. Ecol. Risk Assess. Int. J.* 20 5 1231-48
[11] Wiens J A 2013 *Oil in the environment: legacies and lessons of the Exxon Valdez oil spill* (United Kingdom, Cambridge: Press Cambridge University) 482

IOP Conf. Series: Materials Science and Engineering **976** (2020) 012024 doi:10.1088/1757-899X/976/1/012024

Digital Twin of the Distributed Generation Plant

Yu N Bulatov[1], A V Kryukov[2,3] and K E Korotkova[1]

[1] Bratsk State University, 40, Makarenko st., Bratsk, Russia
[2] Irkutsk State Transport University, 15, Chernyshevsky st., Irkutsk, Russia
[3] Irkutsk National Research Technical University, 83, Lermontova st., Irkutsk, Russia

E-mail: bulatovyura@yandex.ru

Abstract. This article describes the concept of building a digital twin of the distributed generation (DG) plant operating on based on a synchronous generator with a fuzzy auto-tuning unit for automatic voltage regulator and automatic speed regulator. The structure of the digital twin is represented as a hierarchal fuzzy model built using experimental data. Programmatically implemented algorithms for optimization of term aggregations membership functions and numerals from knowledge base rules. The results of experiments on obtaining an optimized neuro-fuzzy model for regulating the generator rotor frequency are presented. The proposed algorithms can be used in further research on the construction of a digital twin of the DG plant, as well as for solving the problem of automatic regulators tuning.

1. Introduction

To increase the reliability of consumers power supply and the formation of Smart Grid networks [1-2], distributed generation (DG) plants are used, including those based on renewable energy [3], as well as electric energy storage units (ESU) [4] and various devices for modes intelligent control. For the optimal solution of the Smart Grid networks management problem, the use of a full mathematical model of the power supply system is required. The contemporary digital technologies allow to build advanced models of power plants that enablė to deliver comprehensive diagnosticating and control solutions. One of the methods to generate such models can be based on digital twin technologies concept, which make it possible to obtain all the necessary data characterizing functioning of a physical object or a system [5-8]. The digital twin includes a detailed mathematical model of the object, the parameters of which are refined with the help of information coming from measuring-and-information systems.

In addition to the concept of "Digital Twin", the term "Digital Shadow" is used, which defines the presence of a system of relationships and dependencies that describe the behavior of a real physical object and contained in excess big data received from a real object using the means of industrial Internet. The digital shadow allows predicting the behavior of a real object, but only under the conditions in which data was collected, which is its main difference from a digital twin. The creation of a digital twin of real power generation equipment or power supply system should be based on the use of various sensors to refine the mathematical model, as well as on the use of the Internet of things, data collection systems for their subsequent processing based on intelligent technologies.

The article presents the main provisions of the concept of building the DG plant's digital twin where plant operates on the basis of a synchronous generator with a fuzzy auto-tuning unit for automatic voltage regulator (AVR) and automatic speed regulator (ASR) [9]. The structure of the digital twin is analyzed as a multiply connected model built using experimental data. The results of experiments on obtaining an optimized neuro-fuzzy model for regulating the generator rotor speed are presented.

IOP Conf. Series: Materials Science and Engineering **976** (2020) 012024 doi:10.1088/1757-899X/976/1/012024

2. Description of the concept and structure of building a the DG plant digital twin

As an object for which the construction of a digital twin is proposed, a small-power turbogenerator plant based on a synchronous generator is considered. The small inertia constant of the rotor of its generator requires taking into account the mutual influence of AVR and ASR in the process of tuning. In addition, optimal control requires adjusting the tuning of AVR and ASR when significant changes are introduced in the operation modes of both DG plant and power systems (PS) in parallel operation modes. These requirements can be met using the use of intelligent control algorithms. In figure 1 a structural diagram of the DG plant under consideration with a fuzzy auto-tuning unit for AVR and ASR [9].

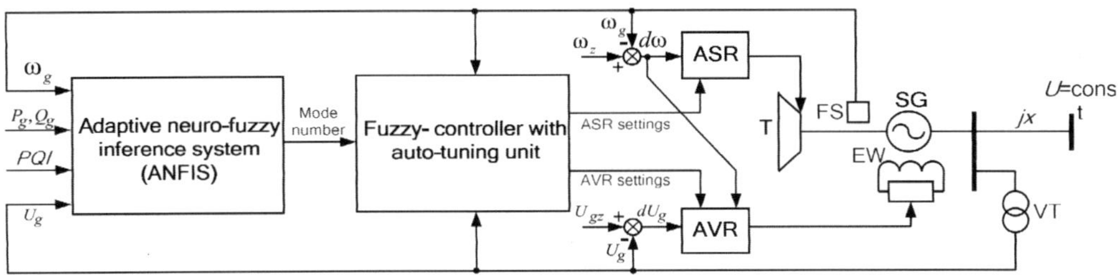

Figure 1. Structural scheme of adaptive system for control AVR and ASR of the DG plant: PQI – power quality indicators; FS – frequency sensor; EW – excitation winding; SG – synchronous generator; T – turbine; VT – voltage transformer.

The operation principle of the control system in question is to identify the operating mode of DG plant generator and to correct the current settings parameters for AVR and ASR when the operation mode is changed. To identify the operating mode, it is proposed to use an adaptive network based on a fuzzy inference system (ANFIS), which is one of the hybrid neuro-fuzzy networks options. In general, ANFIS implements the Sugeno fuzzy inference system in the form of a five-layer feedforward signal neural network. The issue of identifying the operating mode of the DG plant, which resolved by the ANFIS unit, is described in [9].

A fuzzy controller with an automatic tuning unit is a fuzzy logic output system with modules for identifying a digital model of the DG plant harmonized configuring of AVR and ASR [9]. The application of harmonized configuring allows to form the initial fuzzy controller knowledge base. Then, the digital model identification module, based on experimental data, forms a digital shadow for constructing a digital twin of the DG plant.

The proposed concept of building the DG plant's digital twin is based on the representation of the object in the form of a multiply connected structure having a certain set of input and output parameters and relationships formed based on experimental data. The proposed structure of the digital twin of the DG plant under consideration is presented in figure 2. If it is necessary to add other controlled parameters of the DG plant, the proposed structure of the digital twin can be upgraded.

The main elements of the DG plant in question, such as regulators, turbine, excitation system are presented in the form of fuzzy models to identify the dependencies of input and output variables based on fuzzy logic. The particular relations of parameters of the synchronous generator are also presented in separate units of fuzzy logic inference (S1, S2, S3, S4, S5), together forming the generator hierarchical fuzzy model. Thus, the DG plant's elements are presented in a digital twin in the form of fuzzy logic units, that allow to describe adequately the dependence of input and output variables using a small number of rules. In the whole, building of the whole set of relations of the proposed structure of the digital twin is based on the use of hierarchical fuzzy inference systems, which allows to significantly reduce the dimensionality of fuzzy rules base. To reduce the rules bases of the individual units of fuzzy logic inference of elements and relations of the DG plant, it is proposed to use the subtractive clustering method [10] and ANFIS adaptive network, allowing to train neural networks, synthesize and update knowledge bases of fuzzy systems on the basis of the obtained experimental data from the sensors. Under the proposed structure, the sensors must measure all parameters indicated in figure 2.

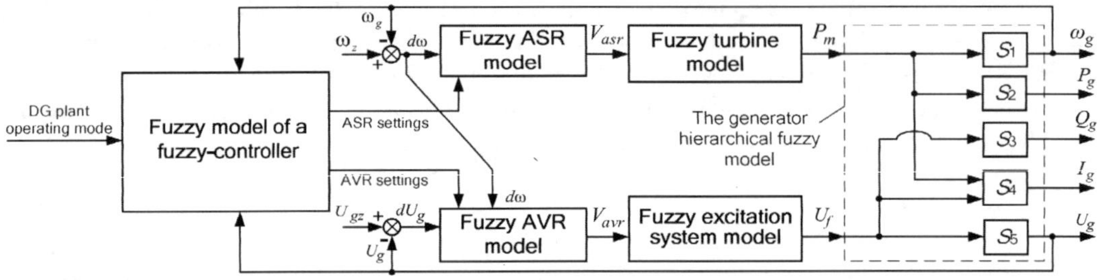

Figure 2. The structure of DG plant digital twin: ω_z, ω_g – accordingly, the set and current values of rotor rotational frequency; $d\omega$ – the deviation of rotor rotational frequency from the set value; U_{gz}, U_g – accordingly, the set and current values of the generator voltage; dU_g – the deviation of the generator voltage from a predetermined value; V_{asr}, V_{avr} – accordingly, the control signals from AVR and ASR; P_m – mechanical power on the turbine shaft; U_f – voltage of the generator excitation winding; P_g, Q_g – the generator active and reactive powers; I_g – generator current of a figure caption.

The measuring system must ensure the identification of power quality indicators. For training and updating fuzzy models the parameters should be measured at the transient process caused by the change of the DG plant operation mode in the course of operation, and when special test signals of low intensity that do not violate the normal operation are applied to the regulators outputs. Updating fuzzy models should be aimed at optimizing the structure and parameters of membership functions of term-aggregations and decreasing the number of fuzzy rules.

Algorithm for building the optimized fuzzy model for a individual relation of the digital twin using the neural networks and genetic algorithm (GA), involves the following steps:

- Training the neural network based on experimental data;
- Translation of experimental parameters to fuzzy variables (fuzzification);
- Building a knowledge base of fuzzy inference system based on trained neural network using the subtractive clustering method;
- The choice of membership functions and the formalization of the knowledge base term-aggregations;
- Determining optimal parameters of the membership functions using GA;
- Determining optimal structure and the number of knowledge base rules of the fuzzy inference system.

The problem of minimizing the number of term-aggregations and the number of fuzzy model rules can be resolved by generating a structure of fuzzy inference system using the subtractive clustering method based on experimental data and subsequent parameter optimization of membership functions using the below-described method.

The method of membership functions parameters optimization of fuzzy terms of the model, allows to minimize the differences between desired (experimental) and model (theoretical) behavior of the object. The symmetrical Gaussian membership function suits best of all to represent terms of knowledge base of the fuzzy inference system:

$$\mu_A(x) = e^{-\left(\frac{x-c}{\sigma}\right)^2}, \tag{1}$$

where c – coordinate of the membership function maximum; σ – mean square deviation, which determines the function width; x – value of variable from the base set.

Optimization of the membership function parameters consists in search through all the variants of the characteristic points and estimating the difference between the reactions of the fuzzy model and the reference values. When using GA, the characteristic points of the membership function (1) are represented on the chromosome by real numbers or binary sequences using one of the known coding methods. After choosing the encoding method, the genetic algorithm procedure is applied to the resulting population of individuals (i.e., to chromosomes containing encoded parameters of the membership functions of the fuzzy system). Thus, the algorithm for optimizing the membership functions of a fuzzy model requires the completion of the following steps:

1. Decoding of each individual in the population, i.e. reading out a set of membership functions and constructing the corresponding fuzzy system:

$$S = (\mathbf{U}_{\mathrm{n}}, \mathbf{Y}_{\mathrm{n}}, \mathbf{C}), \tag{2}$$

where $\mathbf{U_n}$ is the vector of the input linguistic variable values; $\mathbf{Y_n}$ is the vector of the output linguistic variable values; n – the number of lines in the knowledge base; **C** – the vector of parameter values of membership functions.

2. The determination of the fitness function of individuals is carried out according to the difference between the reactions of the fuzzy model and the experimental data using the following quadratic criterion:

$$I = \sum_{i=1}^{n} \left[y_i^{fuzzy}(\mathbf{C}) - y_i \right]^2 \rightarrow \min, \tag{3}$$

where $y_i^{fuzzy}(\mathbf{C})$ – the response of the fuzzy model for various parameters of the membership functions of the terms of the output linguistic variable; y_i – the values of the experimental data output parameters.

3. The use of genetic operators such as selection, crossing and mutation of individuals to determine the minimum of function (3) and the parameters of the membership functions.

To build and optimize a fuzzy model, a program has been developed that has a Windows – oriented interface in which the algorithms described above are implemented.

3. The modeling results

The studies were carried out based on a computer model of a turbogenerator plant with AVR and ASR having a power of 3125 kVA and a voltage of 10 kV, the plant being connected to the power generation system. To study the effect of the proposed algorithm for optimizing the parameters of membership functions on the accuracy of the resulting fuzzy model in the MATLAB system environment, the corresponding experiments were carried out to record characteristics of the input (mechanical power on the turbine P_m shaft) and output (generator rotor speed ω_g) signals when an additional load is connected (figure 3) Based on the obtained data, the neural network was trained and the rules of the knowledge base of the fuzzy logical inference system were formulated. The resulting non-optimized fuzzy model gives the response during the test which is shown in figure 3 b, where a certain deviation from the experimental characteristic is visible.

The procedure for optimizing membership functions and minimizing the number of terms and the number of rules of a fuzzy model allows us to get the response of the model being virtually coinciding with the reference values. A comparative analysis of the oscillograms of the output signal of the constructed optimized model shows acceptable accuracy in the case of using the method of subtractive clustering, which can also significantly reduce the number of terms and the number of the fuzzy model rules (figure 4).

The operability test of the developed program for constructing and optimizing a fuzzy model was carried out using a physical model of an autonomous electrical system. It included a generator (synchronous machine), rotated by a primary engine – a turbine (direct current machine), a power supply of a direct current machine, a synchronous machine exciter, an angular transducer, as well as an active and inductive loads. The generator rotor rotational speed was automatically controlled by changing the voltage at the armature of the DC machine using the developed ASR model (proportional-integral-differential (PID) regulator) using library storage units of Real-Time Windows Target and Simulink of the MATLAB system.

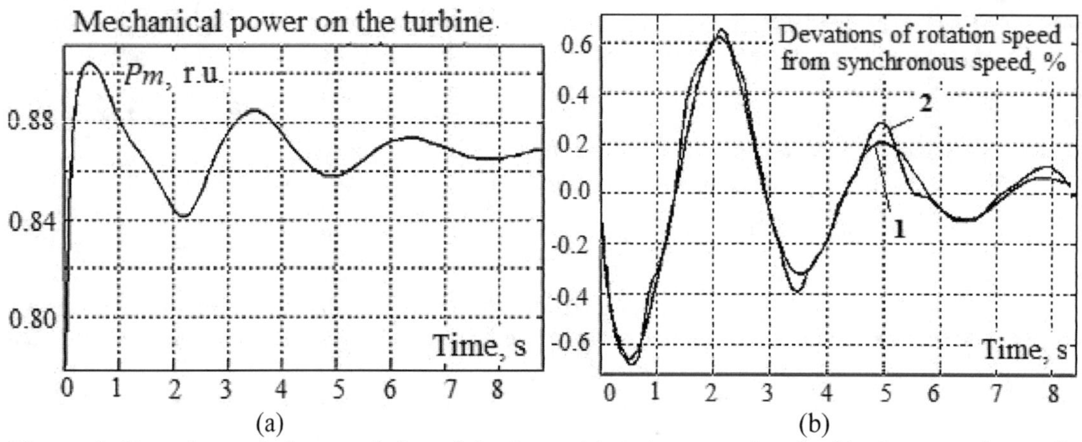

Figure 3. Experimental characteristics of the input (a) and output signals (b): 1 – experimentally obtained (reference) characteristics of the generator rotor frequency; 2 – rotational frequency characteristics generated using a fuzzy inference system.

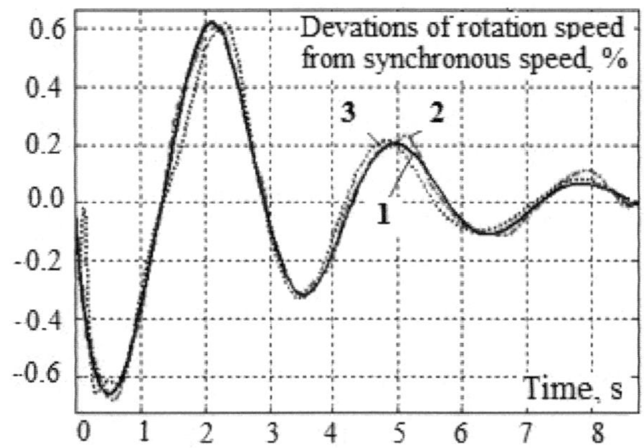

Figure 4. Generator rotor rotational frequency characteristics:
1 – experimental data;
2 – fuzzy optimized model without using the method of subtractive clustering (rule 91);
3 – fuzzy optimized model using the subtractive clustering method (8 rules).

To build a fuzzy model, the experimental characteristics of the generator rotor speed were taken for various groups of settings and tuning coefficients of the PID regulator (figure 5, a). Based on the obtained experimental characteristics, an optimized fuzzy model was built. The results of testing the fuzzy model for random regulator tuning coefficients are shown in figure 5, b. It is necessary to note the obtained accuracy (the root-mean-square deviation is 0.008) and the efficiency of the developed optimization program.

Thus, the algorithm for building and optimizing fuzzy models of individual elements or relations of the synchronous generator of DG plant was developed and programmatically implemented. The results of the research on computer and physical models of the DG plant proved the effectiveness of the use of fuzzy logic and neural networks to build models based on experimental data.

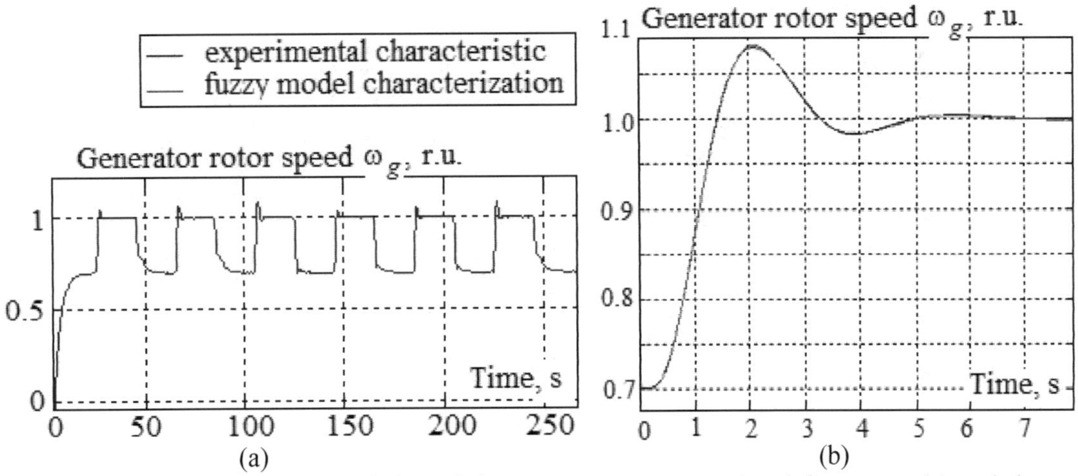

Figure 5. Experimental characteristics of the generator rotor rotational frequency (a) and the test results of the fuzzy model (b).

4. Conclusion

Thus, the structure of the digital twin of the DG plant running on the basis of the synchronous generator with fuzzy automatic tuning unit of AVR and ASR was proposed as a hierarchical fuzzy model, constructed based on experimental data.

Based on the calculation results and simulation, the following conclusions can be formulated:

- The presented algorithms make it possible to obtain and optimize the membership functions of term-aggregations and the fuzzy model number of rules of an individual relation of the DG plant's digital twin based on experimental data;
- The obtained optimized fuzzy model of an individual relation of the DG plant can be used for further studies dedicated to the construction of a digital twin, as well as for solving the problem of automatic generator regulators tuning.

References

[1] Buchholz B M and Styczynski Z A 2014 *Smart Grids – Fundamentals and Technologies in Electricity Networks* (Berlin Heidelberg: Springer - Verlag) 396

[2] Wang J, Huang A Q, Sung W, Liu Y and Baliga B J 2009 Smart Grid Technologies. *IEEE Industrial Electronics Magazine* **2** 16-23

[3] Voropai N I and Stychinsky Z A 2010 *Renewable energy sources: theoretical foundations, technologies, technical characteristics, economics* (Magdeburg: Otto-von-Guericke-Universität) 223

[4] Yoshio Nishi 2001 Lithium ion secondary batteries; past 10 years and the future. *Journal of Power Sources* **100** 101-6

[5] Grieves M W 2014 *Digital Twin: Manufacturing Excellence through Virtual Factory Replication* (Florida: Institute of Technology) 7

[6] Jain P, Poon J, Singh J P, Spanos C, Sanders S and Panda S K 2020 A Digital Twin Approach for Fault Diagnosis in Distributed Photovoltaic System. *IEEE Transactions on Power Electronics* **35** 940-56

[7] Moussa C, Ai-Haddad K, Kedjar B and Merkhouf A 2018 Insights into Digital Twin Based on Finite Element Simulation of a Large Hydro Generator. *IECON 2018 - 44th Annual Conference of the IEEE Industrial Electronics Society Washington DC* 553-8.

[8] Dufour C, Soghomonian Z and Li W 2018 Hardware-in-the-Loop Testing of Modern On-Board Power Systems Using Digital Twins. *International Symposium on Power Electronics,*

IOP Conf. Series: Materials Science and Engineering **976** (2020) 012024 doi:10.1088/1757-899X/976/1/012024

Electrical Drives, Automation and Motion (SPEEDAM) Amalfi 118-23
[9] Bulatov Yu N and Kryukov A V 2018 Neuro Fuzzy Control System for Distributed Generation Plants. *Advances in Intelligent Systems Research* **158** 13-9
[10] Yager R and Filev D 1984 *Essentials of Fuzzy Modeling and Control* (USA: John Wiley & Sons) 387

IOP Conf. Series: Materials Science and Engineering **976** (2020) 012025 doi:10.1088/1757-899X/976/1/012025

Geodynamic ranking of hydrocarbon fields by example of South Cis-Ural

M Yu Nesterenko, Yu R Vladov and A V Tsviak

Orenburg federal research center, Ural Branch of the Russian Academy of Sciences, 29, Naberezhnaja Str., Orenburg, 460014, Russia

E-mail: n_mu@mail.ru

Abstract. In this paper is presented hydro-geodynamics and geodynamics in hydrocarbon fields by example of the Southern Cis-Urals. Seismicity and geodynamics induced by hydrocarbon production by summarizing the published case studies are considered. Changes in hydro-and gas dynamics affect changes in the geodynamics of the crust. Natural and technogenic changes in subsoil of the Earth Crust are usually entail deformations and tectonic movements of earth's surface, which is one of the important factors determining environmental changes and requiring attention in geotechnical survey and field development. The method of classification of the geodynamic state of the developed hydrocarbon fields of the oil and gas basin refers to the field of analysis of geophysical processes, its field of application: identification of the geodynamic state of productive strata and developed hydrocarbon fields of the oil and gas basin, identification of dangerous geodynamic processes, selection of a rational mode development of hydrocarbon fields. The predominant field of application is the classification of geodynamic state of the developed hydrocarbon fields of the oil and gas basin.

1. Introduction

Geodynamics of the Southern Cis-Urals region are poorly understood, since the region is located inside the platform, and platform territories are considered to be geodynamically low-active and low-seismic. Until recently, it was believed that earthquakes rarely appear on the earth's surface by shocks of up to 4-5 Richter scale, and most of the oscillations recorded on the platform are only echoes of strong earthquakes from seismically active regions. The greatest attention was paid to areas with active mountain-folded structures. However, at present, it has become clear that the intra-platform territories are also mobile, especially in areas of tectonic disturbances and near folded areas.

The study relates to the field of analysis of geophysical processes, its field of application: identification of the geodynamic state (GDS) of productive strata (PS) and developed hydrocarbon (HC) deposits of an oil and gas basin (OGB), identification of dangerous geodynamic processes, selection of a rational mode of development of HC deposits of OGB The predominant field of application is the classification of the developed hydrocarbon fields of the OGB according to their geodynamic state.

Intensive oil and gas production in such large regions as Groningen gas field in Netherlands, War-Wink Field in USA, Romashkinskoye field and Orenburg oil and gas condensate field in Russia has disrupted natural environment [1-3], has significantly restructured the hydrogeodynamic and geodynamic processes in the earth's crust at a depth of more than ten kilometers over an area of several tens of thousands of square kilometers. Conditions have been created for the emergence of a number of environmental problems that significantly affect the natural development and quality of life of the region's population.

In large oil and gas regions, dangerous techno-natural processes arise in the upper part of the earth's crust. Such processes develop as a result of an unwanted rock mass response to human technological activity. As a result, seismic activity increases, a hydro geodynamic regime is disrupted and subsidence of an Earth's surface is observed [4-8]. The basis for identifying reasons for increase in seismic activity during hydrocarbon production is based on the results of studies of hydro-gas-dynamic, geodynamic and geophysical processes and water systems dynamics in the regions of developed hydrocarbon fields.

Decrease in pressure in a water system that unites formation waters and surrounding aquifers of a hydrocarbon field, violates natural geodynamic state of the host rocks during hydrocarbons withdrawal and its increase during water injection. It also destabilize dynamic balance between water and a solid part of the Earth's crust [9].

2. The description of the data and methods

There are several methods that solve the problem of assessing the geodynamic activity of oil and gas production areas. For example, method for detecting lithospheric zones of variable geodynamic activity, based on the analysis of data from ionospheric satellite measurements of the magnetic and electric components, therefore to classify the developed deposits with it hydrocarbons on the geodynamic state of the oil and gas basin is not possible.

Method for assessing the influence of factors on the safety of operation of an underground gas storage (UGS) in a porous layer, designed to determine the impact of various natural and man-made processes on the safety of UGS operations.

A method for increasing the reliability and safety of the UGS operation with the determination of the influence of natural-man-made processes, but it does not allow classifying the developed hydrocarbon deposits by the geodynamic state of the entire oil and gas basin, which is its main drawback.

A method for analyzing the results of monitoring at a geodynamic test site built on the basis of a utility model containing a ground test site, a mobile observation device and a center for collecting, registering and processing information. The main disadvantage of the method is the possibility of only studying the territories of oil and gas pipelines, as well as the need for geodetic referencing using a global satellite navigation system.

The closest to the proposed approach is a method for determining the geodynamic activity of the subsoil of a developed hydrocarbon (HC) field by using a seismic stations are combined in a seismic network at the rate of at least three stations for every ten thousands square kilometers, the indicators of the seismic activity of the subsoil of the oil and gas field being developed are summarized, the critical value of the released seismic energy is determined, the data obtained are compared with a given critical energy, and if the critical value is not exceeded, then the summation of the seismic activity indicators continues, and if it is exceeded, then geodynamic zoning of the territory of the developed hydrocarbon fields is performed and areas with anomalously high geodynamic activity are identified. In such areas, a seismological network is built with a density of at least three seismic stations per hundred square kilometers, with seismic stations located at a distance of 3 to 5 km from each other. Taking into account the found seismically active structures of the geological environment, the deformations of the earth's surface in these areas are determined and the level of geodynamic activity of each area is determined according to the aggregated model using normalized partial indicators of geodynamic activity and their weight coefficients, and the choice of particular indicators of geodynamic activity is carried out taking into account the characteristics of the developed HC field from the proposed list, then the found values of the geodynamic activity are assigned to the selected areas, a vector is constructed, the components of which are taken the obtained values of the geodynamic activity of the selected areas, after which the modulus of the vector is determined, normalized by the number of allocated areas, and the magnitude of the vector modulus is judged in the range from 0 to 1 on the geodynamic activity of the bowels of the developed HC field with its surroundings [10].

The common features of the reviewed technical solution and the prototype are the determination of geodynamic parameters characterizing the geodynamic state (GDS) of productive strata (PS) and developed HC fields of an oil and gas basin. However, the prototype method is aimed at determining the geodynamic activity of the subsoil of a single developed HC field, but it is not possible to classify the developed hydrocarbon fields of an oil and gas basin by geodynamic state, which is its main disadvantage.

3. Results and discussions

A method for classifying the geodynamic state of developed hydrocarbon deposits in an oil and gas basin by determining geodynamic parameters characterizing the geodynamic state (GDS) of productive strata (PS) and developed hydrocarbon deposits of an oil and gas basin (OGB), characterized in that four stages are performed sequentially: geodynamic data; determine GDS PS; find the GDS of the developed hydrocarbon fields in the oil and gas basin and build a classification of the GDS of the developed HC fields in the oil and gas basin, and at the first stage: weed out the mothballed HC fields of the oil and gas basin, calculate the total number of PS and PS in each developed hydrocarbon field, as well as the number of developed HC fields in the oil and gas basin open list of geodynamic parameters, for example: average depth of occurrence; oil / gas content area; the average thickness of the reservoir; average net pay; porosity coefficient; oil saturation coefficient; permeability coefficient; net-to-gross ratio; dismemberment factor; oil viscosity coefficient in reservoir conditions; oil density coefficient in reservoir conditions; volumetric ratio of oil; productivity index; drop in reservoir pressure; fund of drilled wells; and so on, determine the values of the selected geodynamic parameters for each PS, find the maximum values of each selected geodynamic parameter among the corresponding parameters of all productive strata of the oil and gas basin, take the basic values of the list parameters, focusing on the maximum values found and normalize the parameter values for each PS, at the second stage they transform an open list of geodynamic parameters into a limited list of geodynamic parameters ranked according to the degree of importance, the corresponding values are selected for the PS of each field from the found normalized values, the weight coefficients of the geodynamic parameters included in the ranked list are found for the PS of each field, the value of the aggregated model is calculated using the ranking results for of each PS, assign the found values of the aggregated models to the corresponding values of the GDS PS, after which, at the third stage, rank the PS in each HC field of the oil and gas basin according to the found GDS, on the corresponding weight coefficients are allocated for the PS in each HC field of the oil and gas basin, the values of the aggregated models for each oil field are determined, the found values of the aggregated models are matched to the corresponding values of the gas flow rate of the developed oil and gas fields, then at the fourth stage to build the desired classification the optimal number of intervals of the gas flow rate of the developed fields is determined HC NGL, find the range of variation of the GDS of the fields, determine the size of the interval, calculate the boundaries of the intervals, determine the quantity with the names of the fields that fall into each interval, build the corresponding histogram and determine the distribution law of the GDS of the developed hydrocarbon fields of the oil and gas basin using the agreement criteria at the selected confidence level, and after each new determination of the values of geodynamic parameters, the steps and operations are repeated in a given sequence.

As an example of the implementation of the method for classifying the GDS of the developed oil and gas fields, let us consider the oil and gas basin in the western part of the Orenburg region with oil and gas fields belonging to the Volga-Ural and Caspian oil and gas provinces. The analysis includes ranking of more than 50 hydrocarbon deposits.

Classification of the developed HC fields of an oil and gas basin, for which the optimal number of intervals of the geodynamic state of the developed HC fields is determined. The optimal number of intervals into which the observed range of variation of a random variable is divided when constructing a histogram of its distribution density is determined by the Sturges rule of thumb. The optimal number of intervals n is determined by:

$$n=1+[\log_2 N] \quad (1)$$

Where N is the total number of observations of the random variable, [ ] - means that you need to take only the integer part of the number.

Substituting our data (N = 56), we get n = 1 + 5 = 6. Next, the range of variation of the GDS of hydrocarbon fields is found by finding the maximum (0.851) and minimum (0.462) values and determining their difference (0.389), determining the value of the interval (0.065) by dividing the range by the optimal number of intervals, calculating the boundaries of the intervals, for example, the boundaries of intervals 2 and 5 are as follows ..., (0.527; 0.592], ..., (0.721; 0.786]; determine the quantity with the names of the fields that fall into each interval, construct the corresponding histogram and determine the distribution law using the goodness of fit criteria at the selected confidence level (figure 1 - a histogram with the number of developed hydrocarbon fields in each of the six intervals, constructed from experimental and theoretical data, in accordance with Pearson's goodness-of-fit criterion.) Moreover, theoretical data with a high 95% confidence probability at three degrees of freedom correspond to the law normal distribution, in particular $X_i = 2{,}221 < X_{i\,kr}(0{,}05; 3) = 7{,}8$.

As a result, a histogram was built with the number of developed hydrocarbon fields that fell into each of the six intervals, built from experimental and theoretical data, in accordance with the Pearson's goodness-of-fit criterion). Moreover, theoretical data with a high 95% confidence level at three degrees of freedom correspond to the law of normal distribution (figure 1).

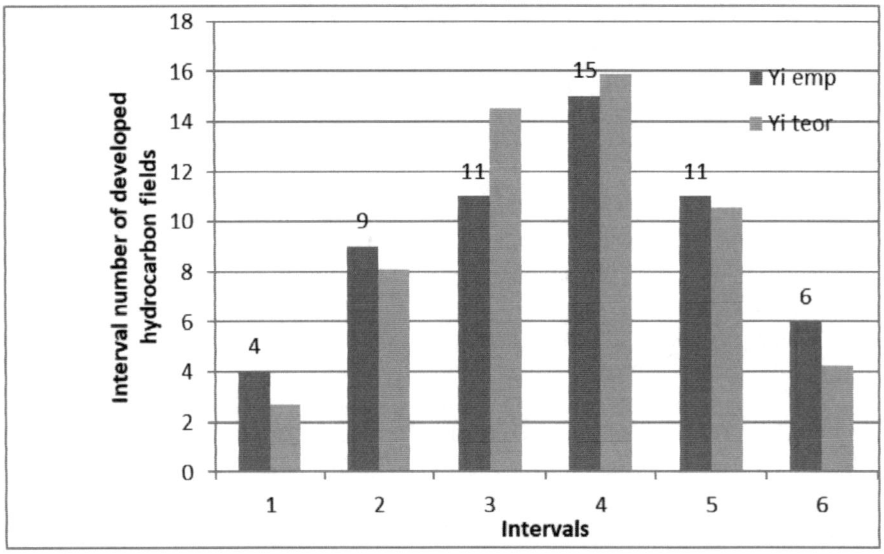

Figure 1. Histogram with the number of developed hydrocarbon fields that fell into each of the six intervals, built according to experimental and theoretical data.

4. Conclusion

The developed method for classifying the geodynamic state of the developed hydrocarbon deposits in an oil and gas basin has significant technical and economic advantages, since it allows with sufficient accuracy to identify the geodynamic states of productive strata and developed hydrocarbon deposits in an oil and gas basin, to identify hazardous geodynamic processes in the depths of the oil and gas basin, to choose rational development modes HC fields and classify the GDS of the developed HC fields. In addition, the method allows you to determine the GDS of the entire oil and gas basin, as well as to compare the oil and gas regions by the GDS values, giving preference to the oil and gas region with the best GDS value.

References

[1] Adushkin V V, Rodionov V N, Turuntaev S D andYudin A E 2000 Seismicity in the oil field. *Oilfield Review* **12** 2-17
[2] Hubbert M and Rubey W 1959 Role of fluid pressure in mechanics of overtrust fauling. *Geological Society of America Bulletin* **70** 115-66
[3] Pennington W, Davis S, Carl-son S, Dupree J and Ewing T 1986 Evolution of seismic barriers and asperities caused by the depressuring of fault planes in oil and gas of South Texas. *Bulletin of the Seismological Society of America* **76** 939-48
[4] Gibowicz S and Lasocki S 2001 Seismicity induced by mining: Ten years later. *Advances in Geophysics* **44** 39-180
[5] Baker M, Doser D and Luo M 1991 Geologic and oil field controls on earthquakes in the War-Wink Field, Delaware Basin. *Bulletin West Texas Geological Society* **31** 5-12
[6] Gan W and Frohlich C 2013 Gas injection may have triggered earthquakes in the Cogdell oil field. *Texas: Proceedings of the National Academy of Sciences of the United States of America* **110** 18786-91
[7] Van Eijs R M H E, Mulders F, Nepveu M, Kenter C J and Scheffers B C 2006 Correlation between hydrocarbon reservoir properties and induced seismicity in the Netherlands. *Engineering Geology* **84** 99-111
[8] Amorese D, Grasso J, Plotnikova L, Nurtaev B and Bossu R 1995 Rupture kinematics of the three gazli major earthquakes from vertical and horizontal displacement data. *Bull. Seism. Soc. Am.* **85** 552-9
[9] Suckale J 2009 Induced seismicity in hydrocarbon fields. *Advances in Geophysics* **51** 55-106
[10] Nesterenko M Y and Nesterenko Y M 2019 Hydro-geodynamic and geodynamic processes in the platform territories of hydrocarbon production. Modern problems of reservoirs and their catchments. *IOP Conf. Series: Earth and Environmental Science* **321** 012004

IOP Conf. Series: Materials Science and Engineering **976** (2020) 012026 doi:10.1088/1757-899X/976/1/012026

First steps of Renewable Technology in Russia

O D Kulkova

Financial University under the Government of the Russian Federation, 49, Leningradsky avenue, Moscow, 125993, Russian Federation

E-mail: kulkova.16.olga@yandex.ru

Abstract. The article gives a brief analysis of some of the benefits provided by Renewable Technology, the ways in which they differ from traditional sources of electrical energy, an overview of the regulatory bodies and the general rules that govern the wholesale electricity market in Russia, the functioning of generation capacity of trading market. Then the mechanisms that exist to encourage market participants to accurately predict supply and demand will be outlined below. Finally, a short description of some other measures that maybe taken in the future to support the growth of the Russian Renewable Energy are to be examined.

1. Introduction

Non-traditional Renewable Energy Technologies ("NRETs") currently play a minor role in Russia's electricity sector, and, in line with global practice and in the interests of diversification, the Russian Government is seeking to promote the use of such technologies in Russia.

For today as opposed to adjustment of oil price by OPEC+ the idea of more focusing on renewable "green energy" is gaining a new momentum in world. The main drivers became global energy companies. The Idea if de-carbonization and "green economy" is in the air within the last months and received a new support from the world leading energy producing giants. After BP which announced in August, 2020 its new priorities on changing its policy from oil- to a new integrated energy company based on "green energy", Shell also announced the similar strategy. BP is seeking to cut down production of carbon resources by 40% until 2030 and to abandon further financing of new projects and closing of the existing projects. The similar view Shell has set out for the future although it outlined only main strategies of transfer to the "green technology" while strategy in detail will be presented in the beginning of the next year. The company plans to significantly reduce the world biggest network of its gasoline stations and also to sell 10 out of 17 oil-refinary plants. Besides, "green bonds" are becoming more popular in the world financial market as a sign of strengthening of this trend. According to Bloomberg, in September 2020 the volume of "green bonds" issued achieved maximum level of 30 billion US dollars. At the same time new discussions of projects on "green hydrogen" are becoming more accrual. Goldman Sachs forecast an increase of this market by 2050 up to 10 trillion US dollars. According to the European Commission volume of generating facilities can increase up to 6 GW by 2024, 40 GW in 2030 and 500 GW by 2050. Total investments can amount to 400 billion euro until the end of 2030. Such huge plans reflect perspectives of significant increase of demand and supply in the near future [1-3].

In regard of this world trend we can try to understand the situation of development of "green energy" in Russia.

2. Materials and methods

The article uses such research methods as abstraction, analysis and synthesis, induction and deduction, and generalization. This study is based on the following information:

- FINAM the Russian financial information portal, Bloomberg Information Agency presented in public sources and media;
- Information of Rosstat (the Russian Statistic Agency) of 2013-2019;
- Analysis of the current Russian legislation laws governing of energy generation.

3. Results

3.1. Energy Consumption Balance.

As a result of population growth and diversification of the economy, an annual growth of approximately 2% in electricity consumption is expected during the foreseeable future, to about 1.492 TWh in 2030. This will require the Russian Federation to increase its installed power capacity by about 160 GW by 2030, at a cost of close to 27 trillion rubles (approx 850 billion US dollars). The Russian Federation has already a low carbon power sector, in which hydroelectric power stations play a major role due to availability of huge hydro resources. However, the anticipated energy demand will require major development of existing power resources. Unfortunately, the Russian Federation currently produces only 0.002 TWh from solar (photovoltaic - PV), wind, marine, geothermal and biomass energy (i.e. non-traditional renewable energy technologies or NRETs) that amounts is 0.28% of total volume of generated electric energy in Russia in 2019. Production of electricity from NRET sources is expected to grow to 5-6 TWh in 2020 and 35-61 TWh in 2030.

3.2. NRETs Available in Russia and Social Benefits Delivered by NRETs.

Despite the small share of investment in the renewable industry so far, when compared to the investment in traditional sources of energy, such major new developments in NRETs as photovoltaic (PV), wind, geothermal, biomass-based generation, marine energy, and small-scale hydroelectric power have been already implemented in Russia.

NRETs deliver significant benefits to Russian society, including the de-carbonization of Russian economy, long-term energy security and diversification of the Russian energy market, comparatively low maintenance costs, positive impact on local employment markets, positive impact on healthcare and environment, compact location and scalability, and attractiveness for investors. It is important that, unlike fossil sources, NRET sources are not vulnerable to fuel price volatility, which may have an additional impact on Russian long-term energy and social policy.

3.3. Specifics of NRET Comparing to the Conventional Sources.

NRET sources have significant differences from traditional sources, which may result in a considerable impact on the Russian energy system. Particular qualities of NRET that need to be accommodated by the regulatory include:

- The long-term costs of electricity generation from fossil-fuels relate to the continuous operation of generation plants' facilities without interruption (fuel costs), while most of the costs for NRET sources are due to capital expenses in equipment;
- NRETs cannot always guarantee a continuous supply of electricity, as generation is only possible when the relevant resource is present and available to generate energy;
- The efficiency of NRET sources is dependent on their geographic position;
- NRETs deliver social and environmental benefits which are not easy to quantify in financial terms, and which are therefore usually not reflected in the market.

In order to accommodate these specific qualities of NRETs, mechanisms may in the future be implemented to compensate for the current biases, and/or the market rules may be changed altogether so that they provide a truly even playing field.

3.4. The wholesale electricity market in Russia – the general situation.
The Russian wholesale electricity market is regulated by the Federal Law "On Electricity Generation " No. 35-FZ dd. 26.03.2003 with amendments made by the latest law - Federal Law No.471-FZ dd.27.12.2019 "On introduction of amendments to the Federal Law "On Electricity Generation " in the part of development of Micro generation ", Government Resolution No. 643 of 27 February 2010, and the Wholesale Market Trading System Accession Contract. Both buyers and sellers of electricity and generation capacity are required to sign this contract in order to participate in the wholesale market. The new amendments were devoted to micro generation that implies NRET.

As in a number of other countries, the wholesale electricity market in Russia includes trading in electricity, and trading in generation capacity. In order to purchase electricity, an entity must first of all purchase the necessary amount of generation capacity. The main participants in the wholesale market are generating companies, importers / exporters of electricity, retail companies, grid companies and major consumers of electricity. The rules governing the wholesale electricity market discussed in this section also refer to Micro generation, and apply to all wholesale buyers and sellers of electricity, including electricity from NRET sources.

3.4.1 Regulatory Bodies. The Ministry of Industry and Energy has overall responsibility for the electric power sector, and the following bodies oversee specific aspects of this market:

- Market Council: a non-profit council of participants in the wholesale market. The market Council supervises the operation of the market, certifies electricity producers and develops the various standard agreements and other documents used in this market;
- System Operator: a 100% state-owned open joint-stock company which oversees the physical operation of the electricity transmission system;
- Administrator of Trading Systems: an open joint-stock company which regulates the wholesale power and capacity market and facilitates electricity trading on standard terms between sellers and purchasers;
- Center for Financial Settlements: a closed joint-stock company which acts as an intermediary for payments on the wholesale market.

3.5. New measures for stimulating the Russian renewable energy sector.
On 28 May 2013 the Russian Government adopted Regulation No. 449, effective from 11 June 2013 (the "RES Standard"). The RES standard includes: (a) rules for calculating RES capacity payments; (b) a methodology for calculating the share of generators' capital expenditures which shall be covered by RES capacity payments; and (c) a methodology for calculating the component of the capacity price that will allow RES generators to cover their capital and maintenance expenses.

The RES standard also introduces changes to: (a) the Qualification Rules for RES generators, as provided in Government Regulations No. 426, dated 3 June 2008; (b) the Regulations of the Ministry of Economical Development No. 438, dated 5 June 2008; and (c) the Rules for the Electricity and Generation Capacity Wholesale Markets, provided in the Government Regulations No. 1172, dated 27 December 2010.

Under the Federal Law "On Electricity Generation ", there are two main methods the Russian Government can use to promote the RES sector: cross-subsidizing electricity prices; and subsidizing generation capacity prices. The RES Standard deals primarily with the second of these methods: selected RES generators may be eligible to receive pre-set capacity payments for up to fifteen years after the launch of new RES generating facilities, to help them cover their capital expenditures related to the construction and maintenance of such facilities, and provide them with a prescribed level of return on their investment. The new RES Standard also provides for certain incentives for localization of RES equipment and for competition between RES projects. The first of the above methods is designed to cover the RES generator's on-going costs and profit, but the details of this method have not yet been fully worked out.

The RES Standard entitles a qualified RES generator to recover a prescribed level of return on its capital investment, which will be calculated using the base level of return (14% for projects selected before 1 January 2015 and 12% for projects selected thereafter) adjusted to reflect the current rate of return on long-term state debt (against a reference rate of 8.5%), and the repayment of the capital itself over a 15-year term [4-6].

4. Discussion

Although a solar energy (PV) is recognized as having the highest level zed cost of energy and as being at present probably the closest to full commercial viability, because of the geographical position of Russia and its large territory wind energy would appear to have the most potential on a national scale. Other NRETs may have greater potential in particular regions of the country. Moreover, speaking about the range of application of the RES Standard, the capacity payment system established by the RES Standard currently applies only to solar, wind and small-scale (less than 25 MW) hydroelectric projects, despite the fact that the Qualification Rules for RES generators (Government Regulations No. 426, dated 3 June 2008) provide for a wide range of RES. Other projects, including biogas projects, are not currently covered by the RES Standard unless they could be referred to Microgeneration that is a general definition in the latest amendments to the Law for NRET [7-8].

5. Conclusion

As mentioned above, the Russian Law "On Energy Generation" contemplates that the Russian Government may establish feed-in tariffs for electricity calculated based on the equilibrium price of electricity plus a regulatory premium which will be established by the Government. However, there are no specific plans to introduce such feed-in tariffs in the nearest future.

The Ministry of Energy, as part of its support for the NRET program, has prepared a draft decree that regulates the retail market for alternative energy. According to this draft, the Federal Tariff Service will make recommendations for setting tariffs in the sector. It will allow suppliers of electricity from renewable energy sources to sell energy to territorial grid organizations (TGO) that are not connected to the Unified National Electric Grid at approved prices within approved price ranges. According to new amendments to the law producers of electricity from NRET sources will be able to declare their own generating capacity, be given a rate and sell the electricity they generate to the TGO at such rate. In this case, the TGO will not have the right to refuse to enter into contracts with suppliers of electricity from NRET sources (who have generation capacity) from NRET sources. The volume of sold electricity will be determined by the guaranteeing supplier. Tariffs will be formed based on the size of investment capital, operating costs, loans and accrued interest, as well as a rate of return of invested capital, fixed at certain percent per annum.

The Russian Government has announced that it is now creating a single NRET development map to better estimate the potential of the country's green energy resources. The map should contain detailed information on the current and future NRETs and an analysis of the territory of the Russian Federation from the perspective of availability of all kinds of renewable resources.

References

[1] *Russian statistical yearbook* 2019 (Moscow: Rosstat) 694
[2] *On Electricity Generation: Federal Law* Retrieved from: http://www.consultant.ru/document/cons_doc_LAW_41502
[3] *Rules for determining the price of capacity of generating facilities operating on the basis of renewable energy sources* Retrieved from: http://www.consultant.ru/document/cons_doc_LAW_146916
[4] Degtyarev K S and Panchenko V A 2019 Development and implemented projects of solar energy in Russia. *Plumbing, heating, air conditioning* **9** 74-9
[5] Nesterenko P E and Kozhanov N T 2015 Alternative energy. *Research and development of young scientists* **3** 104-7

IOP Conf. Series: Materials Science and Engineering **976** (2020) 012026 doi:10.1088/1757-899X/976/1/012026

[6] Belokrylova E A and Kologermanskaya E M 2017 Modern political and legal aspects of the development of renewable energy sources in the Russian Federation. *Bulletin of the Udmurt University. Series "Economics and law"* **2** 85-93

[7] Harmakshanova E V 2020 Measures of state support for the development of alternative energy. *Moscow economic journal* **7** 355-62

[8] *Report of FINAM the Russian financial information portal, Bloomberg Information Agency* Retrieved from: https://www.finam.ru/analysis/marketnews/perspektivy-rublya-v-blizhaiyshie-mesyacy-vyglyadyat-tumanno-20200925-12300

IOP Conf. Series: Materials Science and Engineering **976** (2020) 012027 doi:10.1088/1757-899X/976/1/012027

Market and consumption of oil in China in the first half of 2020: preliminary review

F F Sharipov and O A Timofeev

The State University of Management, 99, Ryazan Avenue, Moscow, 109542, Russia

E-mail: fanissh@rambler.ru

Abstract. China's oil consumption in 2020 faced multifaceted impact of numerous factors, namely COVID-19 global spread, decrease of industrial production and hydrocarbon consumption, global consumption market shrinking etc. Power production is set as one of the sectors, most crucial for implementation two-stage plan for China's socialist modernization by 2050. This strategy's major task is decreasing of coal overwhelming consumption in favor of hydrocarbons. Importing three-quarters of consumed oil, China forms leading trends in global oil transportation, par excellence, for its leading partners: Russia, Saudi Arabia, and Iran. 2020' decrease in oil consumption has had negative impact not only on energy sector per se, but also simultaneously on the wide spectrum of industrial production on a par with transportation, for example, coastal port logistics. Nevertheless, both upstream and downstream oil sectors could rebound post-COVID fast recovery in coming months, which helps to avoid endangering China's long-term plans of development.

1. Introduction

In the first part of 2020, global oil market had a double shock caused by the economic consequences of Coronavirus COVID-19 pandemic and price war between Russia and Saudi Arabia on the subject of review of volumes of production. In this context oil strategy of the largest importers of hydrocarbons raw materials becomes particularly important and first of all China oil imports whose economy recovers faster, than in other developed countries.

2. Impact of the oil market on the long-term development prospects of the energy sector

The XIX congress of the Communist Party of China held in October 2017 adopted a strategy of construction of a modernized socialist state. This strategy is to be implemented in two stages.

In the first phase (2020-2035) on the basis of building an average welfare society [1] it is planned mainly to implement socialist through significant growth of economic and technological strength. By 2035, China should become one of the world's leading innovative economies. At the same time, the environmental situation in the country should be radically improved, which will basically solve the task of creating "beautiful China" [2].

The second stage (2035-2050) yet to make transformation of China into a rich and powerful, democratic and civilized, harmonious and beautiful modernized socialist power. In addition, the level of material, political and spiritual culture as well as the degree of civilization of China's society and ecology must be significantly improved.

In terms of total national power [3] and international influence, China should be among the leading states and generally achieve overall prosperity [4].

IOP Conf. Series: Materials Science and Engineering **976** (2020) 012027 doi:10.1088/1757-899X/976/1/012027

As the energy industry is one of the most strategically important, planning for its development is carried out in full compliance with this strategy. Particularly, in the first stage, a clean, low-carbon and a safe, highly efficient energy system must be broadly formed. At this stage, the level of electrification should be raised. It is assumed that coal and oil consumption will gradually reach peak values and non-fossil energy consumption will be developing steadily. Once the peak is reached, carbon dioxide emissions will begin to decline, there will be a fundamental shift in the protection of the environment, a sharp increase in the efficiency of the consumption of energy will take place while volumes of consumption will be put under strict control.

The second stage will see a comprehensive improvement in the quality of the new energy system, the development of electrification will reach the level of the world's leading countries, the energy structure will be deeply optimised, and non-fossil resources will provide up to half of China's energy consumption by 2045. Coal will only account for one third of the country's energy consumption, while oil and gas will account for other third and non-fossil sources of energy for the rest part. A sharp decline in carbon dioxide emissions will begin, energy efficiency will continue to rise, consumption of energy resources will peak by 2040, cost of energy will continue to decline, and clean energy sources will become widespread. This stage will see China's transition to the status of a "great energy power" [5].

However, the role of oil in China's energy scene will not change significantly, as major changes will affect the coal, gas and renewable energy industries. In 2017, for example, the share of oil in China's energy mix was 19%. In 2050, the share of oil in non-renewable energy consumption is projected to range from 14% (pessimistic scenario) to 16% (figure 1).

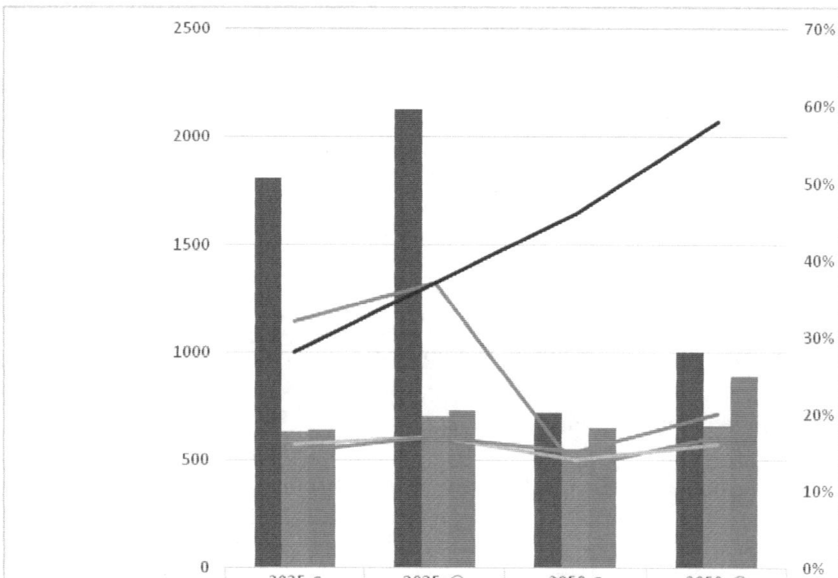

	2035, ☹	2035, ☺	2050, ☹	2050, ☺
Coal, mln tn.	1810	2130	720	1000
Oil, mln tn.	630	700	550	660
Gas, bln, m³	640	730	650	890
Coal, %	32%	37%	13%	17%
Oil, %	16%	17%	14%	16%
Gas, %	15%	17%	15%	20%
Non-fossil, %	28%	37%	46%	58%

Figure 1. Pessimistic and optimistic forecasts of consumption of major energy resources in China.

3. The oil industry - the basis of PRC's global economic strategy

In 2019, China was again ranked second in the world by oil consumption (US 920 million tonnes) and first in terms of imports (figure 2). Imports account for 72% of all oil consumed.

Although China was the sixth largest oil producer in the world in 2019, the country's low oil reserves are responsible for an increasing share of imports in consumption.

In 2019, China imported oil from 48 countries. In 2019, the undisputed leaders were Russia (first place in terms of import volume in 2016-2018) and Saudi Arabia. Riyadh has been particularly aggressive in recent years, seeking to expand its presence in the Chinese market. In particular, monthly exports to China have increased by 84.1% to 7.72 million tonnes since the sanctions were imposed on Iran in June 2020 (in May, 4.7 million tonnes) [6-8].

In 2019, the project to expand the export of American crude oil to China finally failed. For example, in October-November the volume of supplies from the USA fell from 908.4 to 257.2 thousand tonnes.

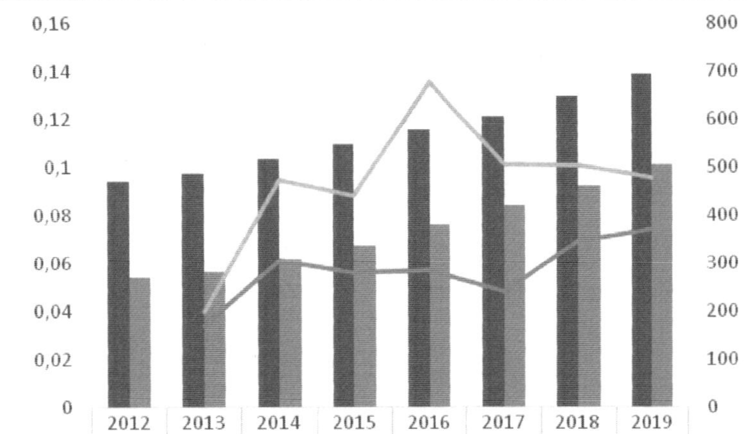

	2012	2013	2014	2015	2016	2017	2018	2019
Consumption, mln, tn	472	488	518	547	578	606	648	696
Imports, mln, tn	271	282	308	336	381	420	462	506
Growth in consumption, %		3,4%	6,1%	5,6%	5,7%	4,8%	6,9%	7,4%
Growth in imports,%		3,95%	9,45%	8,80%	13,56%	10,12%	10,09%	9,55%

Figure 2. Volumes and growth in oil consumption and imports.

In 2016, geologic oil reserves amounted to 125.7 billion tonnes and proven reserves to 30.1 billion tonnes, but technically recoverable oil reserves in 2018 amounted to only 3,573 million tonnes (1.5% of the world total). The 15 largest oil and gas fields account for more than 90% of oil reserves (table 1).

4. Discussion

At the beginning of 2020, the Coronavirus COVID-19 pandemic brought difficult times for the Chinese oil industry (figure 3). Production in January-April increased by 2% compared to the same period in 2019 (to 64.44 million tonnes) and imports by 1.7% (to 167.61 million tonnes), but processing volumes decreased by 3.4% (to 203.48 million tonnes).

IOP Conf. Series: Materials Science and Engineering **976** (2020) 012027 doi:10.1088/1757-899X/976/1/012027

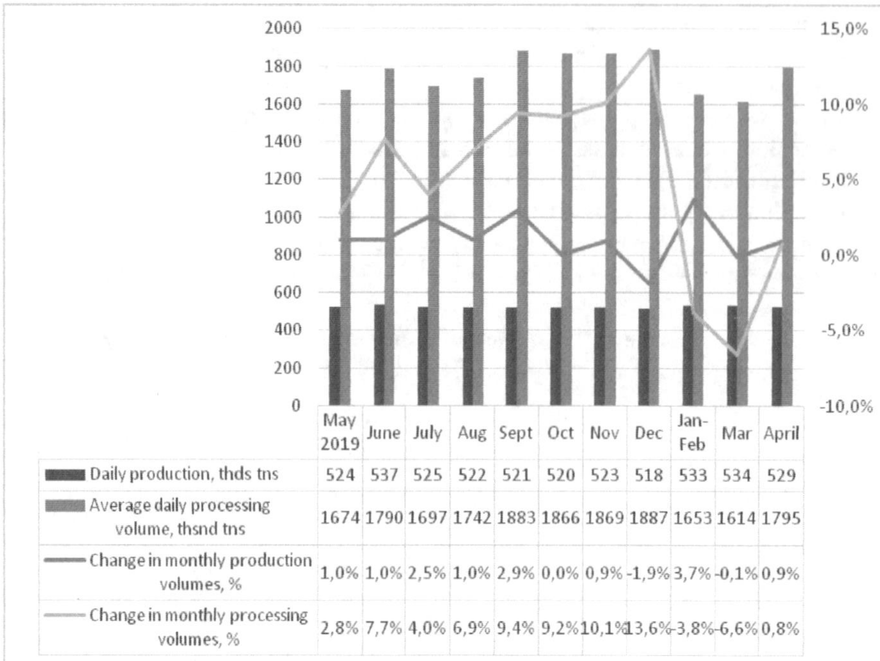

Figure 3. Changes in crude oil production and refining performance in 2019-2020.

Table 1. Main oil fields in China [1-5].

	Geological oil reserves, mln tn	Proven reserves, mln tn
Coastal Zone, Gulf of Bohai	24 287	6558
Ordos	15 838	2672
Sungari-Liaohe Lowland	13 428	3986
Tarim Basin	12 065	2794
Shelf of Bohai Gulf	11 029	2537
Dzungaria	10 673	2202
Zhujiang River mouth	7432	2963
Ciantan Highlands (Northern Tibet)	5095	1121
Tsaidam	3819	624
Gulf of Tonkin	2118	510
Erenhot (Inner Mongolia, Centre part)	1784	267
Sichuan	1611	129
Barkul Kazak Autonomous County (Eastern Xinjiang)	1552	155
Hainan Island Shelf Area	1489	600
Coqên County (SouthTibet)	1104	164
Hailar	1010	201

In April 2020, there was the most significant decline in oil imports from abroad (figure 4). In physical terms, it fell by 7.5%, and in value terms (mainly due to a decline in global oil prices) by 49.3% compared to the same period in 2019.

Despite the rapid development of pipe oil imports, primarily from Russia and Kazakhstan, a significant proportion of foreign raw materials continue to flow through the Chinese seaports. For example, in January 2020, 39.44 million of the 43 million tonnes of imported crude oil were transported by sea. The four largest oil-loading ports - Qingdao, Rizhao (Shandong Province), Ningbo and Zhoushan (Zhejiang Province) - traditionally account for over half of oil imports. However, in February 2020, the situation changed due to the Coronavirus pandemic, of which Zhejiang Province became one of the epicentres (table 2).

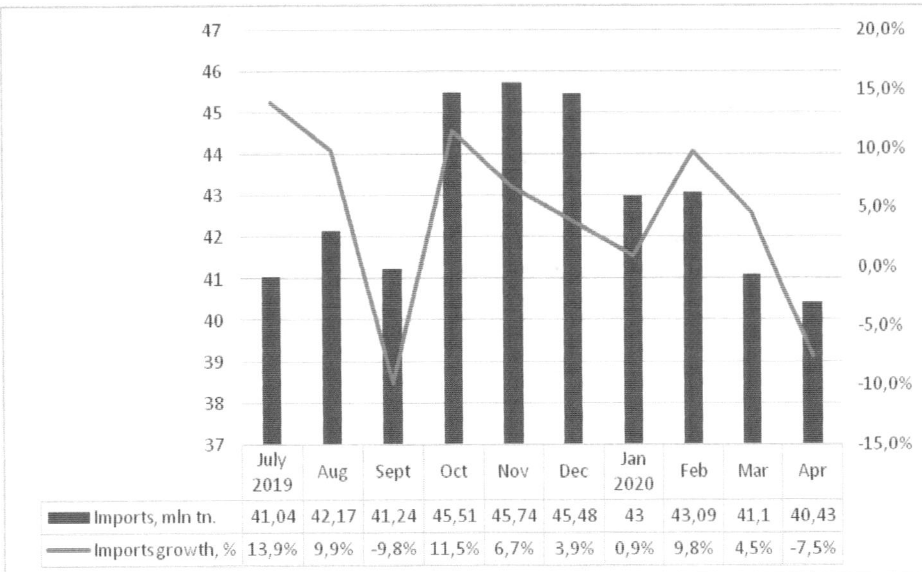

Figure 4. Monthly rates of change in oil imports in 2019-2020.

Table 2. Main ports of oil shipment and delivery to China in February 2020 [1-5].

Ports of shipment / transshipment	Dispatches of tankers	Total deadweight, thousand tonnes
Singapore	17	312.24
Mina Al-Ahmadi (Kuwait)	9	222.98
Ras Tanura (Saudi Arabia)	9	217.94
Kozmino (Russia)	14	193.48
Das Island (UAE)	5	134.81
Ports of delivery	Entries by tankers	Total deadweight, thousand tonnes
Rizhao	22	642.66
Qingdao	21	565.28
Dalian	12	363.41
Dongjiakou (Shandong Province)	12	332.12
Zhangjiang (Guangdong Province)	12	299.18

Due to the huge volume of oil imports, sea transportation is provided mainly by supertankers. VLCC vessels (over 330 m) and above account for 52% of maritime oil imports, Aframax (245 m) for 19% and Suez-Max (285 m) for 14%. Some ports are specialized in unloading certain types of tankers, for example, Qingdao and Rizhao - VLCC, and Dalian and Jinzhou (Liaoning Province) - Handymax and Panamax ships.

5. Conclusion

Despite the short-term negative impact of the Coronavirus pandemic in 2020, China's oil industry can already recover within the full production-consumer cycle: upstream-midstream-downstream in the short term. A minor disruption in the first months of the year will also have little impact on long-term industry plans up to 2050, where oil is expected to play the role of a stabilizer for inter-sectoral shifts in the energy sector.

Acknowledgments
The study was carried out with the financial support of the Russian Foundation for Basic Research within the scientific project № 20-010-00616.

References

[1] An Zhao Zhen 2019 New Situation and New Measures of Sino-Russian Energy Cooperation. *Russian Foreign Economic Bulletin* **1** 7-14
[2] Golobokov A S 2018 Russian energy strategy and new forms of Russian-Chinese energy cooperation on the Far East. *The Territory of New Opportunites. The Herald of Vladivostok State University of Economics and Service* **10(4)** 7-16
[3] *Russian statistical yearbook* 2018 (Moscow: Rosstat) 694
[4] Martyshenko N S and Kletskin A D 2019 Development of cooperation between Russia and China in the oil sector and its impact on the economy of the Primorsky krai. *Bulletin NGIEI* **7(98)** 88-99
[5] Romanova I N, Morkovkin D E, Nezamaikin V N, Gibadullin A A and Ivanova M A 2020 Formation of a policy to ensure environmental safety in modern economic conditions. *IOP Conference Series: Materials Science and Engineering* **734** 012166
[6] Morkovkin D E, Lopatkin D S, Shushunova T N, Sharipov B K and Gibadullin A A 2020 Formation of the conditions for the development of innovation. *Journal of Physics: Conference Series* **1515** 032002
[7] Sharipov F F and Timofeev O A 2020 Coal mining and import in China in 2019: prospects for russian exporters. *Ugol' – Russian Coal Journal* **4** 68-71
[8] Linnik V Yu and Linnik Yu N 2019 Priority directions of innovative development in oil and gas complex at the Siberian Federal District. *Upravlenie* **7** 1 pp 40-9

Key Aspects and Trends of Distributed Generation and Energy Decentralization in Russia

S V Smirnova[1], P V Bolotov[2], D E Petrushin[3,4], M G Popov[3,4] and M V Matrosova[4]

[1]Tula Regional Dispatching Office, a branch of System Operator of the Unified Power System, 99a, Timiryazeva str., Tula, 300012, Russia
[2]Vladimir Regional Dispatching Office, a branch of System Operator of the Unified Power System, 88k, Bolshaya Nizhegorodskaya str., Vladimir, 600016, Russia
[3]Peter the Great St. Petersburg Polytechnic University, 29, Politekhnicheskaya str., Saint Petersburg, 195251, Russia
[4]Petrozavodsk State University, 33, Lenin Ave., Petrozavodsk, 185910, Russia

E-mail: petrushind@petrsu.ru

Abstract. Distributed Generation (DG) is a global trend and it is actively spreading in Russia. In contrast to Western countries, the main incentive for the spread of DG in Russia is the desire of industrial owners to reduce the final cost of electricity. DG in Russia can successfully meet the challenge of supplying electricity to isolated and inaccessible areas and replacing outdated and inefficient generation. The widespread use of DG, including those based on renewable energy sources (RES) makes it necessary to create promising models for the inclusion of DG in the Unified Energy System (UES) of Russia. The paper deals with the efficiency of DG, analyses the application of this technology in Russia formulates the prospects for the development of DG in Russia and gives recommendations for the integration of DG.

1. Introduction

Russian energy system is characterized by a model of centralized generation. It includes 846 power plants with installed capacity of more than 5 MW [1]. This scheme for the development of the Russian power industry is due to the long distances between power sources and consumers, as well as the heterogeneity in the distribution of primary energy resources (oil, gas, coal, large rivers) with energy consumption centers [2]. The global trend is the gradual abandonment of the centralized model and transition to a distributed energy model, as well as a reduction in the share of fossil primary energy resources and an increase in RES [3]. The main aim is to make a forecast of the DG development in Russia until 2035.

2. Materials and methods

Case studies of leading companies were analyzed to determine the initial conditions. A qualitative study was carried out by a questionnaire to make the DG forecast. The questionnaire contained 19 questions. It was required to estimate validity of a statement and estimate the degree of its influence on the DG in Russia. The questionnaire was sent to over 40 organizations. A matrix of criteria was compiled by analysis of the questionnaires. On the basis of this matrix the most probable and significant criteria were identified.

3. Results

3.1. Distributed Generation.

In this paper the DG unit is understood as the power plant, consisting of at least one generation unit. It is connected to the distribution networks or internal networks of power supply at the voltage up to 110 kV. Generating units are as close as possible to the node of power consumption that works in parallel with the utility system or in island (stand-alone) mode. It has total installed capacity to 25 MW. DG used for the production of all types of energy (electrical, heat, cold, etc.) of any primary sources of energy, including RES [4].

According to the report of consulting company Guidehouse Research (USA), the global installed capacity of the DG will be more than 200 GW until 2025. Half of the installed capacity coming from solar power plants (SPP) is using photovoltaic modules (figure 1) [3].

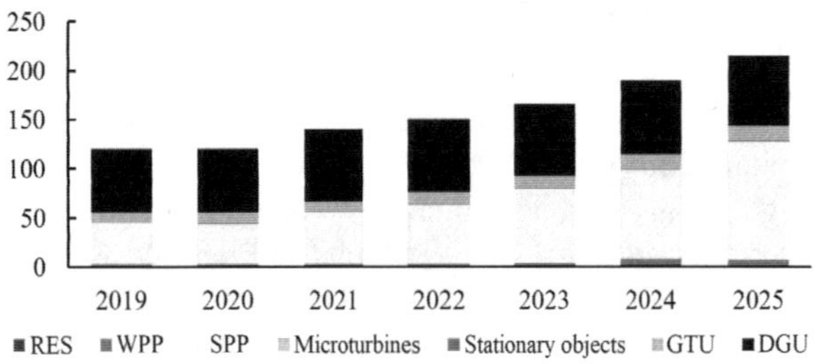

Figure 1. Growth of installed capacity of DG in the world until 2025, MW.

3.2. Relevance of Distributed Generation.

The global energy industry has clearly seen the 3D trend in recent decades: Decarbonization, Decentralization and Digitalization. Structural changes in the energy industry are due to a complex of different reasons. One of the incentives for the "energy transition" for Western countries is the government's response to the request of the citizens, for a transition to a low-carbon economy in general and low-carbon energy in particular.

Using the example of the German energy sector, as the leading economy in the EU and the world, the influence of the 3D technological trend on the structure and volume of the electric power industry can be clearly traced. In 2019, the total amount of electricity produced in Germany was 518 TW·h, with about 40 % of the total volume being generated from RES. Most of renewable energy installations is DG. Significant aspect of the development of the German electric power industry is the final cost of electricity for industrial and domestic use. According to the estimates of NP "Community of Energy Consumers", the final price for industrial enterprises in Germany is one of the highest in the world, due to the high share of installed capacity of generating facilities based on RES (figure 2).

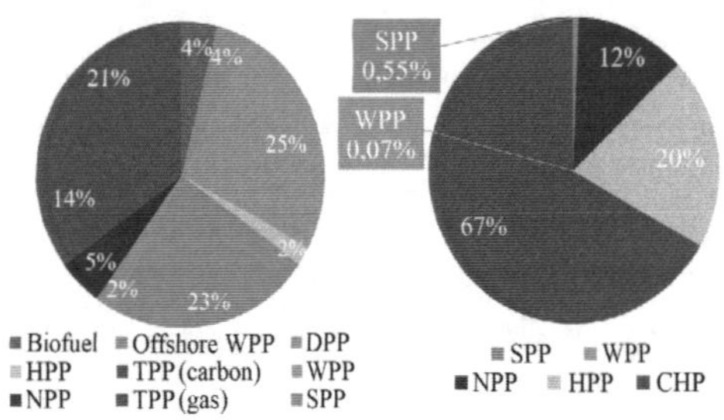

Figure 2. Installed generating capacity structure for 2019 in Germany (left) and Russia [1] (right).

The impact of the 3D trend is weak because of relatively cheap natural gas for Russia. In 2019, most of the electricity was generated by centralized hydrocarbon-fueled power plants. The share of installed capacity of RES power plants is about 1 % of the total installed capacity. According to respondents, the share of DG on hydrocarbon fuel and operating on a cogeneration cycle is 7% of the total installed capacity of the Russian Federation.

IOP Conf. Series: Materials Science and Engineering **976** (2020) 012028 doi:10.1088/1757-899X/976/1/012028

3.3. Current Status of Distributed Generation in Russia.

There is an annual increase in the commissioning of DG facilities in Russia. According to INEI RAS, over the past 10 years, the total installed capacity of 25 MW power plants has increased about 30%. Renewable energy facilities are being introduced to a greater extent in most countries of the world. In Russia, the increase in their number is due to thermal power plants with gas turbine units, diesel generator sets and gas piston units. They are connected to the distribution electric networks or to the internal power supply networks of industrial enterprises. The main incentive for the introduction of DG in Russia is the desire of the industrial enterprises owners to reduce the cost of electricity to increase the operational efficiency of production [5]. The Ministry of economic development predicts the indexation of wholesale gas prices (the main fuel for DG in the Russian Federation) for all categories of consumers (excluding household consumers) by 2-3% per year until 2030 due to rising electricity prices, which means an increase in the attractiveness of DG in the near future.

It is difficult to accurately estimate the current volume of DG in the UES of Russia. The Rosstat and the Ministry of energy of the Russian Federation in their official statistical annual and operational reports do not distinguish this type of generation in the reporting data. The main sources of data are estimates of experts, scientific organizations, indicators of enterprises and analytical work of consulting companies. In addition, in the Russian Federation there is a “gray area” of DG. These units operate without connecting to the UES of Russia. According to an approximate estimate, 487 DG units operated in the UES of Russia with a total installed capacity of 6 GW at the beginning of 2020. The main volume of generating unit is accounted for by industrial enterprise power plants using hydrocarbon fuel or burning production waste: 260 generating units with a total capacity of 3360 MW. According to the calculations of one of the leading engineering companies in the Russian Federation, the ISS group of companies and the Moscow school of management SKOLKOVO, the use of its own generation of industrial enterprise allows you to save up to 5 cents per kWh of electricity, despite the high cost of operating the DG (figure 3).

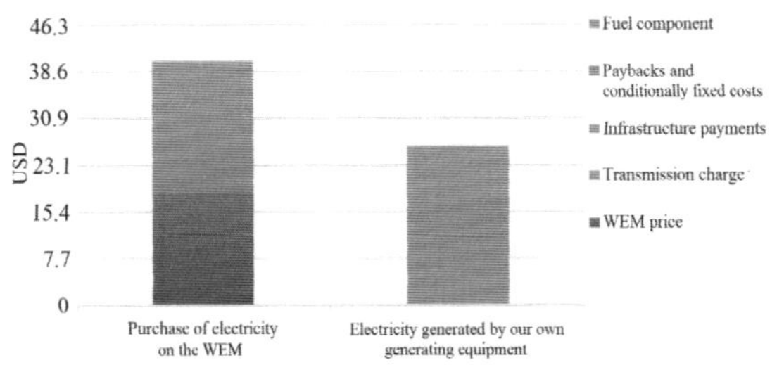

Figure 3. Comparison of the electricity cost for enterprises on the market and using own generation.

The global trend of development of DG based on RES can also be traced in the Russian Federation. Summary values for DG unit in the Russian Federation that operate as part of the UES are shown in table 1.

Table 1. Summary data on DG in UES of Russia.

Type	Indust. PP	Sm. HPP	WPP	SPP	Tidal PP	Bio TPP	TPP
Capacity, MW	3360	480	98	634	1,1	2,4	1395
Number of objects	260	76	11	46	1	1	94

Fifty nine DG facilities based on RES are in operation with a total installed capacity of 735 MW in the Russian Federation in 2020 (excluding small hydroelectric power plants). There are more than 2000 DG facilities located in isolated and inaccessible areas of the Russian Federation and operating on an island regime. Their total installed capacity is estimated at 840.3 MW. The main percentage of installed capacity (63 % – 525 MW) of these, generating facilities is located in four regions of the Russian Federation: The Republic of Sakha (Yakutia), the Kamchatka region, the Krasnoyarsk region and the Yamalo-Nenets Autonomous District. The other 19 regions account for 315 MW of installed capacity. The main type of fuel is diesel, but there are generating facilities that use renewable energy, whose total capacity at the end of 2018 is 16.6 MW (39 units).

The main feature of power generation in isolated and inaccessible areas is that the state subsidies the cost of electricity to compensate for the difference between the economically justified tariff and the established tariffs for households and industrial enterprises. According to estimates of the Analytical Centre under the Government of the Russian Federation, the specific cost of electricity generation reaches 65 cents per kWh.

At present, Russia does not have a legal and regulatory framework for the harmonious development of distributed power generation within the UES of Russia. It is important to note that in recent years the legislative bodies, together with the relevant ministries and the energy community, have been actively engaged in law-making activities. Over the last two years, the Russian Government and the Ministry of Energy have already approved [6] a part of the fundamental documents containing the regulatory and terminological basis for DG, which will ensure stable, consistent development of the energy sector and application of new technologies within the legal framework.

3.4. Technological Trends in Distributed Generation.

The efficient operation of DG within the UES of Russia requires effective mechanisms for the integration, management and regulation of such power sources. The development of information, microprocessor and communication technologies in recent decades, together with modern economic models, has given an active impetus to the development of DG. Due to the large spread of distributed power generation and the need for its effective integration into the UES of Russia, the concept of Virtual Power Plant – an IT system to which distributed power sources are connected, mainly on the basis of RES - has emerged. The blockchain technology can be used in managing the electrical modes, exchange trading in electricity [7].

The digitalization of DG is correlated with the strategic directions of the program of the Ministry of energy "Digital Transformation of the Electric Power industry of the Russian Federation". At the moment, Russia is developing a digital pilot project to create a new system of consumer's relationships on retail electricity market to optimize tariff — AEC [8] — the joint development of SO UPS, JSC, the group of companies STC UPS, JSC, the Ministry of energy and "Energynet". The AEC is a micro-energy system consisting of its own generation, electricity consumers, and network infrastructure. Power generation and consumption is carried out through a controlled smart connection. The economic feasibility of the AEC consists in the ability of each participant to pay the tariff for the maintenance of electric networks within the permitted amount, but not more than its actual consumption from the network, provided that its consumption can be limited to the established level.

An important factor and precondition for the development of DG, which is characteristic of all countries, is the gradual obsolescence of fixed production assets (figure 4), the need to replace obsolete and inefficient generation, and the construction of new generation to cover growing consumption [9].

According to the energy strategy of the Russian Federation, in the period up to 2035, the total installed capacity of power plants in the UES of Russia should achieve 251-264 GW [10], which, taking into account the projected withdrawal of generation unit [10], will require the input of up to 48 GW of new capacity. One of the options to partially cover this amount of capacity is the construction of DG facilities.

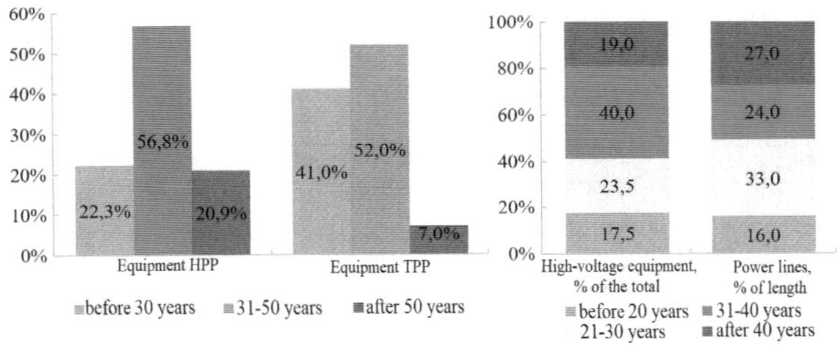

Figure 4. Age structure of generating equipment of the UES (percentage of installed capacity) (left) and power grid equipment of the Unified National Electricity and Distribution Networks (right).

3.5. Trends and technological trends in distributed generation.

High cost of technological connection and electricity transmission and the need to replace worn-out and outdated capacities lead to an increase in tariff.

According to the Alteko, the average annual increase in electricity tariff for domestic consumers will be about 4 % until 2035. At the same time, according to the NP "Community of energy consumers" [11] and the association “NP Market Council”, the price of electricity in the wholesale market for the 1st half of 2019 increased significantly – by 9% compared to December 2018. the price of natural gas for industrial enterprise is projected to increase significantly until 2024 (figure 5) and will average 3% per year [12]. This price ratio is an incentive to build your own generation.

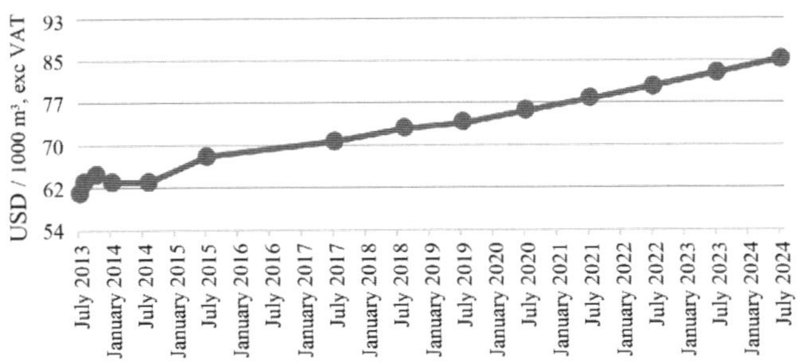

Figure 5. Price of natural gas forecast for industrial enterprises until 2024.

In addition to economic incentives for the development of DG based on hydrocarbons until 2035, an important component of the "Energy Strategy of the Russian Federation for the Period Up to 2035" is to increase the share of DG in the total generation volume and load concentration in regional energy systems. The aim can be achieved by a large-scale upgrade of the Combined Heat and Power by installing heating 250-350 GTU with a capacity of up to 25 MW (total installed capacity of 3.6 GW). It also can be achieved by municipal boiler houses conversion to GTU/GPU. Mini-Combined Heat and Power will total up to 70 GW of newly introduced capacity by 2035. All this creates favorable conditions for the development of industrial enterprises and municipal enterprises generation as well as an increase in the total number of DG on hydrocarbon fuel.

4. Discussion

This study was presented at the Youth Forecast of the Global Energy Development 2020 by SO UPS, JSC team and was awarded the first place. Materials were presented at the conference Electricity Through the Eyes of Youth 2020 by the authors and were awarded the third place. The paper received a positive review from Ph.D., associate professor V.V.Zavalishin.

5. Conclusion

Summing up, it is worth noting that the unconditional advantages of DG will be assessed in the programs for supplying power to isolated and hard-to-reach territories and eliminate the consequences of emergency situations [13]. And within the framework of integration into the UES of Russia, DG is considered not as a competitor to centralized generation, but as a harmonious complement to it.

Acknowledgments
The reported study was funded by RFBR, project number 19-38-90123.

References

[1] *Unified Energy System of Russia* Retrieved from: https://www.so-ups.ru/
[2] Popov M G , Pavlov N G and Petrushin D E 2019 Modern Automatics for Elimination of Asynchronous Operation. *International Ural Conference on Electrical Power Engineering* 75-80
[3] Pritil G 2019 *Global Additions in Distributed Generation Capacity to Increase Significantly* Retrieved from: https://guidehouseinsights.com/news-and-views/global-additions-in-distributed-generation-capacity-to-increase-significantly
[4] Gurevich Yu E and Ilyushin P V 2018 *Peculiarities of Calculating Modes in Power Regions With Distributed Generation* (Nizhny Novgorod: NRU RANEPA) 280
[5] Popov M G and Petrushin D E 2020 Analytical Signals Using for the Power Systems Non-Stationary Modes Analysis. *IEEE Conference of Russian Young Researchers in Electrical and Electronic Engineering* 830-4
[6] *On the Approval of the Rules for the Technological Functioning of Electric Power Systems and on Amending Some Acts of the Government of the Russian Federation* Retrieved from: http://www.consultant.ru/document/cons_doc_LAW_304807/2ca7b483e7f65a8401eea2135f37f80c30c4f304
[7] Bolotov P V 2018 Application of Blockchain Technology in Distributed Generation Based on Renewable Energy Sources. *Electricity in the Eye of Youth* 32-5
[8] *On amendments to some acts of the Government of the Russian Federation on the functioning of active energy complexes* Retrieved from: http://www.consultant.ru/document/cons_doc_LAW_304807/fef1db9e27c611b5b932f67b1ec898f06bc62d38/
[9] Kondrashov M A, Popov M G, Bukharov D S and Petrushin D E 2020 Development of a Cluster Analysis Method for Solving the Problem of Identifying Static Voltage Load Characteristics *Journal of Physics: Conference Series* **1546** 012085
[10] *Energy Strategy of Russia until 2035* Retrieved from: https://minenergo.gov.ru/node/1920
[11] *Electricity for Industry in Russia in 2019 Turned Out to be More Expensive Than in the United States and a Number of EU Countries* Retrieved from: https://www.np-ace.ru/
[12] Zhukov A 2020 *Competitive Model: By 2026, There Will Be Three Times More Own Energy Sources Than Centralized Ones* Retrieved from: https://www.dp.ru/staff
[13] Popov M G, Petrushin D E and Efimov N S 2019 Increasing the emergency control systems efficiency in Kola's and Karelia's power systems. *IEEE Conf. of Russian Young Researchers in Electrical and Electronic Engineering* 1040-3

IOP Conf. Series: Materials Science and Engineering **976** (2020) 012029 doi:10.1088/1757-899X/976/1/012029

Study of the acoustic efficiency of a plate silencer of the original design

D V Chugunkov, E A Zhuravlev, G A Seyfelmlyukova and K C Fomenko

National Research University "MPEI", 14, Krasnokazarmennaya, Moscow, 111250, Russia

E-mail: ChugunkovDV@mpei.ru

Abstract. Energy facilities are sources of noise impact. One of the most powerful sources of constant noise at such facilities is power boilers and draft machines, which emit noise into the surrounding area from air intakes and chimney mouths. To reduce noise from them, dissipative plate-type silencers are mainly used. In many cases, the installation in gas and air ducts of typical dissipative plate noise silencers consisting of plates with flat sidewalls does not provide the required acoustic efficiency with a limited length of the plates and the requirement for low aerodynamic resistance. In this regard, an urgent direction of scientific research is the development of new approaches to improve the acoustic characteristics of dissipative noise mufflers. The article presents the results of field tests of the acoustic efficiency of a dissipative plate noise muffler, consisting of patented noise suppression elements. The comparison of the obtained efficiency of the muffler from the proposed elements of noise suppression with the efficiency of the known designs of plate mufflers is carried out.

1. Introduction

One of the most powerful sources of constant noise at power facilities are power boilers and draft machines. The main ways of propagation of noise from such sources are surfaces of gas and air ducts, air intakes and chimney mouths. The most dangerous sources of noise are chimney mouths, since they are high-rise and affect long distances in the absence of noise shielding and the influence of the terrain. To reduce noise in gas paths of boilers, dissipative plate-type noise silencers are widely used in world practice. A typical plate silencer consists of parallel plates, consisting of cassettes with flat side perforated walls, which are installed in the channels of the gas-air ducts of boilers [1-2] (figure 1).

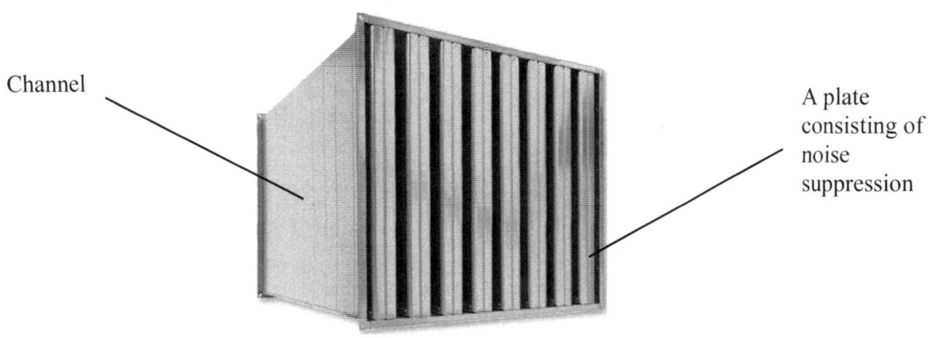

Figure 1. Typical dissipative plate silencer.

The problem of the negative noise impact from chimneys and air intakes of thermal power plants (TPP) on the residential areas of large cities located in the immediate vicinity is especially acute. Figure 2 shows a common case when a thermal power plant is located close to a residential area. In the absence of installed systems for noise suppression from equipment at TPPs, residents of the district are exposed to increased noise.

Figure 2. The close location of the TPP to the residential area.

Silencers are installed in the gas ducts of boilers behind the smoke exhausters in front of the chimneys. The advantages of dissipative silencers are broadband sound attenuation, simplicity of design, long service life and the absence of significant operating costs, as well as low cost of manufacturing and installation [3-8]. Sound attenuation in a plate noise damper can be determined by the formula [1], dB/m:

$$\Delta L = 10\log_{10}\left[\frac{\alpha_{obl} S_{obl}}{2S_{pr}} + 1\right] \quad (1)$$

where S_{obl} – sound absorption area, m^2; S_{pr} – muffler free section, m^2; α_{obl} – sound absorption coefficient.

From formula (1) it follows that the increase in the efficiency of the plate noise mufflers is proportional to the increase in α_{obl} and S_{obl}.

The sound absorption coefficient α_{obl} depends on the properties of the sound-absorbing material (ZPM) of the cassettes of the silencer plates. Certain requirements are imposed on ZPM: fire resistance, hydrophobicity, absence of shrinkage, low density. In the power industry, various types of mineral wool are used as RFPs for noise silencers: glass, stone (basalt) and slag, as well as expanded polystyrene and polyurethane foam.

Basalt wool, which meets the above requirements, is most widely used for stuffing noise-damping elements of dissipative noise mufflers. As the ratio increases α_{obl} the sound attenuation increases. The use of a ZMP with a steel wire frame significantly extends the service life of the lamellar noise mufflers, ensuring a decrease in shrinkage.

Another way to increase the acoustic performance of silencers is to increase the sound absorption area. It is important not only to create a developed sound absorption surface, but also to select a surface shape that will be optimal in terms of production technology.

In work [9] it was shown that, the largest area of sound absorption of the side surfaces of the noise damping elements (cassettes of the NRU "MPEI") of the lamellar mufflers is achieved when using semi-cylindrical concavities.

2. Materials and methods

This article uses the method of experimental research. Full-scale bench acoustic tests of a dissipative noise muffler of an original design in several configurations were carried out.

3. Results

Bench tests were carried out to determine the acoustic performance of the original design of the dissipative silencer.

Figure 3 shows a diagram of a test bench for determining the effectiveness of plate noise mufflers made of original noise suppression elements (NRU "MPEI" cassettes) [10].

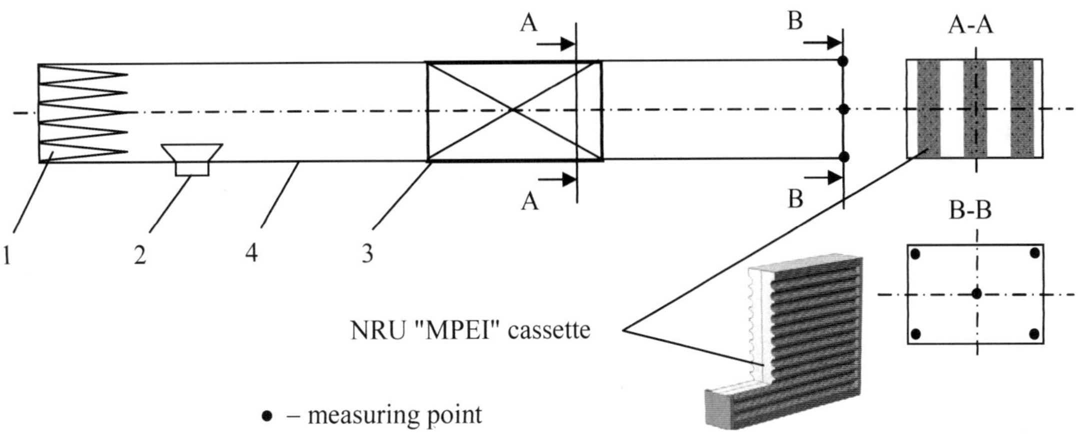

1 - end absorber; 2 - noise source; 3 - a prototype of a noise muffler; 4 - channel of the experimental stand

Figure 3. Test bench layout.

The test bench was a rectangular channel made of sheet steel with dimensions of 1000x1200 mm, from the outside the channel surface was wrapped with a sound-insulating shell, one channel end was plugged, the second channel end was open.

The joints of the stand channel are sealed with a sealant and tightened with a bolted joint. The outer surface is covered with soundproofing material, the joints are soundproofed.

A noise source was installed on one side of the channel; acoustic measurements were carried out at the opposite end of the channel, between them a prototype noise suppressor was installed, which is a parallel plate from the NRU "MPEI" cassette.

The test equipment included a two-channel signal analyzer LDViRTe 3000+ (LD3000 +) used as a noise generator, an acoustic cable, a noise emitter, a noise analyzer SIU Assistant of the 1st accuracy class, which was used for acoustic measurements. The LD3000 + Noise Spectrum Analyzer was used as a reference noise source that generated white noise. White noise is stationary noise, the spectral components of which are evenly distributed over the entire frequency range of the bands. The noise emitter converted the received signal from the LD3000 + into sound waves in the frequency range 50-20000 Hz. Measurements of sound levels (US) and sound pressure levels (SPL) were carried out at several control points with a noise analyzer in octave bands with geometric mean frequencies in the range of 31.5-8000 Hz.

Figure 4 shows the external view of the test bench with the investigated cassettes of the NRU "MPEI".

The tests were carried out in two stages:

- Noise levels were measured at control points when the noise source was operating in a channel without a noise suppressor;
- Noise levels were measured at control points when a noise source was operating in a duct with a noise suppressor in various combinations of location and number of cassettes.

The measurement points along the channel cross-section were chosen from the condition of equal area of the conditionally divided cross-section. The measurements were carried out at the same level of the noise emitter. According to the results of measurements for each arrangement of the muffler, the average values of the SPL were calculated.

The results of full-scale bench tests of the noise muffler are presented in table 1.

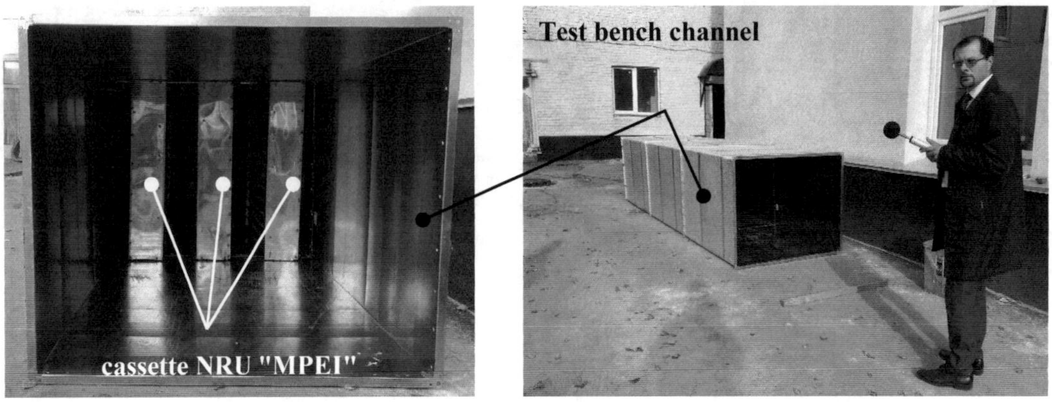

Figure 4. Test bench.

Table 1. Results of full-scale tests of the acoustic efficiency of a prototype of a plate silencer in three configurations at different lengths.

Length, m	Decrease in SPL, dB/m, in octave bands at geometric mean frequencies, Hz							
	63	125	250	500	1000	2000	4000	8000
relative flow area 73%								
1 m	1,0	3,0	4,8	8,2	14,9	13,4	10,3	8,6
2 m	1,5	4,4	9,5	14,6	24,7	19,2	12,2	10,2
3 m	2	5,4	13,1	19,7	28,2	22,3	12,3	10,7
relative flow area 60%								
1 m	1,4	4,1	8,2	14,8	20,3	21,1	15,7	13,0
2 m	2,0	6,0	15,1	24,2	29,8	29,4	22,1	18,2
relative flow area 47%								
1 m	1,4	5,3	12,0	20,4	22,9	23,0	21,4	19,7
2 m	1,8	7,8	17,0	25,3	33,1	31,0	28,2	28,2

It can be seen from the results that with an increase in the length of the muffler, the decrease in noise in octave bands with geometric mean frequencies of 250-2000 Hz increases more intensively.

4. Discussion

To demonstrate the superiority of noise mufflers from MPEI cassettes over noise mufflers from standard cassettes, figure 5 shows a comparison of noise reduction with the same number of plates, their length and distance between the plates.

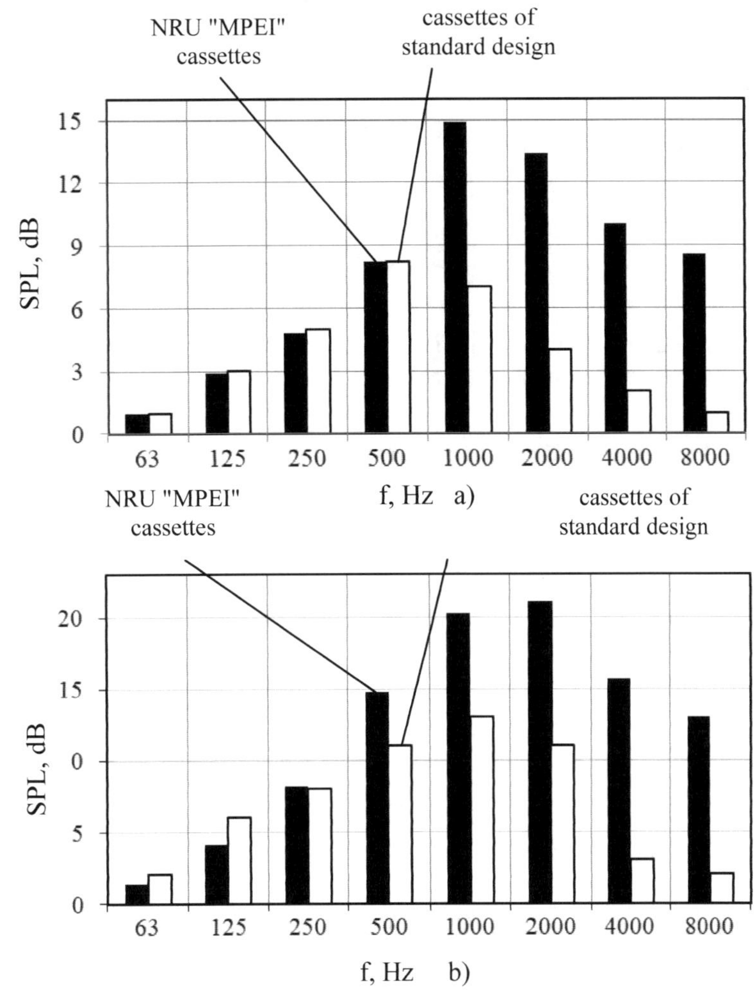

Figure 5. Comparison of the acoustic performance of silencers of two types of cassettes:
a - thickness of cassettes - 200 mm, distance between plates - 400 mm
b - thickness of cassettes - 200 mm, distance between plates - 200 mm

The analysis of figure 5 shows that the acoustic efficiency of the plate noise muffler from the MPEI cassettes is superior to the noise muffler from standard cassettes with flat side walls in octave bands at geometric mean frequencies in the range of 500-8000 Hz, in the low frequency range the noise muffler from the NRU cassettes "MEI" either corresponds in efficiency to a noise muffler made from standard cassettes, or is slightly inferior within the measurement error.

The error in determining the acoustic efficiency of a noise muffler is the sum of the random error in the SPL measurement and the instrumental error, determined by the accuracy of the measuring equipment (noise analyzer).

The results of calculating errors in determining the acoustic efficiency of noise mufflers from cassettes of NRU "MPEI" are presented in table 2.

Additionally, the concavities of the MPEI cassettes contribute to an increase in the passage section of the noise silencer, which reduces the introduced aerodynamic resistance of the noise silencer.

Table 2. The error in determining the acoustic efficiency of a noise muffler from NRU "MPEI" cassettes in several configurations.

Relative flow area, %	Name	SPL, dB/m, in octave bands at geometric mean frequencies, Hz							
		63	125	250	500	1000	2000	4000	8000
73	Acoustic efficiency	1,0	3,0	4,8	8,2	14,9	13,4	10,3	8,6
	Error	±1,7	±1,5	±2,2	±2,2	±2,1	±1,4	±1,4	±1,9
60	Acoustic efficiency	1,4	4,1	8,2	14,8	20,3	21,1	15,7	13,0
	Error	±1,3	±1,9	±1,3	±2,0	±0,9	±0,9	±1,0	±1,0

5. Conclusion

According to the results of bench tests of a dissipative plate noise muffler of the original design, its acoustic efficiency exceeded the efficiency of a noise muffler of a standard design in all similar configurations, especially in the range of medium and high geometric mean frequencies. The results of full-scale tests of the acoustic efficiency of an original design noise muffler, consisting of NRU "MPEI" cassettes, in various configurations are presented in table 1.

Acknowledgments

The study was carried out within the framework of the scientific research program supported by the scholarship of the President of the Russian Federation SP-1944.2019.1.

References

[1] Ivanov N I 2016 *Engineering acoustics. Theory and practice of dealing with noise: textbook* (Moscow: University book) 424

[2] Heckla M and Müller H A 1980 *Handbook of technical acoustics: per. with him. ed* (Leningrad: Sudostroenie) 440

[3] Tupov B V and Medvedev V T 2014 Determination of the lengths of multistage mufflers with different plate thicknesses. *Heat Power Engineering* **3** 66-9

[4] Munjal M L 1987 *Acoustics of Ducts and Mufflers With Application to Exhaust and Ventilation System Design* (NY: Wiley-Interscience) 328

[5] Hood J M and Kidwell S A 1989 Tuned dissipative mufflers for draft fan intakes. *INTER-NOISE-89* **89** 413-6

[6] Ver I and Biker W E 1978 Design of a tuned muffler for large induced-draft fans. *Proceedings, INTER-NOISE-78* **78** 309-16

[7] Tupov V B and Chugunkov D V 2010 Specific features of systems for suppressing noise produced by the gas paths of power-generating boilers. *Thermal Engineering (English translation of Teploenergetika)* **57(12)** 1065-9

[8] Tupov V B, Chugunkov D V and Semin S A Reduction of noise emitted from exhaust paths of gas turbine units equipped with heat recovery boilers. *Thermal Engineering (English translation of Teploenergetika)* **56(1)** 26-30

[9] Zhuravlev E A and Chugunkov D V 2019 Approaches to improving plate silencers for gas-air ducts of a boiler *Conference "Advanced innovative developments. Prospects and experience of use, problems of implementation in production "(October 31, 2019, Kazan): collection of scientific articles on the basis of the ninth international scientific conference. Part 1 Kazan .: Publishing house of LLC "Envelope"* 74-8

[10] Chugunkov D V, Seifelmlyukova G A, Zhuravlev E A, Fomenko K S, Skurikhina A D and Bogdanova A E 2019 Sound-absorbing cassette for silencers of gas-air ducts. *Bul. 9-9* 2018104856

Laboratory approaches to estimating and localizing restrained hydrocarbon reserves in depletion-mode formations

M N Bolshakov, N A Skibitskaya and O O Marutyan

Oil and Gas Research Institute RAS, 3, Gubkin street, Moscow, 119333, Russian Federation

E-mail: rgu2006@mail.ru

Abstract. A significant part of hydrocarbon reserves during development remains in the reservoir in the restrained volumes of the pore space. In depletion-mode formations, bypassed zones of hydrocarbons not affected by displacement can also form. The paper describes laboratory methods that can be used to estimate the amount of remaining restrained hydrocarbons and their distribution in the reservoir.

1. Introduction

Gas production at gas-condensate and oil-gas-condensate fields, as a rule, is carried out using natural reservoir energy - in depletion mode. Reservoir pressure drop during the development triggers the in-situ mechanisms striving to restore the disturbed thermodynamic equilibrium, including the processes of watering due to direct-flow capillary imbibition and the advancement of the bottom or edge waters into the reservoir.

The process of capillary flooding of gas-saturated sediments is accompanied by the formation of restrained volumes of free gas in the pores, removed from the filtration processes in the reservoir pressure drop mode, that reduces the gas-condensate recovery factor.

So, for the Vuktylskoye gas condensate field the recovery factors for dry gas – 0.795, stable condensate – 0.308 [1], for the Orenburgskoe oil-gas-condensate field – 0.71 [2], condensate recovery factor for Astrakhanskoe gas-condensate field – 0.57 [3], the average gas recovery factor is 0.85, and for fields with significant reservoir heterogeneity 0.7-0.8 [4].

In view of this, producing companies, with the aim of a more rational approach to hydrocarbon resources, are striving to increase the recovery factor by using additional methods of increasing recovery, such as hydraulic fracturing, chemical, thermal and other methods. However, we have previously shown that the values of residual (non-recoverable) gas reserves and their distribution in the volume of the reservoir largely depend on the pore space structure (PSS) of the reservoir rock [5].

2. Materials and methods

This paper provides a brief description of two laboratory methods that supplement the standard petrophysical complex of studies with very valuable data. These are direct-flow capillary imbibition (DCI) and aspiration thermomassometry (ATMM) methods. In the experiments, core samples of cylindrical or cubic shape are investigated with the possibility of multiple repetition.

3. Results

The first important point highlighted in this work, which should be considered when designing the development and additional development of the deposit, is the need to build a geological model of the restrained volumes in the pores, as well as the typification of rocks according to the PSS. So, in a large-pore reservoir in large pores, we have a large restrained volume that is not involved in the filtration process, and in a porous subcapillary-pore reservoir, the restrained volume will be small with a high value of the porosity coefficient. In order to extract additional volumes of hydrocarbons, specialists need to carefully select the zones of the reservoir according to the amount of hydrocarbons in the restrained pore volumes and the probability of involving these volumes in development by one or another stimulation method. It should be noted that for an oil reservoir, the amount of restrained oil will correspond to the geometric restrained pore volume, and for a gas reservoir, the amount of gas should be considered based on the equilibrium reservoir pressure at the time of development, since as the pressure drops, part of the gas in the central restrained part of the pores passes into the filter volumes and is displaced. The dynamic volume for an oil reservoir, calculated as the difference between the effective and restrained volumes, determines the maximum recoverable volume of oil in the reservoir without the use of stimulation methods.

The values of the restrained volumes *Kr* are provided by the laboratory method of DCI. The research process consists of the following stages:

- 100% saturation of the rock sample with water;
- Centrifugation of samples, after which there is residual water in the rock (film tightly bound water and water of pore angles). The rest of the pore space is usually called effective; in the productive formation it is occupied by hydrocarbons;
- Direct-flow capillary imbibition with water, by bringing one face of the sample into contact with water. As water is imbibing, gas (air) from the pore space comes out through the open surface of the sample, forming a restrained and dynamic volume in the pores (figure 1). Continuous registration of changes in sample weight over time.

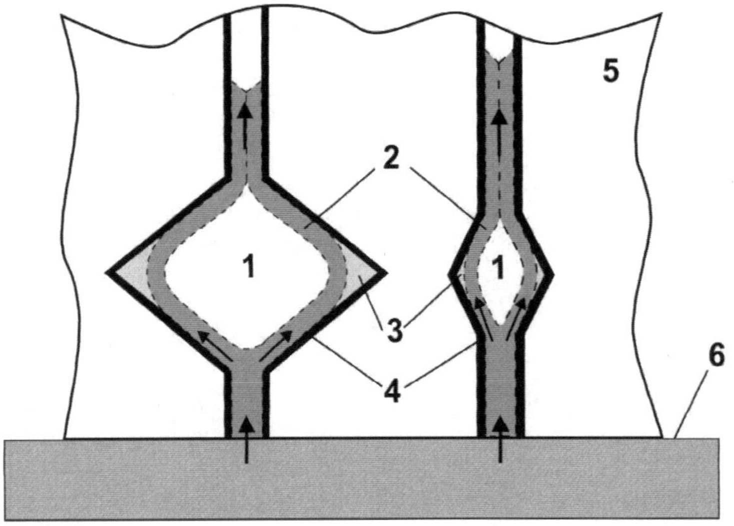

Figure 1. Scheme of the formation of restrained volumes in pores during direct-flow capillary imbibition of hydrophilic rock with water: 1 – restrained pore volume; 2 – dynamic pore volume; 3 – water of pore angles; 4 – film water; 5 – rock sample; 6 – water surface.

Then, in the process of building petrophysical relationships, the values of the restrained volumes are correlated with other parameters, such as open (*Kp*) or effective (*Kp ef*) porosity. For a good correlation, the studied collection of samples must be large and representative. Having a geological model, using the obtained relationships, it is easy to construct a three-dimensional distribution of the

IOP Conf. Series: Materials Science and Engineering **976** (2020) 012030 doi:10.1088/1757-899X/976/1/012030

restrained gas-saturated porosity *WKr* and give recommendations on the most promising zones (figure 2).

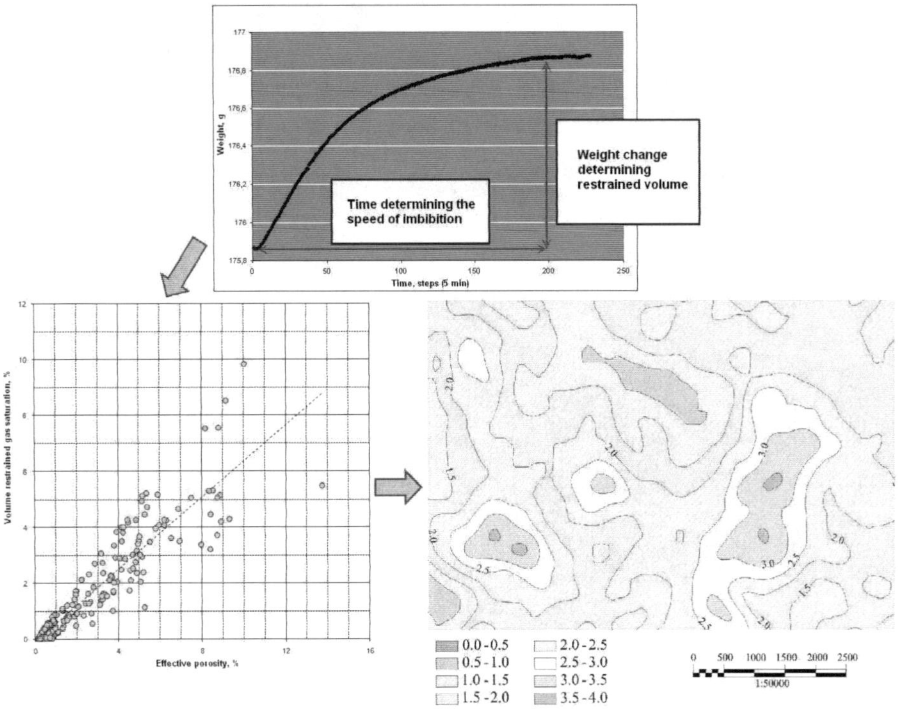

Figure 2. Stages of allocation of zones with maximum concentrations of residual volumes of restrained in the pores gas *WKr=Kr*Kp*:

- Results of direct-flow capillary water imbibition of rock samples in residual water saturation state (DCI experimental curves);
- Petrophysical relationships *WKr=f(Kp ef)*;
- 3D model of the distribution of the restrained gas-saturated porosity and the maps of linear volume values of the restrained gas saturation in the pores *WKr* during watering of the reservoir.

Knowing the distribution of the restrained volumes is important during the development of the reservoir, until it is fully watered, however, studies of the DCI can provide valuable information for watered objects at a late stage of development.

In field practice, the phenomenon of the formation of bypassed zones - areas in the depleted part of the deposit, from which oil remains unrecovered - is widely known [6]. One of the main reasons for the appearance of bypassed zones is the uneven moving of the water-bearing contour, the formation of watering tongues due to capillary forces leading the front of the piston displacement. The formation of oil bypassed zones leads to a decrease in the overall hydrocarbon recovery factor.

The experiment on the DCI with data logging in real time gives us not only the value of the restrained and dynamic volumes, but also characterizes the dynamics of fluid movement due to capillary forces - this is the second important parameter that must be considered when planning additional development of the reservoir. As was shown in [7], the speed of the DCI depends on a number of parameters, such as wettability, porosity, permeability, and can have a different form of relationships for different objects. However, having reliable regularities, 3D models of porosity, permeability and wettability it is also possible to come to creating of a 3D mathematical DCI speed model and then a hydrodynamic model that considers DCI speed (figure 3).

IOP Conf. Series: Materials Science and Engineering **976** (2020) 012030 doi:10.1088/1757-899X/976/1/012030

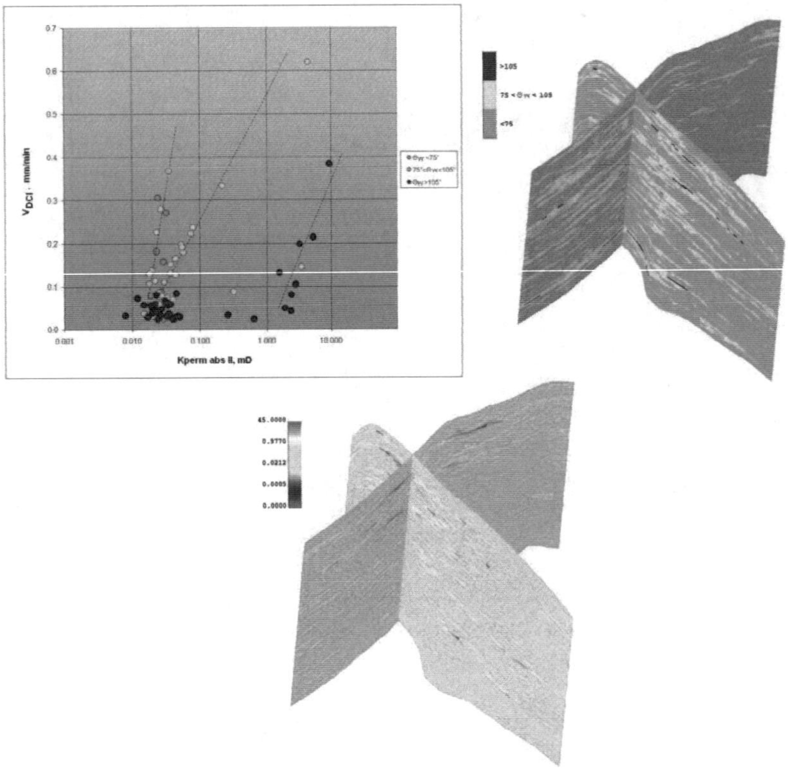

Figure 3. Petrophysical relationships and models as steps for creating a 3D mathematical model of the DCI speed. Left: relationship between the DCI speed and absolute permeability, color gradation for selective wettability [7]; right: 3D model of water wettability; down: 3D model of absolute permeability.

Another laboratory method that can supplement the DCI method with important information is the ATMM method. The laboratory method [8] developed in the 70s of the XX century received insufficient attention, but today its applicability is undeniable. The essence of the method consists in drying a rock sample saturated with liquid (water / hydrocarbons) in an air stream of constant speed and temperature (25 °C). During the experiment, readings of changes in mass, temperature of the sample surface and air are taken.

The ATMM unit improved with the latest technical advances gives a "passport" of the fluid saturation of the rock according to the forms of energy connection with the rock - from 100% saturation to a hygroscopic state. That is, using thermomassometric curves within one experiment, it is possible to distinguish the restrained volume of liquid in the pores, the funicular (dynamic) volume, the fraction of fluid in the corners of the pores, the fraction of fluid in microcapillaries of different radii (figure 4).

4. Discussion

In oil and gas science there is already an understanding of the need to work with effective pore space, speaking of fluid filtration [9,10]. While going along the path of complication, one should also remember that the effective space includes both dynamic bound volumes and volumes that are restrained by the pore structure and the physicochemical properties of the pore surface. The laboratory methods described above provide data to estimate the amount of hydrocarbons in the restrained and free state.

Using the DCI speed in the hydrodynamic model makes it possible to obtain a more reliable picture of the reservoir coverage by waterflooding and to identify possible bypassed zones both at the moment of development and in the predicted future. On the basis of these data, it is possible to make recommendations for the development of the zones of hydrocarbon deposits not affected by flooding.

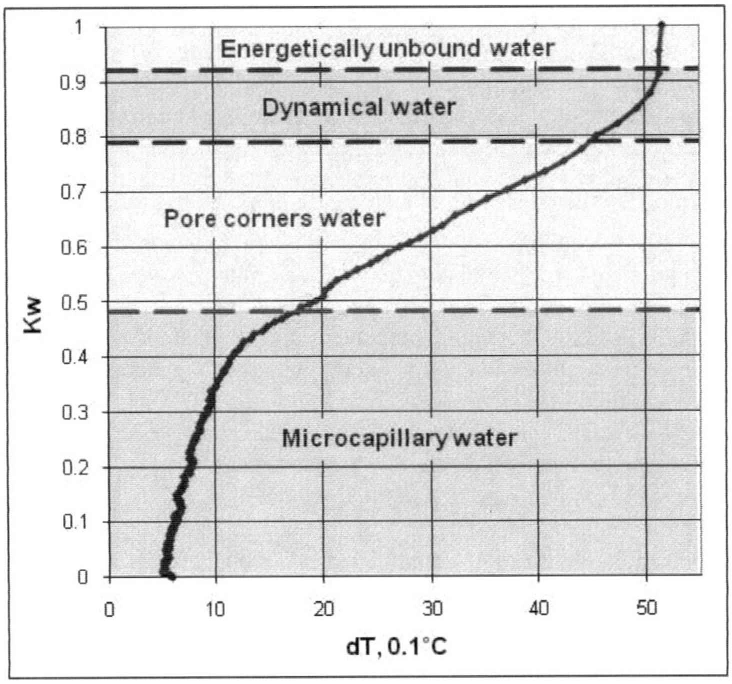

Figure 4. An example of the experimental relationship between the water saturation of the sample (Kw) with the difference in air and sample temperatures (dT).

ATMM method can be an alternative, a control method, or an additional source of information for centrifugation, semipermeable membrane, NMR, DCI methods. An additional advantage of this method is the preservation of the sample in an intact state – in case of centrifugation, a weakly consolidated core can be destroyed.

5. Conclusion

Laboratory petrophysical CDI and ATMM methods are very relevant today, since they allow:

- Estimate hydrocarbon reserves restrained in pores;
- Create a model of the DCI speeds in the reservoir and to localize the potential or existing at the time of development bypassed hydrocarbons zones, due to the capillary movement of water, ahead of the displacement front or the rise of the OWC;
- Obtain fluid-saturation passport of by the forms of the energy connection of fluid with the rock without destroying the sample.
-

Acknowledgments

The results presented in the paper were obtained under the OGRI RAS state research contract, № AAAA-A19-119030690047-6 "Forecast of the resource base state of the Russia oil and gas complex based on system studies of the prospects for the oil and gas potential of natural reservoirs in carbonate, terrigenous and shale formations".

References

[1] Stepanyan G G 1999 *Assessment of the economic efficiency of the development of oil and gas condensate fields at the final stage. Vuktilskoe OGC field as example* (Moscow) 152

[2] *Gazprom Dobycha Orenburg: 40 years of great achievements 2008* Retrieved from: http://www.oren.gubkin.ru/assets/111

[3] Malyukov V P and Apaev Yu S 2013 Influence of geological and physical properties of

complexly constructed deep productive horizons of the Astrakhan gas condensate field on the efficiency of acid treatments of a carbonate formation. *Bulletin of the Peoples' Friendship University of Russia. Series: Engineering Studies* **3** 108-19
[4] Rassokhin G V, Leontiev I A and Petrenko V I 1973 *Influence of flooding of multilayer gas and gas condensate fields on their development* (Moscow: Nedra) 262
[5] Skibitskaya N A, Kuzmin V A, Bolshakov M N and Marutyan O O 2014 Structure-trapped oil-and-gas saturation and its measuring by direct-flow capillary imbibition method. *Oil Industry* **4** 30-2
[6] Mokhov M A, Igrevsky L V and Novik E S 2004 *Oil and gas microencyclopedia* (Moscow: National University of Oil and Gas «Gubkin University») 125
[7] Skibitskaya N A, Kuzmin V A, Bolshakov M N and Marutyan O O 2018 Regularities of the processes of direct-flow capillary imbibition in productive carbonate deposits of the Orenburg oil and gas condensate field. *Actual problems of oil and gas* **3(22)** 2-7
[8] Skibitskaya N A and Morozovich Ya R Declared A method for determining the saturation of rocks with a fluid, such as water. *Inventions. Useful models* **34** 1427329
[9] Zakirov S N, Indrupskiy I M, Zakirov E S and Anikeev D P 2009 Once again about the effective pore space concept. *Oil Industry* **5** 76-80
[10] Indrupskiy I M 2009 About multiphase filtration equations in the concepts of absolute and effective pore space. *Oil Industry* **8** 60-3

IOP Conf. Series: Materials Science and Engineering **976** (2020) 012031 doi:10.1088/1757-899X/976/1/012031

Advantages and Disadvantages of the Production and Using of Liquid Biofuels

V N Pulyaeva, N A Kharitonova and E N Kharitonova

Financial University under the Government of the Russian Federation, 49, Leningradsky avenue, Moscow, 125993, Russia

E-mail: vnpulyaeva@fa.ru

Abstract. This article is devoted to the study of the possibilities of using liquid biofuels to mitigate environmental pollution and stabilize the economy in countries where there is a shortage of natural resources that are energy carriers. To this end, the authors considered the positive and negative aspects of using biofuels as a mixture, including bioethanol or biodiesel. In addition, the article includes an analysis of the situation in Russia regarding the orientation towards replacing traditional fuels.

1. Introduction

The increase in the use of technologies and the growth of environmental pollution associated with this, the depletion of natural resources and the rise in the cost of methods for their extraction have led humanity to search for alternatives to modern fuel. Bioethanol and biodiesel are currently among the most popular alternatives to conventional fuels. Bioethanol is produced from any plant material with a high starch or cellulose content, and as a final product, it is an alcohol with a maximum strength of about 95 degrees. Biodiesel is a mixture of fatty acids that are flammable [1].

Since the issues of environmental protection are currently among the most relevant in the agenda of strategic development of the energy sector in all countries, it seems expedient to study all aspects of the use of liquid biofuels and determine further prospects for its production.

2. Materials and methods

The purpose of this study is to analyze the strengths and weaknesses of the production and using of liquid biofuels and determine the prospects for its further use. In the work, the following tasks were formed:

- Determine the advantages and disadvantages of large-scale production and use of liquid biofuels;
- Identify prospects for further use of liquid biofuels.

The information base of the study was information from government, corporate and analytical organizations. The study used descriptive, comparative, economic-statistical and logical methods; the authors relied on the achievements of such sciences as economics, physics and chemistry.

3. Results

Biofuels are divided into three generations, depending on the production methods. The first generation has the easiest production method and is used in many countries. It is made from crops that are high in sugar and oil. The second generation includes biofuels, which are produced from non-food raw materials, grown specifically as raw materials for the energy industry (figure 1) [2-4].

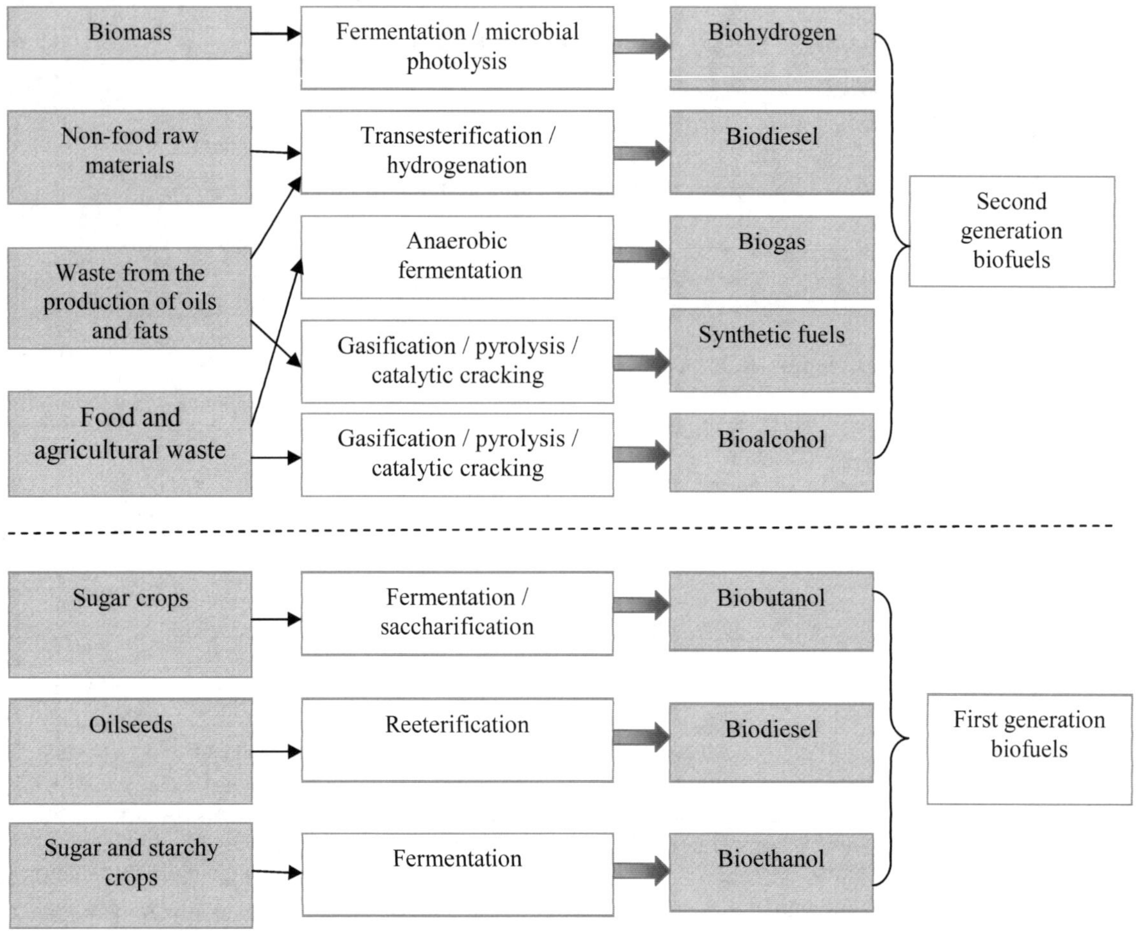

Figure 1. Approaches to First and Second Generation Liquid Biofuel Production.

The fuel of the third generation is the most energy-intensive, since it surpasses the raw materials of previous generations by several times in energy output. The raw materials for this fuel are algae and photosynthetic microorganisms. Unlike fuels of the first generation, the second and third have a very complex production process, since the second requires a large number of chemical transformations and, accordingly, this process carries large energy costs, and the raw materials for the third generation - algae and other microorganisms, require a very complex structure, storage and content. All of the above costs explain the capital intensity of the process of producing second and third generation fuels, respectively, and the choice of first generation biofuels by most countries.

Liquid biofuels have many positive traits [5-7]:

- When using this type of fuel, CO_2 emissions are much lower than with its traditional counterparts.

- If leaked and hit on biological objects for other reasons, biofuel does not harm them and is almost completely subject to water decay.
- The raw material for it is renewable crops.
- The use of biofuels will allow countries that do not have large reserves of fuel to maintain their economic sovereignty.
- High lubricating characteristics, which increases the service life of the equipment.
- High temperature of ignition - 150 degrees, which makes it quite safe.

Despite the fact that liquid fuels made from biological raw materials have so many advantages, at the same time we cannot completely abandon the traditional one. Thus, bioethanol is used with a concentration of 9 to 1 because of its low thermal conductivity, which, as a consequence, leads to a higher consumption of this type of fuel. If we talk about biodiesel, then it is worth paying attention to the fact that it has a sufficiently high viscosity, which, at a high concentration, is very dangerous for equipment, especially in frosty conditions, accordingly, it also cannot be used independently, but only mixed in a ratio of 2 to 8 diesel fuel. Therefore, today it is possible to use only a mixture of traditional and biofuel. Many countries have established ingredient ratios that roughly correspond to the proportions shown above [8].

It would seem that the most rational thing for everyone today is to start using fuel created on the basis of a mixture. But since most countries for objective reasons choose the production of first generation biofuels, for the production of which it is necessary to use high-quality arable land in very impressive volumes, the problem of land distribution between the energy and food industries arises. Very few countries now have sown areas to provide raw materials for each of these industries in the required quantity.

In addition to the problem of the banal lack of such sown areas in many countries, there is another possible problem - the possibility of a massive transition of food crop producers to the production of products used as raw materials for fuel, because this will be much more profitable, the purchase price of fuel companies will be higher. This can lead to a situation where even a country with sufficient land will not have enough raw materials for the food industry, due to the desire of agricultural producers to compete as suppliers for fuel companies. So according to Law of the United States and the National Standard for Renewable Fuels, refineries in the United States must use 50 billion liters of biofuel in the production of gasoline. According to the US Department of Agriculture, in 2017, the corn crop was 268 million tons, which was the lowest since 2011 (the price per ton was more than $ 320). Accordingly, a colossal large share went to the production of bioethanol - 42%. Competition between pastoralists and ethanol producers for the supply of corn was formed in the country, which led to an increase in grain prices and a decrease in the profitability of meat producers. Which led to an insignificant but still food crisis [9].

Another very important risk arises in the production of liquid biofuels - the cultivation of monocultures. For the food industry, agricultural products must be very diverse due to consumer preferences. If a very large plot of land is used to grow raw materials for biofuels, then it will most likely be sown with a certain monoculture due to economic rationality and lack of preferences as such in the use of raw materials. This situation can lead to biological domination by only one type of pest, since their survival rate will increase dramatically, and over time they will develop immunity to pesticides, which will lead to a drop in the harvest, as insects will destroy it. Since only one type of agricultural crop was cultivated by producers, it will not be possible to replace it, which will lead to a crisis of raw materials [10].

Also, in spite of the fact that the consumption of biofuels does not produce carbon dioxide emissions into the atmosphere, we should not forget that while we are using only a mixture, emissions, although reduced, still exist. And since the cultivation of raw materials for biofuels of the first and second generation, which is currently the most profitable from an economic point of view, requires the use of cultivation technology and much more, accordingly, the production of this type of biofuel is not as safe for the environment as it seems.

It should also be understood that a country, having switched partially or completely to second and first generation biofuels, becomes in some way dependent on weather conditions. The volume of possible fuel production in this case directly depends on any changes in weather conditions, which can lead to an extremely acute crisis in the event of natural disasters. So now, the state in the event of a lean year or any natural disasters (fire in forests and fields or flooding) will have to compensate for raw materials not only for the food industry, but also for the fuel industry, which can make its economy very unstable.

4. Discussion

In Russia, there is a sufficient number of cultivated areas for the production of raw materials for biofuel, however, there are other aspects that prevent this. Firstly, in Russia there is no unified state strategy for the use and production of liquid biofuels, therefore there is no systematic support for this industry, which weakens both the coordination of production and its organization. In addition, until recently, the development of this industry was stopped by excise taxes, despite the fact that biofuels were not subject to excise taxes on gasoline, ethanol producers had to pay excise taxes for alcohol-containing products. This form was canceled only recently on November 28, 2019 (according to the Federal Law "On Amendments to the Federal Law" On State Regulation of the Production and Turnover of Ethyl Alcohol, Alcoholic and Alcohol-Containing Products and on Restricting the Consumption (Drinking) of Alcoholic Products" dated November 28, 2018 N 448-FZ (latest edition)), which will undoubtedly increase the profitability of this industry.

In Russia, the ratio of the mixture of biodiesel fuel is regulated: 6% -20% in accordance with state standards that establish requirements for ethanol obtained from plant materials.

In conclusion, we would like to conclude that there are prospects for the development of the biofuel industry in Russia, but there is no urgent need. There is more than enough fossil raw materials for the fuel sector, and CO_2 pollution of the country is not a very acute problem, due to the very large forest areas.

5. Conclusion

Thus, it can be concluded that the production of liquid biofuels has a number of disadvantages. A country that wants to produce alternative fuels must have sufficient acreage to avoid a food crisis, as well as introduce strict control measures, regulating the shares of raw materials supplied to the food and fuel markets. There must also be certain state standards regulating the proportions according to which a mixture of biodiesel with diesel and bioethanol with gasoline will be produced. Do not forget about the need for support and subsidies for the development of the production of second and third generation biofuels, which, despite their capital intensity in production, use a much smaller amount of land and most of them are not arable.

References

[1] Gibadullin A A, Ryabinina E V, Morkovkin D E, Sodikov K A, Trifonov P V, Kirpicheva M A and Kokurina A D 2020 Engineering solutions in the field of digital transformation of the electric power industry. *IOP Conference Series: Materials Science and Engineering* **862** 062055

[2] Tsimbalist A V and Dmitrenko E A 2014 Prospects for the production of bioethanol in Russia. *Science of man: humanitarian research* **3(17)** 57-65

[3] Gibadullin A A, Sadriddinov M I, Kurbonova Z M, Shedko Yu N and Shamraeva V V 2020 Assessment of factors ensuring sustainable development of the electric power industry in the context of transition to renewable energy sources of the national economy. *IOP Conference Series: Earth and Environmental Science* **421** 032051

[4] Morkovkin D E, Gibadullin A A, Safarov B G and Alpatova E A 2020 Definition of factors limiting the growth of industrial production. *IOP Conference Series: Materials Science and Engineering* **862** 042013

[5] Kolnichenko G I, Sirotov A V and Tarlakov Ya V 2010 Liquid biofuel: problems and prospects of creation and use. *Vestnik MGUL Lesnoy Vestnik* **1** 105-7
[6] Gibadullin A A 2020 Assessment of the level of stability and reliability of the electric power complex. *IOP Conference Series: Materials Science and Engineering* **837** 012007
[7] *Biofuel market: problems and prospects. Information and analytical bulletin "Most"* Retrieved from: http://trade.ecoaccord.Org/bridges/0313/3.html
[8] Morkovkin D E, Lopatkin D S, Shushunova T N, Sharipov B K and Gibadullin A A 2020 Formation of the conditions for the development of innovation. *Journal of Physics: Conference Series* **1515** 032002
[9] *Russian statistical yearbook* 2018 (Moscow: Rosstat) 694
[10] Pulyaeva V N, Gibadullin A A, Usmanova T J, Ivanova I A, Rudakova E N and Dmitrieva O A 2020 Formation of modern requirements for the development of human capital in the context of increasing the efficiency of the industrial potential. *Journal of Physics: Conference Series* **1515** 032020

Prospects for projects of the liquefied natural gas production

N A Kharitonova, V N Pulyaeva and E N Kharitonova

Financial University under the Government of the Russian Federation, 49, Leningradsky avenue, Moscow, 125993, Russia

E-mail: vnpulyaeva@fa.ru

Abstract. This research paper examines the current state of the liquefied natural gas industry in Russia, the goals set in the Energy Strategy, as well as the prospects for the development of LNG projects within the framework of market instability, the use of new technologies in the production of liquefied gas. Also the authors assess the competitiveness of these projects. Based on the results of the analytical study, conclusions were formulated. The work can be of value to practitioners and researchers in the industry.

1. Introduction

The gas industry plays a significant role in the modern world economy. Energy transition, modern requirements in terms of ecology and impact on the climate, green energy - all these are factors that reduce the use of petroleum products, as well as any fossil hydrocarbons. Nevertheless, due to its energy qualities and environmental friendliness, natural gas is an exception that meets modern, constantly increasing requirements for industry, energy, and also for use as a motor fuel.

The world market for liquefied natural gas (hereinafter LNG) continues to grow at a significant pace. According to forecasts of leading expert agencies and market participants, this trend may continue until 2040 according to the most pessimistic forecasts. The expansion is planned due to the growth of consumption in the countries of the Asia-Pacific region.

At the moment the estimate of the potential LNG production in the Russian Federation, according to the Ministry of Energy, is more than 80 million tons per year, and taking into account the potentially announced projects, production may amount to about 163 million tons per year. The ongoing large, medium and low-tonnage projects can have a strong impact on the formation of a resource base for the production and export of gas.

Russia is currently in the top of the leading countries in terms of LNG production and transportation capacities. According to analysts' forecasts, by 2035, Russia's share in the world liquefied natural gas market may reach 20% and the country will be able to compete with such market participants as Qatar, the USA, and Australia. Until 2024, the government plans to invest over 2.5 trillion rubles for LNG projects [1].

On April 2, 2020, the government of the Russian Federation approved the draft Energy Strategy of Russia for the period up to 2035. One of the main emphasis of this document is made on the further development of liquefied natural gas production, as well as motivation for the formation of LNG clusters and priority development points in such territories as the Yamal and Gydan peninsulas. To achieve all the ambitious goals of this strategy, it is important to analyze LNG projects and identify their competitiveness in the world market.

IOP Conf. Series: Materials Science and Engineering **976** (2020) 012032 doi:10.1088/1757-899X/976/1/012032

2. Materials and methods

The purpose of this study is to analyze the prospects of projects currently being developed in the Russian Federation and other countries for the production and sale of liquefied natural gas. To do this, it is necessary to solve a number of tasks:

- Conduct an analysis of world experience in the implementation of LNG projects;
- Assess the competitiveness of Russian LNG projects.

As an example of industry development and comparative analysis, the study uses a sample assessment and consideration of medium-tonnage LNG projects, including the use of composite analysis based on world experience.

The information base of the study was information from government, corporate and analytical organizations. The study used descriptive, comparative, economic-statistical and logical methods.

3. Results

For Russia, a completely new direction in the development of the gas industry has become the production of liquefied natural gas, starting in 2009, when the first plant for its production was put into operation.

Today, the Russian Federation is actively expanding the geography of both production and consumption of natural gas, including in the form of liquefied gas. One of the priority areas, according to ES-2035, is to use a gas as an engine fuel. By the beginning of 2019, the number of stationary NGV facilities amounted to more than 400 points, while the volume of sales at gas-filling compressor stations for cars reached more than 650 million cubic meters. Such an increase in indicators, among other things, is due to the existing program for subsidizing certain cost items from the federal budget aimed at acquiring technologies used in the gas engine sector. To date, more than 45 constituent entities of the Russian Federation have and are implementing programs to develop this market.

This energy sector is facing its own specific problems, such as, firstly, an increase in costs for the production and transportation of gas, both to the domestic and global markets. Secondly, it is an oligopoly on the domestic market, which is, among other things, provoked by the creation of the UES gas market as a common one to maintain external competition. An equally significant limiting factor is the still unfinished process of creating a regulatory framework for the functioning of the LNG market within the country.

During 2020, the global LNG market entered a phase where production exceeded consumption. This state, taking into account the commissioning of new liquefaction capacities, may persist for several years. The COVID-19 epidemic has led to a decline in economic activity and demand for energy. As a result, LNG demand is likely to decline in 2020-2021, further increasing pressure on prices.

In a situation of overproduction, the level of costs is of great importance, which determines the long-term competitiveness of the project. Below are the main and considered in the work models for the implementation of LNG projects (table 1).

The Russian Arctic is becoming the world's leading center for LNG production. Several projects are being implemented in different parts of this macro-region. They are located in different segments: from global production to low-tonnage. In the face of increasing competition in the global LNG market (as well as LNG with pipeline gas), there is an oversupply. In such conditions, non-price factors of competition, such as environmental, climatic, geological and political, acquire a special role.

The location of production in the Arctic zone is fraught with significant difficulties in the implementation of the project. This is reflected in increased capital costs throughout the entire production chain and increases the risks of not implementing the project within the agreed time frame within the provided budget. However, external cold accompanies the production of additional products. Large-scale Russian LNG projects are actively using this, increasing their productivity up to 17% in the framework of individual projects, such as Sakhalin-2 [2-4].

Table 1. Models for the implementation of LNG projects [5].

Functions	Models			
	Integrated model	Cheriere model (tolling)	Ownership / Cost Model	Trading / Market Model
LNG buyer	End-user	Aggregators and end users	Various	Various
LNG seller	National or international vertically integrated companies	Project owner	Various, according to the distribution of shares	Various
Financing	Whole production chain or LNG plant	LNG plant only or separate gas pipelines	Various, supported by larger owners or long-term commitments	Market based, no risk of LNG sales
Ownership structure	Closed, often as an SPV	Closed to maximize sales margin	Depends on the owner, rather with maximizing sales margins	Depends on the owner, rather with maximizing sales margins
Project scope	Onshore large-scale projects	Expansion of existing production facilities and construction at new sites	Expansion of existing production and construction at new sites, FLNG and medium-sized projects	Expansion of existing production facilities and construction at new sites
Gas supply	Included in the project outline	LNG plant only	Combined	LNG plant only
Gas operator	National or international vertically integrated companies	An independent company without ownership of production assets	An independent company without ownership of production assets	An independent company without ownership of production assets
Gas source	Fields with proven gas reserves	Gas is purchased on the market	Gas is purchased on the market	Gas is purchased on the market

In other countries, manufacturers seek to improve their efficiency in other ways. Thus, the Japanese company Chiyoda Corporation is starting a project using artificial intelligence or AI to increase the productivity of the RT Donggi-Senoro LNG plant located in Indonesia. Such technologies can also be actively applied in the Arctic, which will further significantly increase productivity (table 2).

Table 2. Artificial intelligence vs cold climat [5-6].

Project	Installed capacity, million tons	Actual capacity, million tons	Production start year	Year of fixing the result	Cause
Sakhalin-2	2x4,8	11,5	2009	2018	Climatic features, technological improvements
Yamal LNG	3x5,5	16,5+7-10%	2017	2018	Climatic features
PT Donggi-Senore LNG	2	No data	2015	2019 (project start)	Artificial Intelligence

High quality and a large resource base allow companies to increase production by building additional LNG production lines in industrial parks. The use of the existing infrastructure and its modernization at the existing sites greatly affects the efficiency of the entire LNG production complex. For example, thus, the placement of the T4 production line ensured a cumulative synergistic effect and a significant increase in the efficiency of the Yamal LNG project from Gazprom by 3-4%.

4. Discussion

For the development of the LNG industry in Russia, a set of basic measures is envisaged, including legislative, infrastructural, regulatory, and technological ones. First, it provides for the development and improvement of a mechanism to ensure legislative, as well as regulatory and legal conditions, such as measures of tax and customs tariff incentives and regulation, focus on cost-effective improvement at all stages of the LNG industry: production, transportation, storage, sale and use of LNG.

Secondly, a set of measures for the liberalization of LNG exports, the creation of a mechanism to control balanced competition in external markets for natural pipeline gas and LNG.

It also provides for the modernization and construction of auxiliary and additional infrastructure, such as port, transport, electricity and others on the principles of public-private partnership, including within the framework of the creation of an LNG cluster on the Yamal Peninsula and the Gydan Peninsula. In the Arctic zone of the Russian Federation, it is planned to create specialized hubs for transshipment, storage and trade of LNG, the implementation of projects for the construction of terminals in Kamchatka and Murmansk, as well as the development of the production of low-tonnage LNG and the formation on its basis of the internal LNG market as a tool to ensure the energy security of territories remote from unified gas supply system

Today in Russia there are several companies engaged in the distribution of LNG abroad: Gazprom, Gazprom export, Rosneft, Yamal LNG, Arctic LNG-1, Arctic LNG-2, Arctic LNG-3 - which, in turn, are subsidiaries of PJSC Novatek "(figure 1).

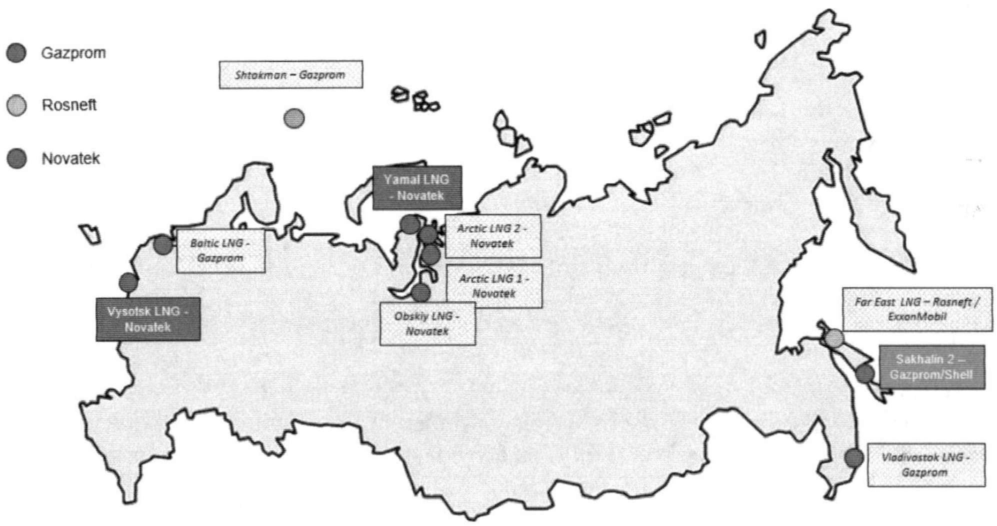

Figure 1. Map of Russian LNG projects [5].

According to the 2020 market reviews by IGU and GIIGNL, Russia ranks 4th in terms of LNG exports for 2019, exporting 29.3 million tons of LNG (8% of the global market). 51% of LNG exported by Russia comes from Europe and 46% to the Asian market (figure 2).

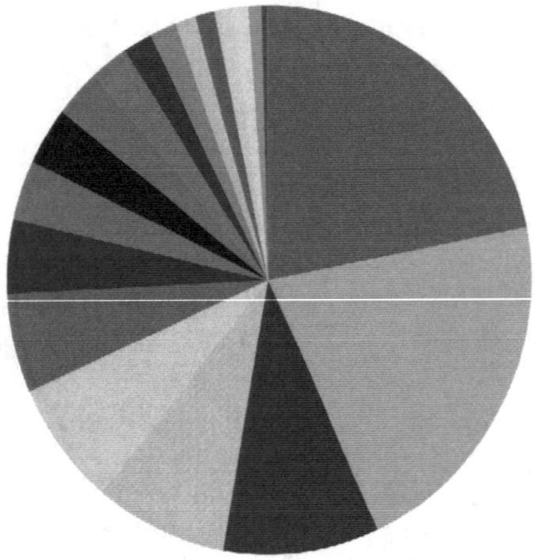

Figure 2. Export structure and market share [5].

The quantitative benchmark for the LNG industry is production of 80-140 million tons by 2035. In a number of publications in previous years, a target level for the Russian industry of 140 million tons was determined, which is confirmed by the statements of the Minister of Energy of the Russian Federation Alexander Novak and the Chairman of the Management Board of PAO NOVATEK. The role of the LNG industry in the Russian gas industry should grow significantly: from 4.2% in 2018 to 22.4% in 2035.

LNG is an essential element of Russian energy exports. This applies to both low-tonnage (up to 80 thousand tons per year) and medium-tonnage (up to 2 million tons per year) segments, as well as large-scale projects in the Arctic, the Baltic and Asia [7-9].

Taking into account the projects implemented under the Production Sharing Agreement, in the period up to 2035 more than 75% of the planned LNG production in Russia will be exported under exceptions from the Federal Law "On Gas Export", which means without the participation of Gazprom. Thus, this law has actually lost its significance and role for "maximizing gas rent" in Russia [10-11].

It is no coincidence that the liberalization of LNG exports is one of the elements of Russia's energy strategy until 2035, which makes it possible to fully unleash the potential of LNG production in Russia.

5. Conclusion

For research on this topic, a huge array of various information from authoritative sources was taken, comparing and analyzing which, the authors analyzed and identified potential prospects for the development of Russian LNG projects, in accordance with world experience and based on statistically confirmed materials and knowledge bases on the effectiveness of LNG enterprises, both in Russia and in other countries. The scientific value of this work lies in the presence of a value judgment of the above information and the formation of certain conclusions made by the author of this research work. Based on the research carried out within the framework of the designated topic, conclusions were made on the current state of the LNG component of the fuel and energy complex in the Russian Federation, the main aspects and directions of increasing stability were considered, on the basis and within the framework of the Energy Strategy 2035, studies were carried out and conclusions were drawn based on qualitative and quantitative indicators provided by the largest companies representing LNG production in the Russian Federation. The main models used in LNG projects were considered, their advantages, disadvantages and characteristic features. The main priorities for the development of the Russian direction of the LNG industry were also identified; The prospects and advantages of introducing AI into production in the Arctic are considered.

Development of production and consumption of LNG, the desire to enter the Russian Federation in the medium term among the world leaders in its production and export.

References

[1] Gibadullin A, Morkovkin D, Shmanev S and Ilminskaya S 2020 Ensuring a balance of production and consumption of electric energy in order to preserve the environment. *E3S Web of Conferences* **175** 14011
[2] *Small-scale LNG: New Opportunities for Russian Gas in Europe* Retrieved from: http://www.konoplyanik.ru/ru/publications/07-787(2019)_small.pdf
[3] Gibadullin A A, Gilts N E, Romanova Ju A, Romanova Ir N and Bahretdinova Kh A 2019 The development strategy of the environmental safety of the electric power complex. *IOP Conference Series: Materials Science and Engineering* **537** 042065
[4] *Prospects and potential of LNG exports from Russi* Retrieved from: http://www.ngv.ru/magazines/article/perspektivy-i-potentsial-eksporta-spg-iz-rossii/?sphrase_id=1898605
[5] *NewPlayers, New Models* Retrieved from: https://www.oxfordenergy.org/wpcms/wp-content/uploads/2019/04/New-Players-New-Models.pdf
[6] *Shell LNG Outlook 2019* Retrieved from: https://www.shell.com/energy-and-innovation/natural-gas/liquefied-natural-gas-lng/lng-outlook-2019.html
[7] Lopatkin D S, Shushunova T N, Shaldina G E, Gibadullin A A and Smirnova I L 2019 Renewable and small energy development management. *Journal of Physics: Conference Series* **1399** 033061
[8] Gibadullin A A, Pulyaeva V N, Kharitonova E N and Kharitonova N A 2019 Convergence of the Russian power industry in a period of declining reliability and sustainability of the industry. *Journal of Physics: Conference Series* **1399** 033034
[9] *Russian statistical yearbook* 2018 (Moscow: Rosstat) 694
[10] Gibadullin A, Abanina I, Medvedeva M, Shmaneva L, Alekseeva G and Morkovkin D 2020 Assessment of the global energy potential and existing areas of innovative development in the context of environmental safety. *E3S Web of Conferences* **175** 14012
[11] Sadriddinov M I, Mezina T V, Morkovkin D E, Romanova Ju A and Gibadullin A A 2020 Assessment of technological development and economic sustainability of domestic industry in modern conditions. *IOP Conference Series: Materials Science and Engineering* **734** 012051

IOP Conf. Series: Materials Science and Engineering **976** (2020) 012033 doi:10.1088/1757-899X/976/1/012033

Formation of a system for achieving sustainable development of the electric power complex

A A Gibadullin

State University of Management, 99, Ryazan Avenue, Moscow, 109542, Russia

E-mail: 11117899@mail.ru

Abstract. The article analyzes the influence of various factors on the electric power complex. The study showed that the internal factors associated with the condition of the equipment in operation have the greatest influence. The analyzed indicators indicate that during the periods under consideration, there are practically no changes associated with a decrease or increase in the number of accidents at stations and networks. At the end of the work, recommendations were presented to ensure the stability of the electric power complex, which should include long-term and short-term plans.

1. Introduction

The issues of achieving sustainable development of the electric power complex are relevant for any state and industrial system, since it is the electric power facilities that provide power supply to enterprises and organizations, preserve the country's defense capability and ensure comfortable life for the population. Any negative changes associated with the influence of external and internal factors on the electric power complex have a negative impact on all technological, operational and organizational processes occurring at electric power enterprises [1-2].

Any processes affecting the electric power complex can be divided into two components - these are external and internal factors. External factors include changes in the economy, politics, regulatory changes, the emergence of innovative and new engineering solutions, techniques and technologies, and more. Internal factors include various changes occurring within the system, for example, changes in the principles of management of electric power facilities, physical and moral obsolescence of production facilities, increased operating costs, reduced energy efficiency of production, lack of a policy to modernize and update production facilities, and much more [3-5].

In our study, we will not consider the factors that affect the electric power industry from the outside, since almost any change in the external environment, such as changes in the economic policy of the state, the introduction of new regulatory and legal rules for the functioning of the industry, pandemics and crises, cannot significantly affect for the electric power complex. At the same time, changes within the energy system always affect the electric power complex and may lead to the destruction of the built system within the entire industry or individual electric power enterprises. Thus, we consider it necessary to limit our research only to internal factors and consider their influence on the state of the energy system [6].

2. Materials and methods

In this study, the aim of the work is to formulate a policy to ensure the sustainable development of the electric power complex. The work set the following tasks:

- Analyze the indicators of the accident rate of the electric power complex;
- To formulate measures to achieve the stability of the electric power complex.

The study used scientific methods and approaches, and as an information base, statistical indicators, annual reports of government agencies and analytical reports of various organizations were used.

3. Results

The policy in the field of sustainable development of the electric power complex in Russia is formed at the national level and for each energy system its priorities are determined to ensure the supply of electricity to consumers, the reliability and uninterrupted operation of electric power facilities, the possibility of electricity flows between power systems, the operation of the wholesale and retail market and the policy in the field of modernization and innovative development of the electric power industry. At the corporate level, specific programs are being developed to implement the goals and objectives of the power industry development set at the federal level. But speaking about the policy in the field of sustainable development, it is necessary to take into account all the factors that determine this direction of development. Among the main groups of factors, it is advisable to highlight the following [5-8]:

- Financial stability;
- Resource and raw material supply;
- State and dynamics of development of the electric energy market;
- Share of intersystem flows;
- Level of environmental safety;
- The size of the export potential;
- The level of reproduction of fixed assets.

These groups of factors can be divided into a large number of components, which will ultimately lead to the determination of the causes of the emergence of certain negative factors that hinder and restrain the stable operation of the electric power complex. In our opinion, it is advisable to consider the last group of factors associated with the level of reproduction of fixed assets. It is advisable to include the following components in this group of factors, so we refer to the conditions for ensuring sustainable functioning [8-10]:

- Level of interchangeability;
- The level of physical wear and tear of fixed production assets;
- Volumes of capacity reservation;
- Equipment accident rate.

The factors that affect sustainable development include:

- The amount of state support;
- Dynamics of electricity consumption;
- The level of obsolescence of fixed assets;
- Dynamics of the introduction of innovative and digital technologies in the electric power complex.

Undoubtedly, the level of physical and obsolescence of equipment, the volume of capacity redundancy, the level of equipment accidents, and the amount of state support, the dynamics of the introduction of innovative and digital technologies in the electric power complex is influenced by the condition of the equipment and the proportion of stops or equipment failures. In this regard, let us consider the volume of accidents at power plants and networks (figures 1 and 2) [11].

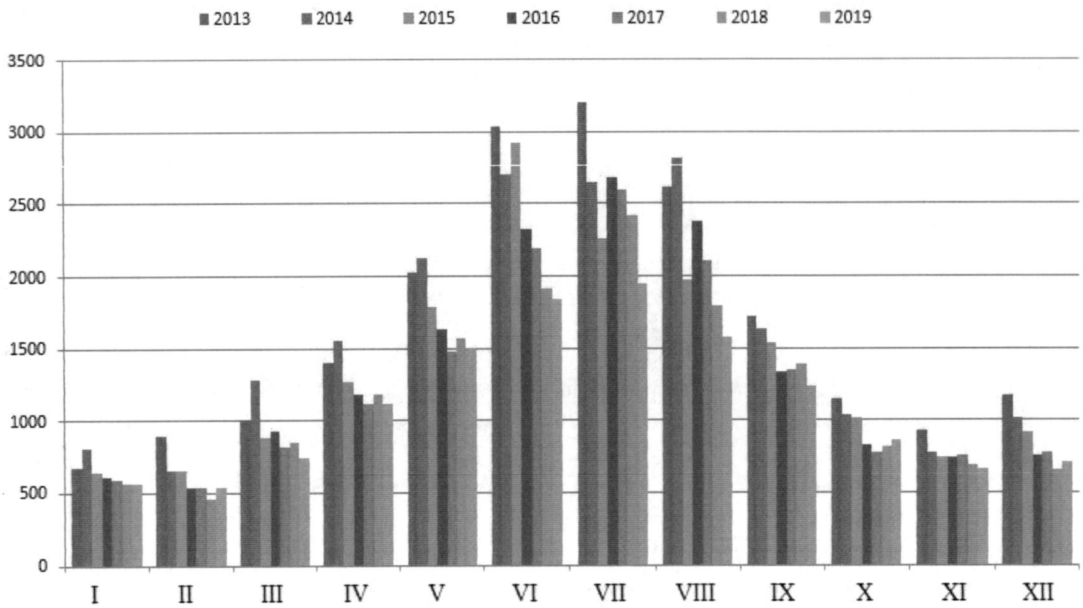

Figure 1. Emergency in electrical networks of 110 kV and above.

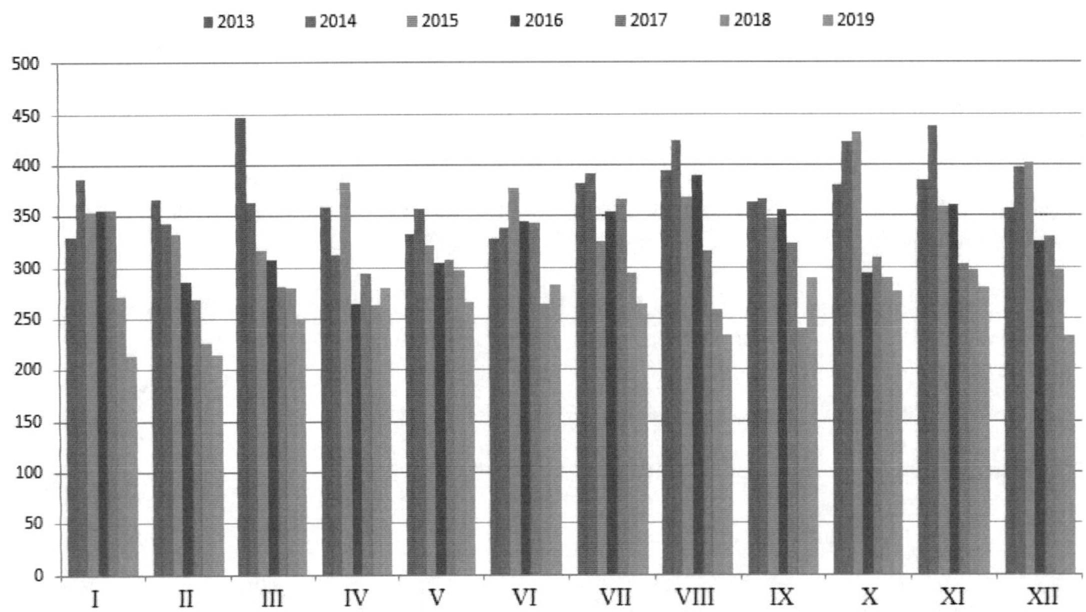

Figure 2. Emergency at power plants with an installed capacity of 25 MW and above.

The presented figures show that the number of accidents annually and monthly is approximately at the same level. This situation indicates that there is no policy at the state, regional and corporate level to ensure the sustainable operation and development of the electric power complex.

If we analyze the level of depreciation, commissioning and disposal of fixed assets in the electric power industry, then we can conclude that depreciation reaches 60%, and some production facilities are operated outside the park resource, which indicates an extension of the standard period of use of these capacities. If we consider in interrelation the indicators related to wear, retirement and commissioning, we can conclude that there is no renewal of production capacities in the industry, and the commissioning of new capacities occurs only due to the development of the energy system in new territories and the emergence of new consumers.

In this regard, we consider it necessary to propose mechanisms aimed at ensuring sustainable development in the long and short term.

4. Discussion

At the facilities of the electric power industry, measures are constantly being taken to maintain equipment in working order, however, we believe that at the state level it is necessary to develop certain criteria that should be applied at all energy facilities and be monitored by state supervisory organizations. It is advisable to divide such requirements into short-term and long-term perspectives.

In the short, term [6-9; 12-13]:

- Implementation of a policy on online monitoring of the technical and technological state of the main production facilities;
- Elimination of worn out and energy wasting capacities;
- Implementation of a policy to assess the condition of production facilities;
- Implementation of planned current, medium and major repairs;
- Reduction in the use of equipment in critical condition;
- Implementation of a policy to prevent the occurrence of shutdowns, failures and accidents at power facilities.

Long term [6-9; 12-13]:

- Develop a phased policy for the modernization and renewal of production facilities;
- To propose the conditions for redundancy and power flows in order to remove worn-out and energy-wasting capacities from operation;
- Development of a policy to increase the installed capacity utilization factor at efficient enterprises and to reduce the specific consumption of equivalent fuel at energy enterprises;
- Introduction of a policy to prevent physical and moral deterioration of production facilities;
- Carrying out a policy of transition to environmentally friendly and renewable energy sources;
- Development and implementation of innovative and digital technologies in the electric power complex.

Thus, we can conclude that ensuring the sustainable operation and development of the electric power complex must be carried out in the framework of the short and long term.

5. Conclusion

As part of the study, it was revealed that the electric power complex affects all areas of activity, because of which, ensuring its sustainability is a necessary task for the industry. The work revealed that there are external and internal factors that negatively affect the electric power complex, however, it was found that internal factors have the greatest impact on the electric power industry. The study presented a group of factors that ensure the stability of the electric power complex. It was revealed that most of the factors, in one way or another, arise due to the deterioration and accident rate of equipment, because of which, the indicators of accident rate at stations and in networks were analyzed. The presented indicators indicate that the accident rate does not change for several years and months, therefore, mechanisms were proposed in the work to ensure the stability of the electric power industry in the short and long term.

References

[1] Romanova I N, Morkovkin D E, Nezamaikin V N, Gibadullin A A and Ivanova M A 2020 Formation of a policy to ensure environmental safety in modern economic conditions. *IOP Conference Series: Materials Science and Engineering* **734** 012166.

[2] Kobtseva O N, Novoselova N N, Novoselov S N, Shichiyakh R A, Morkovkin D E and Sidorchukova E V 2017 Organizational and economic features of import substitution formation and realization in the conditions of spatial restrictions. *International Journal of Applied Business and Economic Research* **15(23)** 25-35

[3] Morkovkin D E, Lopatkin D S, Shushunova T N, Sharipov B K and Gibadullin A A 2020 Formation of the conditions for the development of innovation. *Journal of Physics: Conference Series* **1515** 032002

[4] Gibadullin A A, Kerimova Ch V, Chorshanbiev S R, Shmanev S V, Romanova Ju A and Sozontov A A 2020 Management of mechanisms ensuring sustainable functioning and economic development of the electric power complex. *IOP Conference Series: Materials Science and Engineering* **734** 012170

[5] Linnik V Yu and Linnik Yu N 2019 Priority directions of innovative development in oil and gas complex at the Siberian Federal District *Upravlenie* **7** 1 pp 40-9

[6] Buevich A, Karamova O and Sumarokov E 2019 Improvement of the institutional structure of the real sector under the conditions of the digital economy. *Studies in Computational Intelligence* **826** 675-86

[7] Kulyasova E V et al 2019 *J. Phys.: Conf. Ser.* **1260** 032024

[8] Sharipov F F, Krotenko T Y and Dyakonova M A 2021 Transdisciplinary strategy of continuing engineering education. *Lecture Notes in Networks and Systems* **139** 480-8

[9] Dudin M N, Zasko V N, Dontsova O I, Osokina I V and Berman A M 2018 Renewable energy sources as an instrument to support the competitiveness of agro-industrial enterprises and reduce their costs. *International Journal of Energy Economics and Policy* **8(2)** 162-7

[10] Nikolaeva E K, Yudina N A, Dunaeva T U, Livshits S A and Enikeeva S R 2019 Economic indicators improvement due to the introduction of energy-efficient technologies. *E3S Web of Conferences* **124** 05045

[11] *Ministry of Energy of the Russian Federation* Retrieved from: https://minenergo.gov.ru/

[12] Yudina N A et al 2019 *IOP Conference Series: Earth and Environmental Science* **288** 012072

[13] Sharipov F F, Krotenko T Y and Dyakonova M A 2021 Digital Potential of Economic Education: Information Technologies in a Management University. *Lecture Notes in Networks and Systems* **133** 561-72

Trends in the development of the Smart home system and its integration into modern society

E M Deyanov and N A Kharitonova

Financial University under the Government of the Russian Federation, 49, Leningradsky Prospekt, Moscow, 125993, Russia

E-mail: egor.deyanov@mail.ru

Abstract. The stages of development of smart home systems and integration of these systems into the city infrastructure are considered. The main advantages and disadvantages of these solutions are identified, and problems associated with their installation and use are identified. The research on the volume of markets and target audience is analyzed, and the trends of recent years in the areas of both "smart home" and "smart city in General" are considered. Priority directions of development of the entire field of solutions are indicated. Specific examples of using smart systems in different configurations and with different functions are considered. Conclusions are drawn about the current situation on the market, as well as about the future prospects for implementing smart home systems in the lives of users.

1. Introduction

The term "smart home" appeared in the early 2000s and is inextricably linked with the concept of "smart city", the implementation of which becomes a target for the development of energy companies and involves the use of its infrastructure to improve the efficiency of urban infrastructure use. Since 2015, the degree of distribution of various smart home technologies in Russia, as well as public awareness of modern solutions in this area, has increased significantly [1]. However, the understanding of the concept of "smart home" in Russia and in the rest of the world is slightly different. While in the West, experts have been implementing various solutions and integrating them into the city's infrastructure for quite a long time, becoming partners of governments, in the Russian market, marketers have instilled in the population a not entirely correct idea of "smart" technologies. In the view of most Russians, "smart" is considered to be almost any equipment with a few advanced functionalities, for example, "smart" kettles, water heaters, refrigerators, etc. At the same time, not everyone understands the difference between a refrigerator that can be used to access the Internet, and a "smart" water supply system in the house, which is able to control not only water leaks, but also pipe breaks, as well as quickly call a repair team to fix the problem [2-4].

However, at the current stage of technology development, more and more construction companies are beginning to cooperate with manufacturers of equipment and systems for smart homes, which suggests great prospects for the development of this industry in the Russian market.

2. Materials and methods

Speaking about the "smart home" system, it is worth considering not only the solutions themselves, but also their application in real life, as well as the target audience that may be interested in these technologies. Initially, the smart home system was positioned as a way to reduce energy and water consumption, which reduces operating costs. Experts say that using such solutions can reduce energy consumption by 30-40%, water consumption by 20-30%, and operating costs by 10-30%. In addition, there are three stages of integration of "smart" systems in the urban environment. The first stage is considered to be the development of an efficient infrastructure, then we are talking about primary digitalization and, finally, the last stage is the digital ecosystem (Smart Sustainable City). It should be emphasized that the degree of distribution of each stage in the Russian market differs significantly for Moscow, millionaire cities and the rest of Russia [3-4].

The second advantage of "smart home" is security, which is understood not only to protect against illegal entry into the house, but also to prevent accidents in the presence of various failures and breakdowns. At the moment, there are already solutions on the market that allow you to install special sensors in the house/apartment that monitor the situation in the room around the clock, which allows you to inform the user about failures and problems in a timely manner. For example, some smart systems send a notification to the owner in the event of a water leak or gas leak and then call special services to fix the leaks themselves. Some systems are able to eliminate leaks themselves by shutting off the water or gas supply. There are also sensors that monitor Windows and doors. In case of illegal entry, the user immediately receives a notification from the security system, which can be configured so that the camera distinguishes a cat from a person and reacts depending on the situation. Fully automated security systems are certainly capital-intensive, which indicates that they are not available to the average user, and is considered as the main drawback that prevents their widespread use. For example, to supply a standard two-room apartment with "smart" equipment, you will need about 5 thousand dollars (if you do not take into account the most expensive and modern options) [2]. Thus, investments in the "smart home" will pay off in at least 30 years, which, of course, few people are satisfied with. Average users prefer cheap solutions that do not create a common ecosystem of "smart home", but only help to optimize some processes in the home.

Real and full-fledged smart home systems are found mainly in the capital. For example, when creating a project for the reconstruction of the Narkomfin house in Moscow, which is a monument of avant-garde and constructivism, architects and developers decided to implement "smart" systems in the house. The apartments have special touch panels that allow you to control the climate, lighting and curtains. These control interfaces are connected to a common centralized KNX apartment system, which allows you to program thermostats, panels, and sockets. The programmable smart home system allows you to record scenarios, turn on heating and sockets according to a schedule, and turn off all electrical appliances as necessary.

Most often, the integration of the "smart home" system into the city's infrastructure is associated with the creation of a centralized management system for modern residential complexes (smart intercoms are installed in entrances, electric car chargers are installed in Parking lots, and video surveillance is conducted around the perimeter of the complex, which is integrated into the video surveillance system of the entire city). At the same time, the investment associated with the creation of such systems also varies significantly. If the most common "smart" sockets are currently not so expensive, then to fully integrate the "smart" system into each individual apartment, you need, among other things, a design project, which is likely to include special design solutions. Despite the fact that the cost of "smart" systems has significantly decreased in recent years, however, it is still quite high. For example, the cheapest options for a "smart" system that includes lighting, temperature, curtains and blinds, as well as energy management, will cost the user 35-100 thousand rubles. Such systems have a number of advantages: various scenarios, notifications, remote control, and schedules. Scripts allow the user to program the system for a specific set of actions. For example, when coming home in the evening, the user can turn on the "Evening" scenario using voice control, which involves closing curtains and blinds, setting the light to warm tones, dimming the light, turning on pleasant music, turning on the kettle. Remote monitoring allows the user to monitor the status of various indicators in the house in real time – from electrical appliances to video surveillance cameras. It should be noted that online monitoring of video surveillance cameras is already quite widely used by ordinary users as an effective security measure; however, integrated video surveillance systems are still expensive.

Notifications are also an important part of the smart home system, as they allow you to quickly get information about certain incidents. For example, if there is a power outage in an apartment, the system can automatically switch to the UPS (uninterruptible power supply) and at the same time send a notification to the user about the power outage. Now, there are already relatively inexpensive solutions for security systems that can notify the user in the event of illegal entry into a residential premise, as well as call out-of-house security. If we talk about centralized control systems, we can give an example of the KNX system, which just allows you to program equipment for various scenarios and connect all smart systems of a house or apartment to a single application in a smartphone [5-6]. However, it should be noted that due to lack of investment, lack of all necessary technologies and poor public awareness, such systems are still very poorly distributed in Russia.

3. Results

Speaking about the trends in the development of the smart home market, it is worth considering both the whole world and Russia in particular. Over the past few years, the global market for smart home solutions has been growing rapidly (figure 1), while experts predict that, the trend will continue.

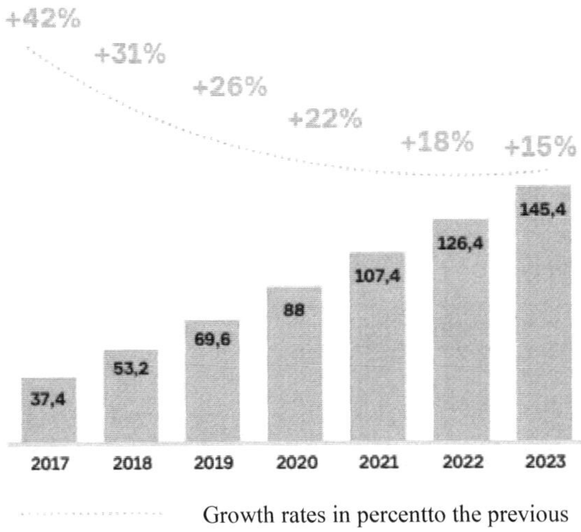

Figure 1. Global smart home solutions market volume, $ billion [1].

It should be emphasized that the Russian market lags behind the Western one in terms of development by about 5 years, although the situation has significantly improved over the past two years, and by 2023, it is expected to grow by more than three times the previously achieved level (figure 2).

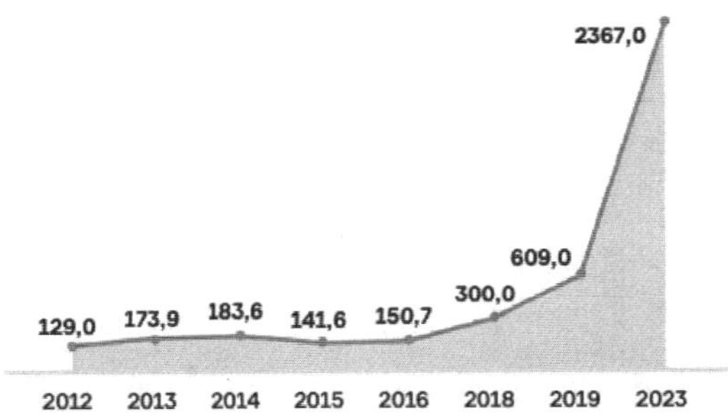

Figure 2. The volume of the market of solutions for smart houses in Russia, $ billion.

At the moment, the market leader is the United States. Their share is approximately 44.4 per cent. The second position is occupied by China with a market share of 13.5%. Germany closes the top three with a share of 5.8% [7-8].

Without a doubt, the volume of the market largely depends on the activity of end users. After analyzing Yandex's research on the audience profile of smart homes, we can draw several conclusions. First, users are poorly aware of the market for these technologies – 60% of users are looking for only one brand – Xiaomi. Second, potential users of smart home systems are mostly men aged 25 to 44 (68% of the audience is in this age group), 65% of the audience has an above-average income, and 64% of search queries about "smart home" are in millionaire cities. In particular, for Moscow and the Moscow region -36%, Saint Petersburg and the Leningrad region -10%, millionaire cities (except Moscow and Leningrad) – 18%, the rest of Russia – 35%). It should also be emphasized that the main audience of smart homes is young, affluent, modern urban people [9-10]. Third, interest in smart homes increased by 31% from 2017 to 2018. At the same time, users are more interested in this topic with the onset of autumn, by the summer interest decreases, which is due to a large number of vacations. In addition, it is obvious that climatic conditions in some regions of Russia are pushing users to find ways to optimize energy consumption and improve the convenience of various home control systems.

World practice shows that most smart home technologies do not immediately enter the free B2C market. In most cases, solutions are first purchased by companies, and then come to the consumer market. We can't say that now most offices and technological premises are equipped with smart systems, but every year more and more companies use different solutions. For example, many processes in the fuel and energy sector have long been automated –the indicators of devices are monitored by special sensors that are programmed for both normal monitoring and a set of actions in emergency situations. Without a doubt, the introduction of such technologies has significantly reduced the percentage of accidents at work, since it is possible to exclude the "human factor". Elements of a smart office system can include motion sensors that monitor power consumption, humidity and temperature sensors, for example, in conference rooms, climate control in different parts of the building, and so on.

After the introduction of such technologies in business and their successful verification, solutions for the consumer sector often begin to be developed. In most cases, solutions are moving from the business to the premium construction segment. For example, a lighting system that allows you to run various scenarios will cost the user 7000-20000 rubles. Temperature and floor heating control systems will require from 7,000 to 25,000 rubles. Depending on the configuration. Solutions for curtains or blinds are offered at a price of 7000-15000 rubles, and security systems with leak sensors and cameras cost about 7000-30000 rubles. Naturally, for this amount, the user will not get a full home equipment, so each additional element, whether it is a sensor or a camera, will cost even more. Systems for energy management are priced on the market from 7000 to 20000 rubles, and solutions for voice control cost up to 15000 rubles [10].

It should be noted that the above offers include a minimum number of elements, so when calculating the cost of equipping a house/apartment, the entire system will cost much more. It should also be emphasized that for proper configuration of all mass-implemented equipment, specialists in "smart homes" are needed, which are currently not so many. Thus, high prices for installation, configuration and maintenance are due to the fact that smart home systems are still rarely used by the population and very few companies have the necessary qualifications to work with them.

4. Discussion

According to the road map of Moscow until 2030, "smart home" is one of the directions of development and implementation of the "smart city" concept. Already, the Moscow government is actively developing infrastructure that allows not only to reduce the number of personnel who provide certain functions, but also to increase the security of the city as a whole. Currently, Moscow has one of the most effective video surveillance systems in the world, which allows you to reduce the number of accidents and other offenses, as well as monitor the overall situation in the city online. In addition, many urban infrastructure management processes are translated into an online format due to the integration of various information developments into the city's ecosystem, which is ensured by the work of companies that produce solutions for "smart" apartments, houses and courtyards, 49% of which are based in Moscow [7].

The growing interest of investors in the problem contributes to more active development of new solutions, and secondly, to reduce their overall cost. According to experts, in the next 5 years, the price of "smart" gadgets and "smart home" systems in General will decrease by 1.5–5 times. To date, the high price for solutions is set due to the lack of necessary experience in this field. First-time firms need to create such systems according to established standards, localize them for each country separately, and develop strategies to inform the public, create the necessary base of experience in the field of maintenance and reduce production costs. In addition, at the moment there are no systems that allow you to completely manage the entire house from a single device, so new solutions are constantly being developed. In 2014, Apple introduced Apple Homekit, an application that combines more than 50 different devices, but this ecosystem also has its drawbacks. When the technologies are sufficiently developed, it will be possible to talk about reducing their cost in the market [7].

Every year, the number of venture capital investments in smart home technologies is growing, which indicates the investment attractiveness of the market. Manufacturers of household appliances are increasingly implementing smart technologies in their products, so in the near future the number of users of "smart homes" will steadily grow, which will lead to the development of the overall urban infrastructure [8].

5. Conclusion

Summing up, it should be noted that the field of "smart homes" has huge prospects. Investors are already investing heavily in the development of "smart" technologies, which makes it possible to reduce the cost of relevant solutions from year to year. At the same time, it should be emphasized that at the moment there are no full –fledged smart home systems that combine the management of all components and suitable for the price category of average residents, however, the latest developments will soon make the Russian market of smart home technologies more accessible due to the development of healthy competition between manufacturers of modern IT solutions.

Thus, for the end user, the development of the market has only positive aspects: greater awareness of the population, lower price and improved functionality that allows not only to meet certain user requirements, but also to save water and electricity consumption. With the advent of cheap and centralized management systems, the smart technology industry will move from the "unreliable" category to the "ubiquitous" category.

References

[1] Krykova A A and Schmatok K O 2019 Features of the development of the "Smart home" concept: Russian and foreign experience. *Azimuth of scientific research: science and management* **3(28)** 397-9

[2] Chibisova I S and Chibisov E B 2018 Technologies of the smart house in Russia. *The age of science. Technical science* **16** 206-7

[3] Lobikova O M and Lobikova N V 2019 Problems of building smart homes. *Greater Eurasia: Development, safety, cooperation* 1000-3

[4] *Smart home in Russia: fantasies, expectations, reality and facts* Retrieved from: https://vc.ru/kauri_iot/90305-umnyy-dom-v-rossii-fantazii-ozhidaniya-realnost-i-fakty

[5] *Smart home. Smart systems market overview 2019* Retrieved from: https://innoagency.ru/files/Smart_home_%2015_02%2019.pdf

[6] Morkovkin D E, Lopatkin D S, Shushunova T N, Sharipov B K and Gibadullin A A 2020 Formation of the conditions for the development of innovation. *Journal of Physics: Conference Series* **1515** 032002

[7] Gibadullin A A, Kerimova Ch V, Chorshanbiev S R, Shmanev S V, Romanova Ju A and Sozontov A A 2020 Management of mechanisms ensuring sustainable functioning and economic development of the electric power complex. *IOP Conference Series: Materials*

Science and Engineering **734** 012170
[8] *Dumping in the smart home market* Retrieved from: http://www.smartrussia.ru/katalog/obie/text1/
[9] Gibadullin A A, Pulyaeva V N and Yerygin Y V 2018 The Need for a Digital Substation during the Digitalization of Energy. *International Youth Scientific and Technical Conference Relay Protection and Automation RPA* 8537223
[10] *Who is looking for smart homes on Yandex. Portrait of the audience 2018* Retrieved from: https://yandex.com/support/direct/technologies-and-services/audience-segments-in-direct.html

IOP Conf. Series: Materials Science and Engineering **976** (2020) 012035 doi:10.1088/1757-899X/976/1/012035

Development of liquefied natural gas transport and creation of energy infrastructure for its functioning in Russia

N A Miagkov and N A Kharitonova

Financial University under the Government of the Russian Federation, 49, Leningradsky Prospekt, Moscow, 125993, Russia

E-mail: kharitonovana1951@gmail.com

Abstract. This is as a monitoring result of the actual scientific literature in the field of using liquefied natural gas as a motor fuel. The ecological and technological advantages of using liquefied natural gas in this area are considered. Based on this, conclusions are drawn about the competitiveness of liquefied natural gas relative to other types of motor fuel. Analyzed the current energy strategy of the Government of the Russian Federation in the field of promoting and popularization of liquefied natural gas as a motor fuel and the development of energy and filling infrastructure. Based on the identified shortcomings of the strategy, an alternative option is proposed that relies on local gas filling systems, mini-gas liquefaction plants, and mobile gas stations. An assessment of the technical and economic parameters of the project for the creation of such an infrastructure was carried out because of which a conclusion was made about its overall efficiency. Options for implementing the concept of alternative development of the energy and refuelling infrastructure of liquefied natural gas and its integration into the state energy policy are proposed.

1. Introduction

In the conditions of the actively gaining pace of development of the sixth technological order and the beginning of the fourth energy transition in terms of energy consumption, energy efficiency and environmental friendliness of the resources used are in the first place with the possibility of replacing classical energy carriers with more efficient ones. In particular, the use of liquefied natural gas (hereinafter LNG) as an automobile and ship fuel, the feasibility of which is confirmed by numerous studies. Despite the fact that LNG is not a product of renewable energy and its use does not fit into the classical ideas of the fourth energy transition, the lack of an objective possibility of widespread introduction of electric vehicles, the high cost of technological solutions in this area and a number of other factors make LNG one of the few alternatives allowing to provide at the same time a sharp decrease in the level of emissions into the environment, as well as economic benefits for transport owners due to the availability of relatively simple technical solutions for its widespread use.

However, despite all the advantages, the introduction of LNG-fueled transport in Russia is hampered by the complete absence of a filling infrastructure, the appearance of which is advisable when a critical level of consumers is formed [1-2]. Thus, the lack of infrastructure makes the rapid development of LNG transport impossible.

2. Materials and methods

Based on the study of Russian and foreign publications, the authors concluded that the vast majority of researchers note the exceptional advantages of this type of fuel in all respects over traditional petroleum products and the availability of ready-made technical developments for trucks and cars on LNG [3-5], which made it possible to conclude about the need to create a network of cryo-filling stations for vehicles powered by LNG. According to the state program «Development of Energy» (as revised on August 26, 2020), it is planned to subsidize the creation of 130 filling stations in the amount of 3 billion rubles by 2024 and by 2030 it is planned to reach the target of 200 operating stations [6], which, according to authors, for a country of such a scale as Russia, is a half measure, since it will allow transferring to LNG a small part of public transport.

For this reason, the authors made an attempt to analyze the factors influencing the decision of the carriers to switch to LNG and propose an alternative version of the state strategy. Based on the results of the analysis of materials from open sources, the authors were able to propose a project for a mini-plant for liquefaction of gas and a filling network with a special territorial positioning.

3. Results

Currently, the main areas of search for alternative fuels and improvement of existing types are:

- Deeper refining of oil products and stricter fuel quality standards;
- Creation of a transport operating on electric energy or having a hybrid engine;
- Wider distribution of biological fuels [2-7].

All three areas have a solid history of scientific research from a technical, environmental and economic point of view, as well as extensive practical experience in their development [4]. In this regard, the development of transport on LNG is a new practice, which has received scientific and practical development only in the last 5-10 years [2]; however, the exceptional characteristics of LNG make this technical direction very attractive. First, it should be noted the excellent performance of LNG from an environmental point of view (figure 1).

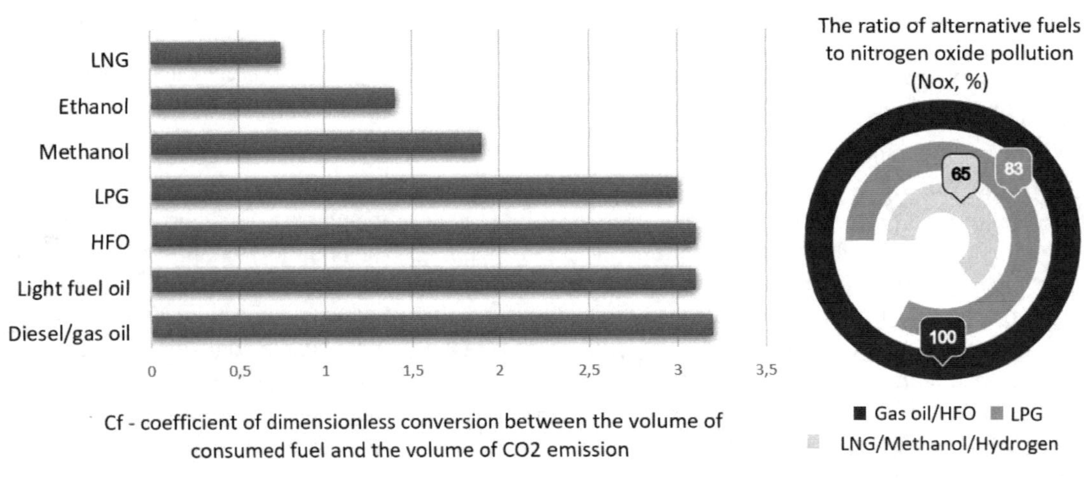

Figure 1. Environmental characteristics of LNG as a type of motor fuel relative to some other types of motor fuel [2].

Despite the fact that environmental standards for petroleum products are tightening every year, the potential for improving gasoline and diesel fuel is technologically limited and significantly affects their cost. While there are technologies that increase the environmental characteristics of natural gas [8], which suggests that, there is a potential for improving the environmental characteristics of LNG. At the same time, one of the main advantages of LNG as a fuel is its low cost (figure 2).

IOP Conf. Series: Materials Science and Engineering **976** (2020) 012035 doi:10.1088/1757-899X/976/1/012035

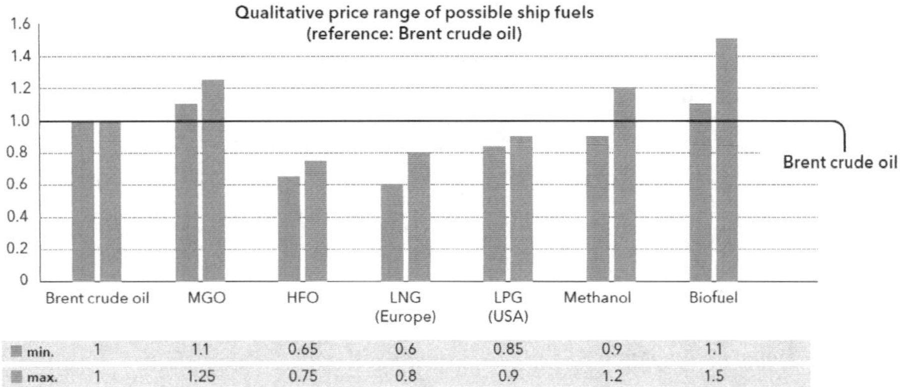

Figure 2. Comparison of average world prices for various types of fuel for the period 2005-2016 [2].

Data from Russian sources indicate that the cost of a kilometer for a truck running on LNG is approximately 8.9 rubles versus 17.8 rubles for a diesel vehicle [2]. In addition, experts note that the use of LNG reduces the wear of engine mechanisms [9]. At the same time, vehicle owners can receive additional benefits from tax incentives: now, in 21 subjects of the Russian Federation, a tax deduction is provided for average of 50% of the transport tax for vehicles with installed gas equipment [8]. Taking into account the energy policy of the Russian state [3], in the future it can be expected that the system of benefits will be introduced in all regions.

However, as noted earlier, the transfer of transport to LNG limits the infrastructure factor. It is not advisable for transport owners to switch to a new type of fuel until a wide refueling infrastructure has been created throughout the territory of the intended use of transport. At the same time, the created infrastructure will be unprofitable until a certain minimum volume of consumption occurs. The Russian Government sees a way out of this situation in subsidizing the construction of cryo-filling stations and the production of vehicles powered by LNG, as well as a wider distribution of tax incentives for owners of LNG vehicles [3; 10]. It should be emphasized that most of the projects in this area are expected to be implemented by PJSC «Gazprom» and PJSC «Novatek» [2]. At the same time, the companies' own vehicle fleet [11] will provide the initial demand for cryo-filling stations.

An alternative option for the development of infrastructure, contributing to the widespread use of LNG transport, is the use of mobile filling systems (figure 3).

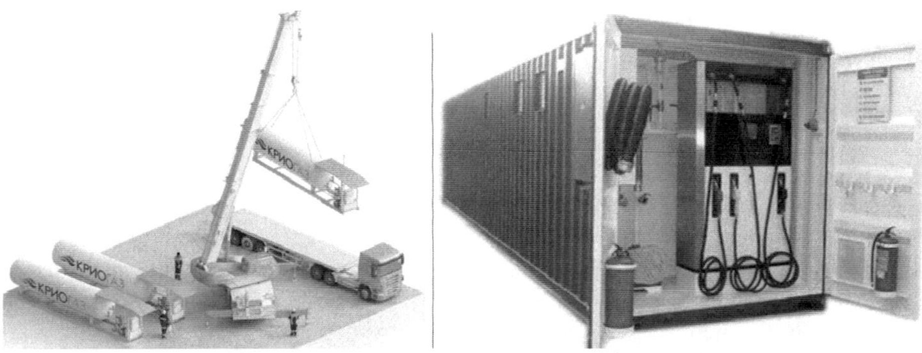

Figure 3. Examples of mobile filling systems [12].

Placement of a mobile filling system on the example of a specific commercial proposal, including a tank for storing LNG, a device for sale (filling station), a fire extinguishing system and auxiliary

IOP Conf. Series: Materials Science and Engineering **976** (2020) 012035 doi:10.1088/1757-899X/976/1/012035

elements (pumps, etc.), with a total cost of $ 36,000 [13] can be carried out on existing gas stations under a lease agreement.

In turn, the gas liquefaction infrastructure is proposed to be implemented in the form of a gas liquefaction mini-plant, which supplies the nearest gas stations. The capacity of such a plant can be increased through additional investment as demand from vehicle owners grows.

According to the authors, the «anchor» consumers for LNG infrastructure could be public transport fleets, river and seaports, given the tightening of fuel standards for ships [1].

Using a public commercial offer for the sale of equipment for a cryogenic filling station [14], as well as a publicly available approximate estimate for the construction of a mini-plant for liquefying gas [15], the authors formed a financial model of a private LNG-filling infrastructure project from a mini-liquefaction plant, 20 petrol stations and two port petrol terminals. The main characteristics of the project are presented below (figure 4).

THE COMPONENTS OF THE LPG INFRASTRUCTURE AND THEIR COST

№	Equipment and costs	Value thsd RUB
I	Inlet Gas Drying Equipment	453 868
	Compressor equipment	
	Liquefaction unit	
	Refrigerating machine	
	Tank for cryostorage of capacity 63 cubic mete 63м3	
	Control and measuring devices and automation + fittings	
	Gas power plant (2 x 500 kW)	
	Equipment for purification from CO2, S, N2 and other impurities	
II	The project with the examination of industrial safety	31 963
	Installation supervision, commissioning	6 393
III	Design and construction of structures and utilities, installation of cryogenic and fire-fighting equipment	181 547
IV	Staff training (2-3 persons)	2 740
	Unforeseen expenses	22 100
V	Carrier (2 x 56 cubic meters)	2 740
	Carrier (2 x 8 cubic meters)	22 100
	Containers x 4 (40 ft)	24 839
	Mobile gas filling stations x 20	54 720
	Refueling ship terminal x 2	94 600
	Total	897 609

ECONOMIC EFFICIENCY

Indicators	Value	
Project implementation period	20	years
Initial investment	897 609 190	RUB
Net value	3 067 852 705	RUB
Profitability Index	3,42	RUB
Payback period	8	years
Discount rate	6,34	%
Net present value	782 170 762	RUB
Discounted profitability index	1,99	RUB
Discounted payback period	10	years
Internal rate of return	20,41	%

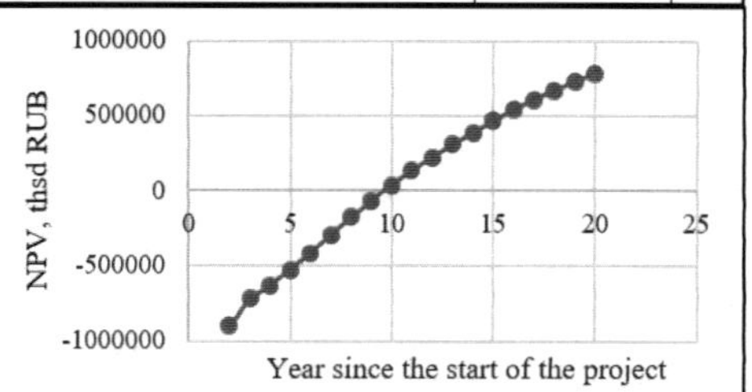

Figure 4. Technical and economic characteristics of the energy infrastructure project for LNG.

The approbation of the financial model showed that the approach proposed by the authors to the development of infrastructure for wider distribution of LNG as a transport fuel in Russia would not only create a filling network, but also ensure its profitability and an acceptable payback period.

4. Discussion

The study showed that it is possible to create a cost-effective local filling infrastructure due to the low initial production capacity, reliance on «anchor» consumers and the use of mobile filling stations instead of standard cryo-filling stations. The concept of infrastructure development proposed by the authors can be integrated into the current state policy for the development of the internal LNG market or considered as a complete alternative to the current state strategy. In general, the proposed concept would require from the federal and local authorities less financial support and more assistance in organizing partnerships between factories and government-owned fleets [16-17].

If earlier the scientific «mainstream» in the application of LNG in transport was a discussion of the advantages and disadvantages of this type of fuel and the general prospects for implementation [18], at the moment the focus of the scientific community is on the issues of practical implementation of fuel in the automobile [19], marine [20] and railway [21] transport, as well as directions of infrastructure development in these areas [22].

5. Conclusion

In order to summarize the ideas it is worth to declare that LNG has exceptional environmental and economic characteristics, significantly superior to gasoline, diesel and fuel oil, which makes it possible to consider it as a new promising type of automotive, ship and train fuel. At the same time, the necessary technological base has already been created for the use of LNG in the indicated directions. The main obstacle to the introduction of LNG in transport now is the problem of creating a cost-effective infrastructure. The government of the Russian Federation is subsidizing the creation of filling stations on the main transport routes, which should allow creating the necessary minimum infrastructure and making it possible to use the first LNG vehicles.

At the same time, the authors proposed an alternative concept for the development of infrastructure, which allows creating a wider filling network, providing filling terminals for river and sea transport, reducing the amount of necessary investments and increasing economic efficiency, and based on the use of mini-liquefaction plants and mobile filling stations as well as strategies for their placement and business, taking into account the regional needs of the country's economy. The results of the research carried out may be of interest to both government agencies and businesses.

References

[1] *Assessment of selected alternative fuels and technologies* Retrieved from: https://www.dnvgl.com/maritime/publications/alternative-fuel-assessment-download.html
[2] *«Gazprom» starts construction of Russia's first cryogenic filling* Retrieved from: https://news.rambler.ru/other/40925790-gazprom-nachal-stroitelstvo-pervoy-v-rossii-kriozapravki/?updated
[3] Rozhko D Ya 2020 Alternative and transitional energy sources for urban transport. *Journal of the Siberian Federal University: Engineering and Technology* **5** 586-96
[4] *Scientific electronic library «Cyberleninka»* Retrieved from: https://cyberleninka.ru/
[5] *Western carmakers asked for unprecedented benefits for LNG* Retrieved from: https://www.vedomosti.ru/business/articles/2020/08/23/837479-avtokontserni-poprosili
[6] *Decree of the Government of the Russian Federation of 15.04.2014 N 321 «On approval of the state program of the Russian Federation «Development of energy»»* Retrieved from: http://www.consultant.ru/document/cons_doc_LAW_162194/
[7] *CO2 Emissions from International Maritime Shipping* Retrieved from: http://www.unepdtu.org/-/media/Sites/Uneprisoe/Working-Papers/2017/Working-Paper-4_Emissions-fromShipping.ashx?la=da&hash=F8FC98CB8712757219146CEBD-6B651EA5E0051D4
[8] *Equinor`s low-carbon solutions: keeping CO2 out of the atmosphere* Retrieved from: https://www.equinor.com/en/what-we-do/low-carbon-solutions-in-equinor.html
[9] Kulaev K B 2018 Prospects for development of the market of gas engine fuel in Russia. *Innovation and Investment* **5** 367-70
[10] *Official website of the Federal Tax Service of the Russian Federation* Retrieved from: https://www.nalog.ru/rn77/
[11] *Cryoblock "Gazprom Gazomotornoye Toplivo" provides liquefied natural gas to the fleet of "Automobile Company - Mostrans"* Retrieved from: http://gazprom-gmt.ru/press-center/news/Krioblok_Gazprom_gazomotornoe_toplivo_obespechivaet_sjijennyim_prirodnyim_gazom_avtopark_Avtomobilnoy_kompanii__Mostrans
[12] *Online marketplace «Alibaba.com»* Retrieved from: https://russian.alibaba.com/product-detail/autoware-mobile-lng-petrol-gas-diesel-portable-filling-station-62280633969.html?spm=a2700.8699010.normalList.10.1cd174d58rWmh4&s=p
[13] *Official website of the Bank of Russia* Retrieved from: https://cbr.ru/
[14] *FAS LNG filling station* Retrieved from: https://fasenergo.ru/
[15] *Preliminary estimation of the cost of the main equipment, design, construction and installation works, carriers and unforeseen costs for the construction of a mini-LNG plant with a*

capacity of 1.5 t/h Retrieved from: http://www.transgasindustry.com/1/1-5t-spg.shtml

[16] Gibadullin A A, Pulyaeva V N, Kharitonova E N and Kharitonova N A *2019* Convergence of the Russian power industry in a period of declining reliability and sustainability of the industry. *Journal of Physics: Conference Series* **1399** 033034

[17] Gibadullin A A, Pulyaeva V N, Kharitonova E N, Erygin Yu V and Kharitonova N A 2019 Formation of a mechanism to ensure the stability of the electric power complex. *IOP Conf. Series: Materials Science and Engineering* **537** 042031

[18] Pronin E N 2012 Prospects of LNG on transport. *Transport on alternative fuel* **2** 47

[19] Shchendrigin A V, Zhuravsky M A, Lyugai S V and Evstifeev A A 2018 Current state and prospects for using LNG as a motor fuel. *Gas Industry* **9** 100-5

[20] Khoroshev V G, Popov L N and Gatin R I 2019 Prospects for using alternative fuels in shipping power plants. *Proceedings of the Krylov State Scientific Center* **4** 200-8

[21] Kondratenko S E 2018 Liquefied natural gas as an alternative to diesel fuel by railway transport. *Gas industry* **1** 24-8

[22] Lyugay S V 2018 Scientific and technical tasks in the field of using natural gas as a motor fuel. *Gas industry* **1** 14-23

IOP Conf. Series: Materials Science and Engineering **976** (2020) 012036 doi:10.1088/1757-899X/976/1/012036

Assessment of prospects and directions of digital transformation of oil and gas companies

A S Flaksman[1], D I Kokurin[2], D K Khodzhaev[3], M A Ekaterinovskaya[4], O V Orusova[4] and A V Vlasov[5]

[1]State University of Management, 99, Ryazan Avenue, Moscow, 109542, Russia
[2]Russian state Academy of intellectual property, 55A, Miklukho-Maklaya str., Moscow, 117279, Russia
[3]The Tajik state university of finance and economics, 64/14, st. Nakhimov, Dushanbe, 734067, Republic of Tajikistan
[4]Financial University under the Government of the Russian Federation, 49, Leningradsky avenue, Moscow, 125993, Russia
[5]Russian University of Transport, 9b9, Obrazcova Street, Moscow, 127994, Russia

E-mail: flax75@mail.ru

Abstract. The article is devoted to the prospects for digital transformation of oil producing companies using the example of Gazprom Neft PJSC. For this, the article analyzes the current state of digital technologies in industries, as well as an analysis of the risks of potential implementation of digital technologies in various industries. The main technological and economic processes in the company of the oil and gas complex were investigated, taking into account, which the assessment of the possibility of introducing digital technologies into the business processes of the oil and gas company was carried out. In the final part of the article, an assessment of the results of the introduction of digital technologies at the enterprises of the oil production complex is carried out.

1. Introduction

Russian oil and gas companies, along with foreign ones, make significant investments in the most attractive and potential projects that can increase the efficiency of oil production, refining and distribution, while simultaneously contributing to an increase in sales profitability. The first priority is to improve the efficiency of management and staff training. Augmented and virtual reality technologies, as well as systemic digitalization of fields can solve most of the problems.

According to the Strategy for the Development of the Information Society in the Russian Federation, activities "in which the key factors of production are data presented in digital form, and their processing and use in large volumes allows, in comparison with traditional forms of management, to significantly increase the efficiency, quality and productivity in various types of production ...", characterizes the digital economy [1]. The program "Digital Economy of the Russian Federation" [2] speaks of the "alternative value" of data, which becomes a "new asset" as they are used "for new purposes and use to implement new ideas". Consequently, it becomes clear that at the state level there is an understanding of the decisive role of digital technologies in significantly strengthening the competitiveness of the national economy. In this regard, the problem of the formation and development of the digital economy from the point of view of innovative development is important for both theory and practice of management.

Accelerated implementation, the use of digital technologies and the creation of truly new investment and business models are possible with the expansion of digitalization opportunities, infrastructure solutions, reducing the cost of computing equipment and improving the software architecture. At the same time, it should be noted that with the growth of the volume of generated global information, there is no proportional development of technologies capable of processing and storing such a volume of data.

For companies, having access to a new digital or information and communication technology is a powerful driver of increasing business competitiveness, but it is worth noting that its distribution is limited among competitors due to the high cost and lack of highly qualified specialists.

The commercialization of digital technologies goes far beyond the virtual environment. Industry / Industry 4.0 and other new production and management systems based on digital technologies - such as Big Data, artificial intelligence - act because of digitalization and integration of vertical and horizontal value chains, digitalization of offered goods and services and the emergence of completely new digital business models and modern platforms for interaction with customers. It should be noted that such an important production indicator as low production costs loses its significance in ensuring the company's competitiveness in the world arena.

According to the World Economic Forum, digitalization of the oil and gas complex alone will bring additional income of $ 1.6 trillion by 2026 [3].

2. Materials and methods

The purpose of this study is to analyze the prospects for digital transformation at the enterprises of the oil and gas production complex.

To achieve this goal, the following tasks are being solved:

- Analyze the current state of digital technologies in industries;
- Conduct a risk analysis of the potential introduction of digital technologies in various sectors of the economy and industry;
- Consider the best practices in the use of digital technologies in domestic and foreign companies in the fuel and energy complex;
- Study the main technological and economic processes in the oil and gas company;
- Assess the possibilities of introducing digital technologies into the business processes of an oil and gas company;
- Assess the results of the introduction of digital technologies at oil and gas enterprises.

Within the framework of scientific research, methods of descriptive, statistical, comparative, economic analysis, the method of expert assessments are used.

The information base of the study was the materials of various studies, the analytical report "Digital Transformation in Russia" based on the results of a survey of Russian companies for 2018 from Team-A Management LLC [4], Strategies for the development of the information society in the Russian Federation [1], annual reports and the investment program of PJSC Gazprom Neft.

3. Results

One of the breakthrough trends of the last decade is the integration of digital and cloud technologies into real economic processes. "Industry 4.0", "Big Data", artificial intelligence and others are already becoming realities of everyday life. This caused a serious transformation in many spheres of social and economic life and necessitated a revision of the principles of managing the innovative development of enterprises and the national economy as a whole. Expanding digital opportunities and infrastructure solutions while lowering costs are accelerating digital adoption and enabling new business models.

The oil and gas industry includes a large number of industrial, economic business processes, a number of excerpts follow from this:

- A large number of units that poorly interact with each other;
- Bureaucratic processes;
- A lot of technologies that can be modernized by transferring to digital technologies.

According to experts, the massive use of IT technologies will help oil companies to raise the oil recovery factor by 2-7%, as well as a possible reduction in operating costs by 25%.

Vygon Consulting analyzed the potential implementation or conversion of technological processes of an oil and gas complex in Russia to digital technologies by 2030 and concluded that these innovations will help increase oil production by 155 million tons, which will help compensate for lost production with depleted long service life wells [1-4].

Figure 1 shows the production potential in Russia for 2030.

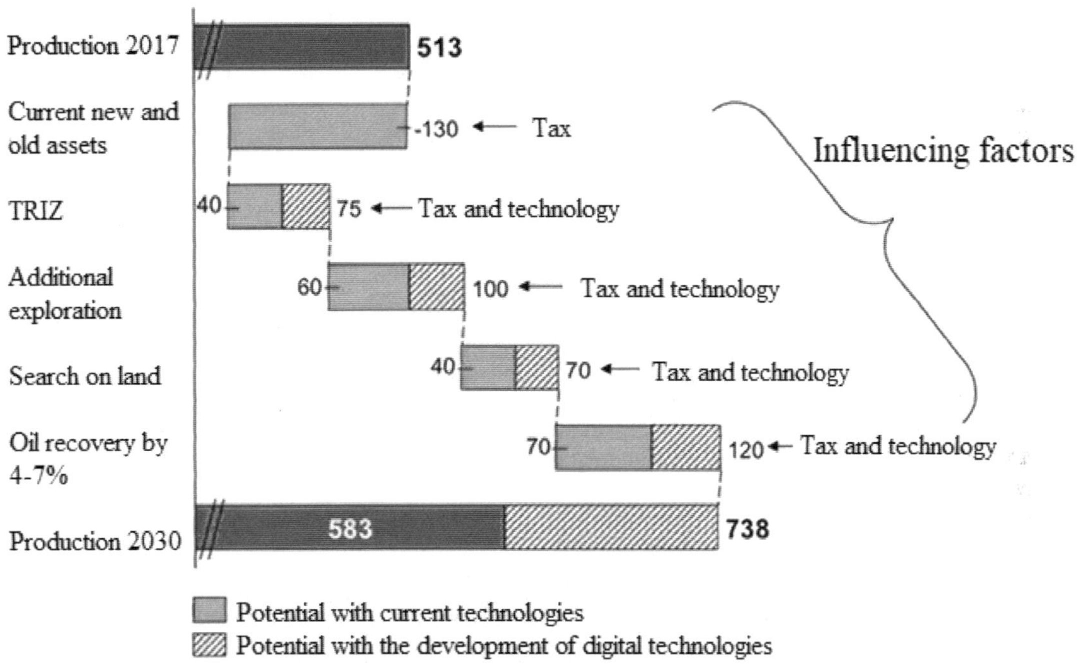

Figure 1. Theoretical production potential in Russia for 2030 in scenarios with current technologies and with the development of digital technologies.

In addition to the obvious advantages of transferring business processes to digital technologies, namely, such as:

- Increase in profit;
- Increase in production volumes;
- Reduction of costs at the stages of preparation.

The transition to digital technologies gives a number of non-obvious advantages, namely, such as:

- Strengthening of positions in the market;
- The reliability of the company, by reducing the likelihood of potential accidents;
- Speed up all processes, from development to sales.

Accenture, after conducting its research, concluded that 36% of the world's oil producing companies has already implemented BigData technology, and 38% are going to introduce it in the next 5 years. Despite the fact that the oil price has collapsed, the majority of companies in the oil and gas industry are not going to cancel plans for the introduction of digital technologies, as well as to reduce the flow of investments aimed at this area. There has also been an increase in demand for the work of service companies that process large, massive data that interpret the results of seismic studies. This service is very expensive, but such an investment pays off and also helps to bring additional benefits. Using targeted drilling, with the help of geophysical guidance, the accuracy of work becomes many times higher. Thanks to 3D seismic exploration, oil companies can save about 5-7% for every dollar invested. The Institute of Oil and Gas Problems of the Russian Academy of Sciences (RAS) has been trying for many years to create recommendations for the implementation of digital technologies in the oil and gas industry.

One of the main issues in the transition of the oil and gas industry to digital use is the automation of refining. All this suggests that oil companies have all the opportunities, prerequisites, and most importantly, the need for digital technologies. In Russia, the advanced process control system or APC (Advance Process Control) has become widespread, which allows you to adhere to the established regulations and predefined target functions, as well as to make timely and correct decisions in crisis and non-standard situations. Thanks to the use of on-line analyzers, it becomes possible to establish monitoring of raw materials and products in real time. Adjacent installations are usually combined into an integrated system, this helps to synchronize technical processes that are controlled from a common center. The use of the latest analysis methods in the processing of petroleum products helps to increase profit indicators. Using as an example, one of the projects in Russia that started using the Industrial Internet of Things (IIoT), also known as the Industrial Internet of Things, EY analysts demonstrated the advantages of implementing this technology. Its essence lies in the fact that data modeling takes place using neural networks and machine learning methods. At refineries that use atmospheric and vacuum distillation in their installations, the introduction of IIoT helped to raise the forecasting quality by 15%, as well as improve the accuracy of the final result, or rather, deviations from the expected result decreased by 90%.

Market leaders have already created or have begun to create digital programs (Rosneft, Lukoil, Gazprom Neft), which suggests that we are facing a serious modernization in the oil and gas industry. Confidence in digital technology is growing stronger after reports of successful pilot projects from companies that have already tried out the technology.

All this suggests that oil and gas companies have all the opportunities, prerequisites, and most importantly, the need for digital transformation.

After analyzing the strategies of the largest international companies that use digital technologies, we can conclude about the main trend groups [5-8]:

- There is a period of global introduction of technologies that have a strong impact on the success of oil and gas companies;
- Major players are increasingly beginning to invest in technology at the initial stages of development, using corporate venture funds for this;
- IT departments are being transformed into growth and development units and are increasingly moving away from the cost centralization model. Many companies are moving to a model where the IT department is directly headed by the CEO, thanks to this, the status of the CIO (Chief Information Officer), or, in other words, the head of the information technology department, is increased;

- One of the most important issues is the choice between developing competencies within the company or trusting outsourcing. World experience shows that both development models have a right to exist. For example, some global companies use outsourcing to improve the efficiency of cost management (British Petroleum), but there are also opposite successful examples of company development that leave technological development within their own business, not only in order to control operations as much as possible, but also to preservation of intellectual property for inventions (Shell);
- Many digital technologies are created in symbiosis between vertically integrated oil companies (VICs) and oil service companies (for example, GE and BP, Baker Huges and Shell);
- However, even despite this, a large investment flow occurs in the development of competencies within the vertically integrated oil companies;
- Oilfield service companies are trying to collaborate with IT companies in order to create fully integrated solutions (for example, Schlumberger and IBM, Atos and Siemens).

Based on the results of the study, the following technologies can be identified for Russian oil and gas companies, into the depths of which can be introduced in the near future:

- Compression of processes;
- BigData services;
- Robotic Process Auto mation (RPA);
- Internet I ndustrial o f Things (IIoT).

4. Discussion

It should be noted that the introduction of digital technology has a number of advantages, both material and non-material [9-10]:

- Reduced costs;
- Increased income;
- Inflow of information;
- Perfect monitoring;
- Prevention of emergency incidents.

However, along with these advantages, there are risks in the development and functioning of business processes based on digital technologies, which include the following [8]:

- Low level of infrastructure;
- Lack of qualified personnel;
- Significant costs in an unstable market;
- Cybersecurity;
- Lack of a refined regulatory framework.

5. Conclusion

The study notes that oil and gas companies are always at risk, as this industry is influenced by many more factors than others are. For many companies, digital technologies are seen as what can become an "anchor" of stability in difficult conditions of existence in the market. Many companies have already or are adopting digital technologies. Other companies are already building their competitive strategies with the assumption that they will have to use digital technologies in order not to lose market share.

References

[1] *On the Strategy for the Development of the Information Society in the Russian Federation for 2017-2030* Retrieved from: http://kremlin.ru/acts/bank/41919

[2] *Official site of ANO Digital Economy of the Russian Federation* Retrieved from: https://data-economy.ru/organization

[3] *Digital revolution* Retrieved from: https://oilcapital.ru/article/general/05-12-2017/tsifrovaya-revolyutsiya-91a53a31-8a30-4ea7-a680-8d0c195751eb

[4] *Digital transformation in Russia: analytical report based on the results of a survey of Russian companies in 2018* Retrieved from: http://www.interface.ru/iarticle/files/39873_50679725.pdf.

[5] Gibadullin A A, Ryabinina E V, Morkovkin D E, Sodikov K A, Trifonov P V, Kirpicheva M A and Kokurina A D 2020 Engineering solutions in the field of digital transformation of the electric power industry. *IOP Conference Series: Materials Science and Engineering* **862** 062055

[6] Gibadullin A A, Sadriddinov M I, Kurbonova Z M, Shedko Yu N and Shamraeva V V 2020 Assessment of factors ensuring sustainable development of the electric power industry in the context of transition to renewable energy sources of the national economy. *IOP Conference Series: Earth and Environmental Science* **421** 032051

[7] Gibadullin A A 2020 Assessment of the level of stability and reliability of the electric power complex. *IOP Conference Series: Materials Science and Engineering* **837** 012007

[8] *Grigory Vygon on the future of "difficult" oil in Russia* Retrieved from: http://vygon.consulting/pressroom/our-publications/1190/

[9] Gululyan A G 2014 On the assessment of the economic efficiency of the introduction of technologies for "smart" fields. *Problems of economics and management of the oil and gas complex* **06** 16-20

[10] Linnik V Yu and Linnik Yu N 2019 Priority directions of innovative development in oil and gas complex at the Siberian Federal District *Upravlenie* **7** 1 pp 40-9

Development of a technology for non-reagent deironing of water for power facilities

A Yu Kurbatov, E N Kuzin, Yu M Averina and M A Vetrova

Mendeleev University of Chemical Technology, 9, Miusskaya square, Moscow, 125047, Russia

E-mail: kurbatov@muctr.ru

Abstract. As part of the study of the possibility of using hydrodynamic water treatment for preliminary deferrization of artesian water before feeding it to a deep osmotic treatment unit. It was found that in the process of water treatment in a hydrodynamic device with constant air ejection, the complete oxidation of iron (II) compounds occurs. The influence of the volume of consumed air and the number of processing cycles on the efficiency of the deferrization process has been studied. Data were obtained on the distribution (size / ratio) of air bubbles in the fluid flow during its hydrodynamic treatment. Based on the results obtained, it was concluded that when hydrodynamic treatment is combined with ceramic ultra filtration membranes, the iron removal rate is 99%, while hardness salts, hydrogen sulfide and other pollutants are partially removed from the water, and complete disinfection occurs.

1. Introduction

The most widespread method of water treatment for the needs of energy facilities is reverse osmosis. Unfortunately, artesian or river water contains significant concentrations of iron compounds, which negatively affects the service life of reverse osmosis membranes. A prerequisite for the efficient operation of reverse osmosis post-treatment systems is the maximum removal of iron compounds from water [1].

One of the possible options for water purification from iron compounds is its reagent treatment with solutions of strong oxidants (hypochlorite, peroxide, potassium permanganate); however, the use of these reagents before the reverse osmosis system is undesirable due to the negative effect of the oxidant on the reverse osmosis membrane [2].

Today, the most common and at the same time simple, but not sufficiently effective method of purifying natural waters from ions of dissolved iron is based on simplified aeration and oxidation of iron by atmospheric air, followed by the removal of the formed insoluble iron compounds on mechanical filters (polymer, bulk, mechanical) [3].

The main disadvantage of this purification method is the low rate of the oxidation process and, as a consequence, an increase in the process time with a concomitant complication of the purification apparatus. The increase in contact time is due to the specificity of the interaction of iron ions with air oxygen dissolved in water. The oxidation reaction occurs at the interface between the air bubble and water, which means it, is limited by the contact area. The solubility of oxygen in water under normal conditions is about 9.0 mg/l [4], and, therefore, the maximum concentration is usually reached at the very beginning of the aeration process. In this regard, practically the only available way to intensify the process of oxidation of iron ions with air oxygen is a significant increase in the contact surface of the phases at the water - air bubble interface.

In practice, a significant increase in the contact surface of the liquid-gas phases can be obtained either by creating sufficiently small liquid droplets (the so-called water fog), or by creating sufficiently small gas bubbles in the liquid. Taking into account the technological design of the process, as well as its economic component, saturation of the liquid with finely dispersed gas bubbles is more rational.

Methods such as air ejection or bubbling (for example, through finely porous ceramics) can further reduce the size of air bubbles in order to further increase the contact surface of the water-air phases, however, even for such methods there is a limitation on the attainable minimum bubble size (about 280 micrometer) [5].

To further increase the degree of dispersion of the gas in the liquid and reduce the bubble size, an effective way is the imposition of vibration influences or acoustic vibrations propagated in the processed multiphase system. In this case, the greatest effect can be achieved with a certain combination of parameters of the impact on the multiphase medium and the dynamic properties of the medium itself [6-7]. The optimal combination of the geometry of the reservoir, the physicochemical properties of the water-air mixture and the hydrodynamic parameters of the fluid movement provides a sharp increase in the degree of air dispersion, and, as a consequence, the intensification of the oxidation process of ferrous iron.

Literature sources [8] describe in detail methods for calculating the natural frequencies of a multiphase liquid-gas system in vessels of various shapes, when they coincide with the frequencies of an external influence, the rate of mass transfer processes increases significantly, dispersion increases, an intense directional movement of gas bubbles occurs, which outwardly resembles a violent "boiling" in the entire volume of the vessel.

2. Materials and methods

Determination of the content of iron (II) ions was carried out on a DR 2800 spectrophotometer (HACH USA), and the determination of the total content of metals on an atomic emission spectrometer with a magnetic plasma "SpectroSky" (Russia) [9].

To assess the degree of air dispersion and the possibility of oxidation of iron compounds in the process of hydrodynamic treatment, we used an installation (figure 1) assembled at the Department of Innovative Materials and Corrosion Protection of the Russian Chemical Technology University named after I. DI Mendeleev [10]. The hydrodynamic device is made of stainless steel 12X18H9T and allows air to be ejected directly into the working chamber by imparting a rotational velocity component to the flow, the presence of which leads to the emergence of centrifugal mass forces in the flow and the formation of a radial static pressure gradient. In this case, the velocity of the liquid at the wall of the working chamber significantly exceeds its average value, and at the axis, there is a sharp drop in pressure and the release of the smallest air bubbles from the liquid, caused by the cavitations phenomenon, contributing to the intensive dispersion of the gas. The flow rate of air ejected into the hydrodynamic device was controlled by rotameter.

Model and real artesian treated in a hydroacoustic device and assessed the size of the formed dispersed air bubbles in distilled water, and also measured the sizes of Fe (II) and Fe (III) forms during hydrodynamic treatment and after filtration through a ceramic membrane filter with a pore size 0.05-0.5 micrometer.

The bubble size was determined using the "Imagescope" software by analyzing photographs obtained with a microscope.

3. Results

At the exit from the hydrodynamic device, it was possible to achieve a sharp increase in the specific surface area of phase contact and a high degree of homogenization due to the impact on the gas-liquid mixture of the field of centrifugal forces, which contribute to the intensive fragmentation of gas bubbles and renewal of the interphase surface (figure 2).

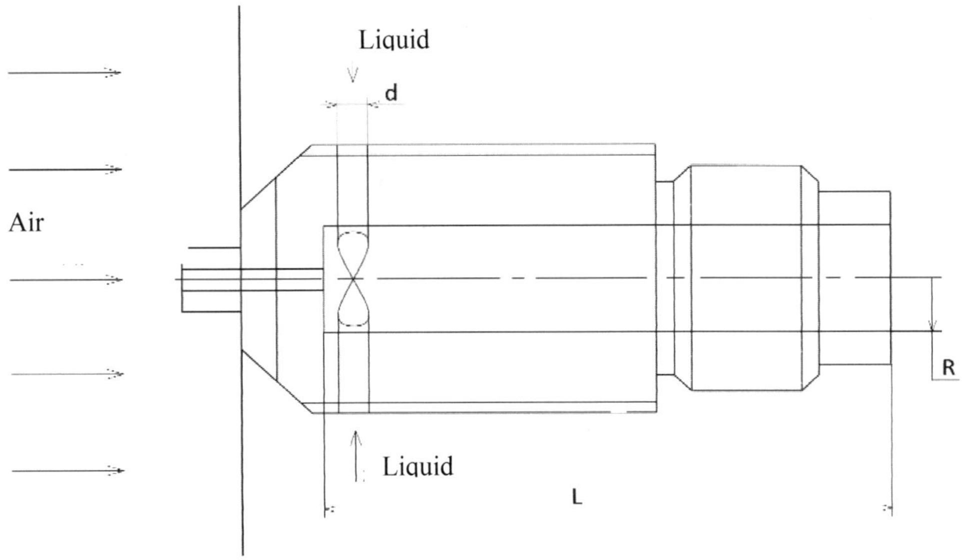

Figure 1. Schematic diagram of the operation of the hydrodynamic device: d - diameter of the inlet channel cameras (RK); R is the radius of the RK; L is the length of the RK.

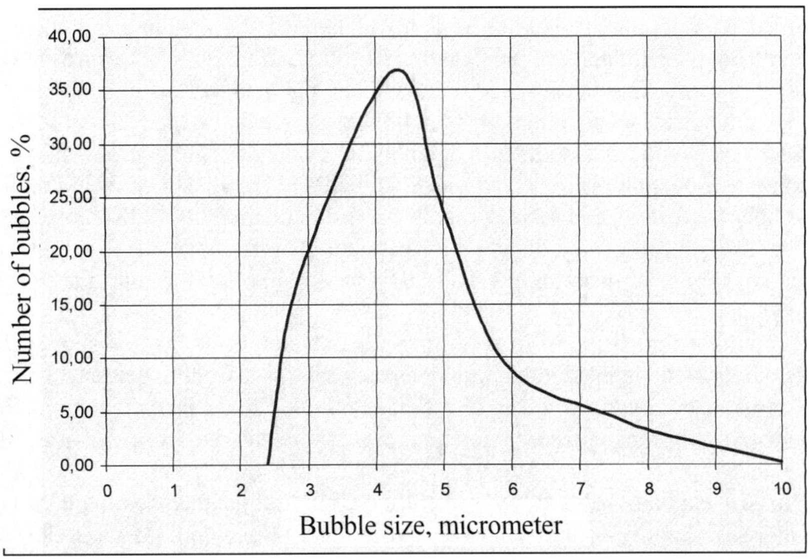

Figure 2. Dispersed air bubble size distribution.

From the data in figure 2, it can be seen that the sizes of air bubbles at the outlet from the hydrodynamic device are in the range of 2.4-10.0 microns. Most of the bubbles (up to 40% of the total) had a size of 5 microns; the average size is 5 microns. The calculated average bubble ascent velocity was approximately 0.1 mm/s. Based on these data, it can be assumed that the processes of heterogeneous oxidation of iron compounds at the air bubble - water phase boundary will occur much faster due to an increase in the contact area.

The effect of the number of water treatment cycles without oxygen ejection is shown in table 1.

From the data in table 1, it can be seen that a single passage of water through a hydrodynamic device allows a 68% decrease in the concentration of dissolved iron (II), the second cycle of treatment reduces the concentration of iron (II) to the MPC standard, the third cycle reduces the concentration of iron (II) to the level of the detection limit techniques. It should be noted that the residence time of the treated water in the hydrodynamic device is 1-3 seconds, and the linear velocity of the water is in the range of 25-32 m/s.

Table 2 shows the dependence of the residual content of iron (II) ions depending on the volume of ejected air into the working chamber.

Table 1. Efficiency of removal of iron and manganese compounds from the model water solution.

Index	Residual concentration, mg/l					
	Feed water	Cycle 1	Cycle 2	Cycle 3	Cycle 4	Cycle 5
Iron (II)	0,65	0,21	0,09	0,01	0,00	0,00

Table 2. Efficiency of removal of iron and manganese compounds from the model water solution.

About air, Vg/Vzh, %	Source water	0,0	1,5	3,0	4,5	6,0	7,5	9,0	10,5
Iron content, mg/l	0,65	0,21	0,13	0,09	0,05	0,18	0,23	0,31	0,37

From the data in table 2 it can be seen that with a volume of ejected air equal to 4.5%, the maximum possible oxidation efficiency of iron (II) compounds is achieved. An increase in the volume of ejected air in excess of this amount leads to a sharp change in the hydrodynamic treatment of the water flow, which in turn significantly reduces the efficiency of oxidation of iron (II) ions.

4. Conclusion

According to the results of the experiments, it was found that the use of hydrodynamic treatment of natural water makes it possible not only to increase the degree of air dispersion, which is confirmed by a significant decrease in the size of the formed dispersed air bubbles, but also to lead to a significant intensification of the process of heterogeneous oxidation at the bubble-water interface. This process is due to the formation of cavitations phenomena in the processed natural water during its processing.

The characteristics of the air dispersion process have been studied - in particular, data on the size distribution of dispersed air bubbles formed as a result of hydrodynamic treatment have been obtained. It was found that most of the air bubbles formed during the treatment of water have a size of 2.5 to 6 microns, which, on average, is an order of magnitude smaller than when bubbling through fine-porous ceramic membranes is used [11].

It was found that for the complete oxidation of iron compounds in water, only 3 processing cycles are sufficient, while the optimal volume of ejected air was determined - 4.5 deironing percent. Taking into account the design features of the hydrodynamic device, it is possible to replace the air ejection with ozone, which will not only further increase the cleaning efficiency, but also enhance the disinfecting effect.

The data obtained will further expand the scope of application of hydrodynamic treatment, which has recently been gaining more and more popularity, in view of the fact that hydrodynamic treatment of natural waters does not require additional use of chemical reagents, and, therefore, the absence of additional effluents, and at the same time has a fairly simple and reliable hardware design [12-16]. The above factors, in turn, contribute to the improvement of the ecological component of the water treatment process as a whole.

Acknowledgments

This work was supported by the D. Mendeleev University of Chemical Technology of Russia through the Internal Bridging Grants for Young Science Educators Contest, proposal no. K-2020-015.

References

[1] Kundenok S B, Popova T Yu and Medved P V 2018 Individual technology of water treatment of artesian water with a complex composition of impurities. *Bulletin of the Engineering School of the Far Eastern Federal University* **3(36)** 118-27

[2] Frog B N and Pervov A G 2015 *Water treatment* (Moscow) 255
[3] Lekomtsev V R 2017 Review of the main methods of iron removal of water. *Young scientist* **29(163)** 17-20
[4] Tolstoy M Yu, Shishelova T I and Shestov R A Studies of oxygen solubility. *Izvestiya vuzov. Applied Chemistry and Biotechnology* **1(12)** 86-90
[5] Averina Yu M, Terpugov G V, Pavlov D V and Varaksin S O 2010 An integrated approach to the problem of water deferrization. *Water treatment* **12** 23
[6] Romankov P G and Kurochkin M I 1982 *Hydromechanical processes of chemical technology* (Leningrad: Chemistry) 190
[7] Kobasko R F and Kulik N I 1960 *Oscillatory phenomena in multiphase media and their use in technology* (Kiev: Technics) 118
[8] Fedotkin I M 2001 *Physical effects during hydrodynamic cavitation and their practical application* (Kiev) 134
[9] Kuchumov V A and Shumkin S S 2017 Analysis of the chemical composition of the initial alloy in the production of permanent magnets from alloys of the Sm-Co system. *Scientific and technical bulletin of SPbSPU* **23(1)** 219-25
[10] Avduevsky V S, Ganiev R F, Kalashnikov G A and Kostrov S A 1994 Hydrodynamic oscillator *Bul.* **13** 2015749
[11] Menshikov V V, Kurbatov A Yu and Averina Yu M 2015 Study of the technology of oxidation of iron dissolved in natural waters by forced aeration using ceramic fine-pored tubular membranes. *Eurasian Union of Scientists* **6-6(15)** 6-10
[12] Hydrodynamic oscillation generator 1994 *Bull.* **13** 2015749
[13] Ukrainskiy L E 2006 *Dynamic fundamentals of wave technology* (Moscow) 239
[14] Arrojo S and Benito Y 2008 A theoretical study of hydrodynamic cavitation. *Ultrasonsonochem* **15** 203-11
[15] Dindar E An 2016 Overview of the Application of Hydrodinamic Cavitation for the Intensification of Wastewater Treatment Applications. *InnovEnerRes* **5** 137
[16] Sun X, Liu J, Ji L, Wang G, Zhao S, Yoon J Y and Chen S 2020 A review on hydrodynamic cavitation disinfection: The current state of knowledge. *Science of The Total Environment* 139606

Assessment of oil and gas resources of northern Afghanistan and their impact on energy security in the country

Ahmad Tamim Mehrad, V P Zvolinski, D O Kapralova and Milad Ahmad Niazmand

People's Friendship University of Russia (RUDN University), 6, Miklokha Maklaya, Moscow, 117198, Russian Federation

E-mail: ahmadtam32@gmail.com

Abstract. Developing and utilization from hydrocarbon resources of northern Afghanistan is very important for ensuring the energy security and development of the least developed country in Asia. The estimation made by the USGS and Afghanistan geological survey revealed substantial oil and gas resources in the Amu Darya and Afghan-Tajik basins of northern Afghanistan that created an optimistically bright landscape for ensuring the energy availability and development of the country. This study evaluates the Amu Darya and Afghan-Tajik hydrocarbon basins and their development prospective. In 2020, field trips were arranged to Sheberghan and Sar-e-Pul oil and gas fields for collecting information, conducting interviews, and evaluating the infrastructures of oil and gas fields. The findings described that the infrastructures for producing oil and gas are old, poorly maintained, and have limited production capacity. There are serious obstacles, such as security problems, undeveloped infrastructures, and technical issues that are needed to be considered.

1. Introduction

The economic development of any country is directly related to the sustainable meeting of energy demands, especially oil and gas. Currently, oil and gas are the dominant sources of energy in the world, and countries with significant energy reserves not only have a stable and strong economy but also play an active role in international issues. Oil and gas reserves have been the main sources of economic development in the majority of the gulf and north African states. Countries like The USA, Canada, Norway, Iran, Nigeria, The Russian Federation, and the CIS have also reached different stages of development largely due to their rich oil and gas reserves.

Afghanistan is situated along the energy corridor of Central and Southwest Asia with the vast endowment of natural resources including, rich hydrocarbon reserves [1-2]. The presence of hydrocarbon reserves is a panacea for Afghanistan, and the development of these resources can help the poorest country in Asia to resolve part of its economic problems, create work opportunities and take crucial steps toward reaching stability. Oil and gas resources in Afghanistan have been known for a long time, but they were exploited in limited quantities by the government and some influential war-lord leaders [3]. Most of the oil and gas resources of Afghanistan are situated in the north of the country within the Amu Darya and the Afghan-Tajik basins, and it is believed that these basins are the continuation of the great oil and gas fields of Central Asia, which is situated in Turkmenistan, Uzbekistan and Iran [4].

According to intergovernmental agreements concluded between the Government of Afghanistan and the Soviet Union, from the second half of 1958, topographic and geological-geophysical surveys were conducted in the territory of northern Afghanistan to study the geological structure of this region and setting up exploratory drilling for oil and gas. As a result, from 1960-1983, several oil and gas fields were discovered in the northern provinces of Sheberghan, Sar-e-Pul, and Faryab [5].

Before the withdrawal of the Soviet troops in 1989, about the 80 percent of natural gas produced at the Shebergan gas fields, was exporting to the Soviet Union, supporting the activities of the Northern fertilizer plant, the 36,000 kW thermal power station, and the textile factory in Mazar-e-Sharif [6]. In 2005, by using a series of Soviet's old charts, available data, and modern technics, the USGS conducted a comprehensive scientific survey for identifying minerals in Afghanistan [7]. The survey revealed that two northern basins of Amu-Darya and Afghan-Tajik could contain as much as 1.6 billion barrels of oil and 15.6 trillion cubic feet of natural gas respectively [8]. In 2011, the Afghan Ministry of Mines and Petroleum estimated that northern Afghanistan has about 1.95 billion barrels of undiscovered technically recoverable crude oil reserves and 16.2 trillion cubic feet of discovered natural gas reserves [9].

Afghanistan's development has been affected by decades of political turmoil and conflict. The country's economic growth, which was 9.3% between 2003 and 2014, fell to 2.9 percent in 2019 [10]. It was mainly influenced by the withdrawal of international forces, as assistance and activities related to their participation supported the period of economic expansion. The cutback in aid, accompanied by extensive corruption, created a gap in the Afghan economy and an overall catastrophe in the country.

2. Oil and gas basins of northern Afghanistan

The oil and gas resources of northern Afghanistan are accumulated in two proliferous geological basins of The Amu-Darya and Afghan-Tajik figure 1.

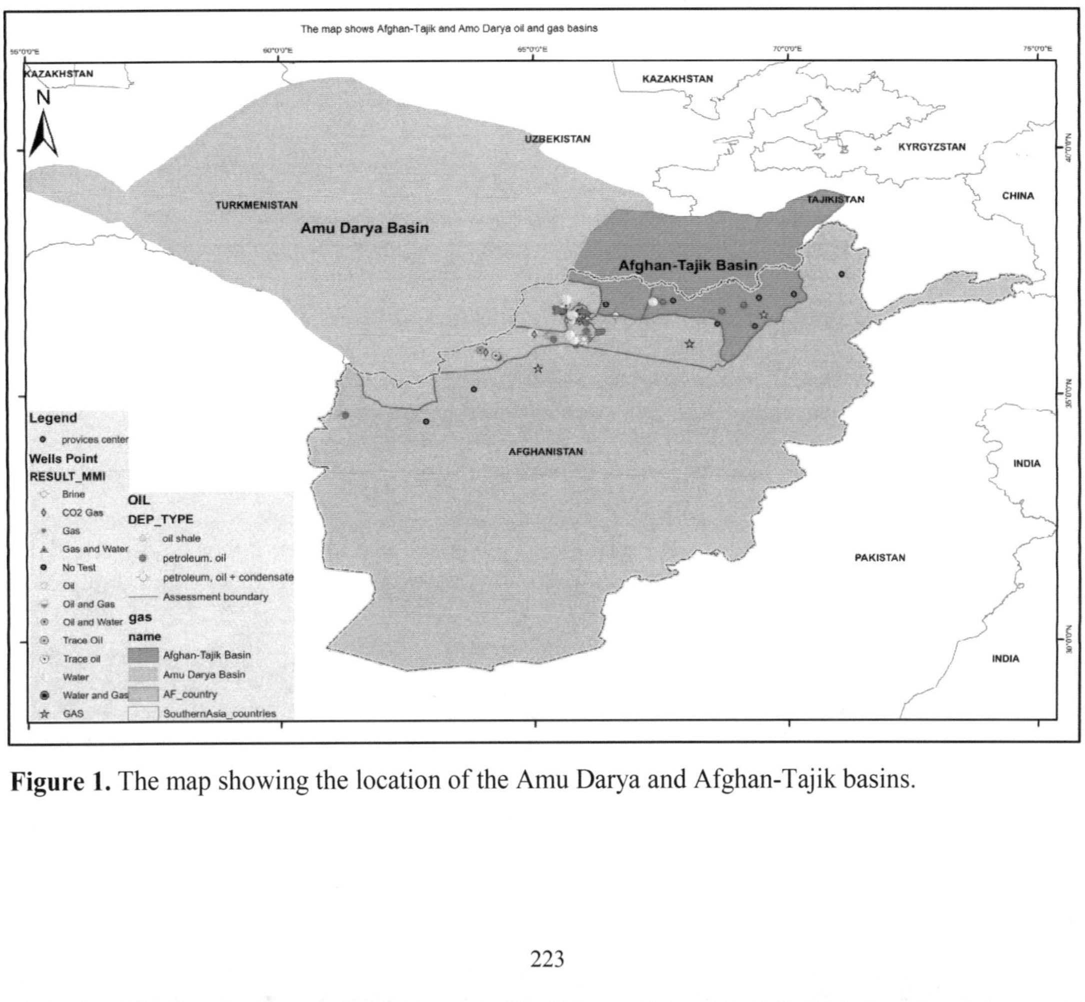

Figure 1. The map showing the location of the Amu Darya and Afghan-Tajik basins.

2.1. The Amu Darya (Amu River) oil and gas fields.

The Amu Darya basin is located in Faryab and Sar-e-Pul provinces in the northwest of Afghanistan. This basin is associated with the foredeep of the southeastern part of the Turan Plate, extending far beyond Afghanistan [11]. The part of the basin which is located in the Afghan territory extends from the north to the south. The northern part includes the North Afghan and Maimana basement protrusions, Daulatabad troughs, and the Kushkin uplift zone, whereas the south is limited by the outcropping of the basement in the mountain-folded structures of Afghanistan (table 1) [12].

Table 1. The estimated mean volumes of undiscovered, technically recoverable, conventional oil and gas resources for the Amu Darya basin province.

Crude oil	942 million barrel
Natural gas	52 025 billion cubic feet
Associated and dissolved natural gas	2 321 billion cubic feet
Non-associated natural gas	49 704 billion cubic feet
Natural gas liquids	582 million barrel
Natural gas liquids in oil accumulation	55 million barrel
Total liquids in non associated gas accumulation	527 million barrel

2.2. Afghan-Tajik oil and gas fields.

The Afghan-Tajik basin is located in the north and northeast of Afghanistan and is extended to Turkmenistan and Tajikistan. This region extends along with the northern and northeastern parts of the country from Takhar to the Jawzjan provinces. From the east, it reaches to the Hindu Kush mountain, in the west and north to the Amu Darya, and in the south to The Alborz mountains [13]. The Afghan-Tajik basin has formed by the northward protrusion of the Pamir-Badakhshan block that began in the late Oligocene and continuing to the present [14]. The Afghan-Tajik basin has significant potential for the accumulation of crude oil, as the basin, is situated in the area with Cretaceous-Paleocene carbonate reservoir rocks that are associated with thrust faulting and folding of northern regions of Afghanistan (table 2) [15].

Table 2. The estimated mean volumes of undiscovered, technically recoverable, conventional oil and gas resources for the Afghan-Tajik basin province.

Crude oil	996 million barrel
Natural gas	7 072 billion cubic feet
Associated and dissolved natural gas	225 billion cubic feet
Non-associated natural gas	8 847 billion cubic feet
Natural gas liquids	85 million barrel
Natural gas liquids in oil accumulation	4 million barrel
Total liquids in non associated gas accumulation	81 million barrel

3. Methods

The main objective of this paper is to evaluate the potential and challenges of developing oil and gas resources of northern Afghanistan and their impact on the economic development and stability of the country. This paper is designed based on:

- Area observation
- Interviews were conducted with relevant government officials, academics of Kabul, Polytechnic, Balkh, and Jawzjan universities, experts, and representatives of private companies who linked with this study.

- Analysis of relevant data obtained from the Afghan Ministry of Mines and Petroleum, Soviet sources, USGS, the World Bank, Afghanistan Geological Survey, Sheberghan oil and gas authority, and Northern oil and gas authority.

4. Discussion

4.1. Afghanistan's energy status.

Afghanistan's energy comprises both renewable (hydro, solar, wind) and non-renewable (natural gas, petroleum, coal, wood, charcoal). In 2019, wood comprised nearly half (44%) of Afghanistan's energy consumption, while oil and gas comprised almost (8%) and (17%) of Afghanistan's energy consumption basket [15]. Oil is widely used in transportation and power generation sectors, whereas gas for heating and cooking purposes. In the last two decades, Afghanistan's energy consumption has increased tremendously, mainly due to population growth and changing life patterns (figure 2).

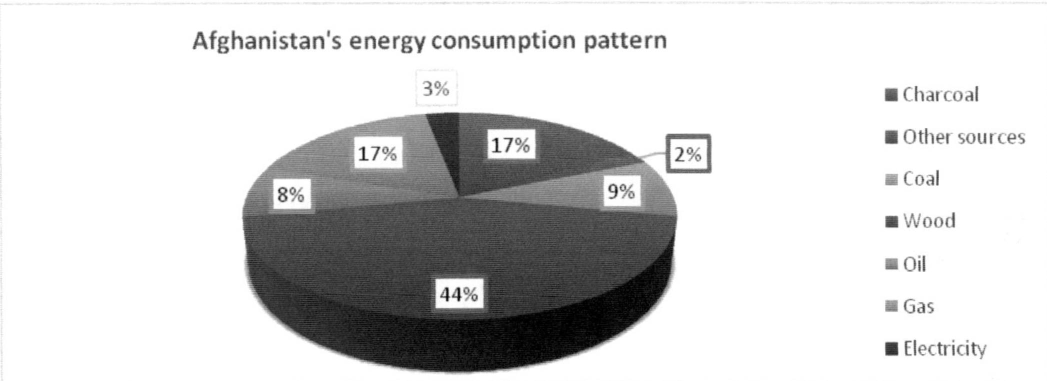

Figure 2. Afghanistan's energy consumption pattern 2019 (designed based on NSIA data).

4.2. Afghanistan's oil and gas production, consumption, and imports.

Afghanistan's oil and gas consumption has been increasing steadily for the past two decades, and it is expected to have a similar trend in the future. The country is a net importer of oil and gas, as it produces only 5 % of its total domestic need figure 3. Since 2017, natural gas and crude oil production have decreased significantly due to the country's instability and disputes with contracted companies, while the import has increased.

Afghanistan needs almost 4.5 million tons of oil and 656 thousand tons of natural gas annually, which is mostly imported from Russia, Iran, and Central Asian countries. Importing oil, gas, and electricity costs about 1.85 to 2.5 billion dollars for Afghanistan every year that is roughly 20-25 percent of the country's total imports. In 2019, Afghanistan's import reached 7.4 billion dollars, in which petroleum and electricity shared 24 percent of the total import (figure 4).

4.3. Development potentials and economic impact.

Despite having substantial untapped minerals and hydrocarbon reserves that are estimated to worth more than 3 trillion dollars, Afghanistan is still the poorest country in Asia and one of the poorest in the world. The Afghan economy is extensively dependent on foreign aids and, every year needs around 10.1 billion dollars for expenditure. Considering the current trend that Afghanistan's revenue barely reaches 2.5 billion dollars per year, about 7 billion dollars of aid is needed annually to grow the country's economy and ensure the national budget. On the other hand, Afghanistan has an annual population growth rate of 2.3, which indicates that there will be more demand for oil and gas in the future. The assessment of the current population growth rate and energy consumption models reveals that by 2050, Afghanistan will have an estimated population of 59.8 million and will need to nearly 18 million tons of oil and gas (figure 5).

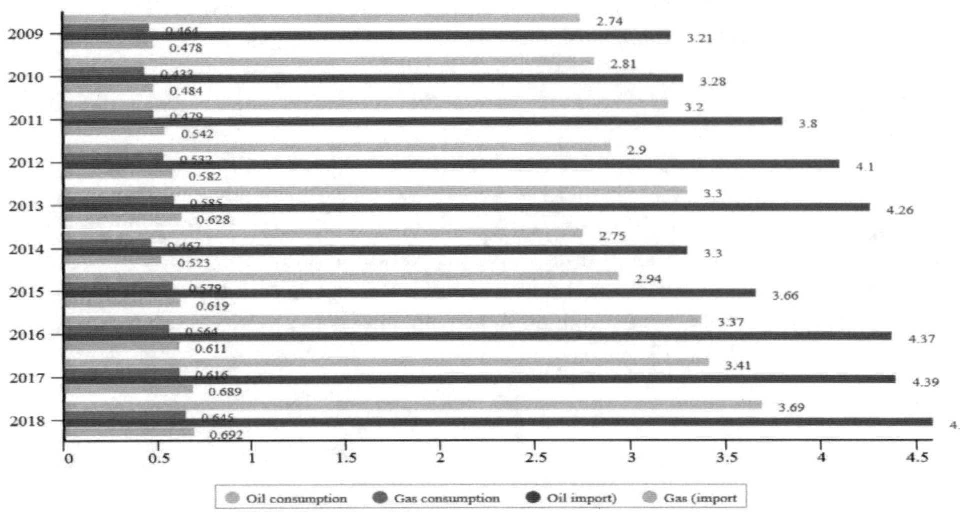

Figure 3. Afghanistan's oil and gas consumption and imports from 2009-2018 (designed based on NSIA data).

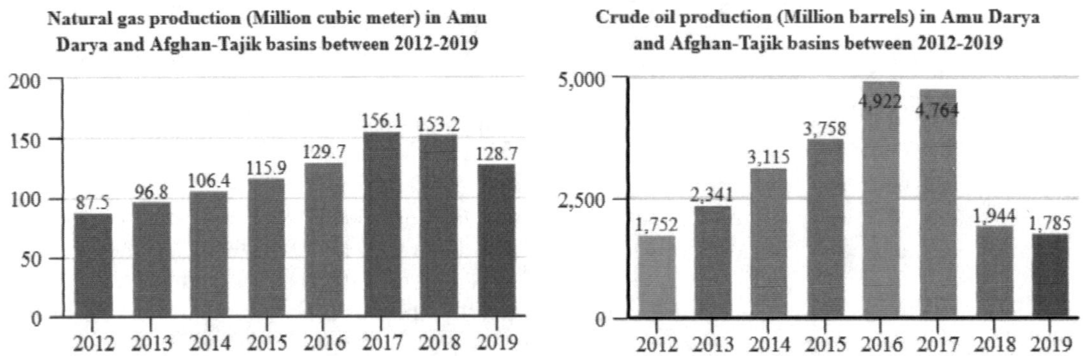

Figure 4. Oil and gas production in the Amu Darya and Afghan-Tajik basins between 2012-2019 (designed based on NSIA data).

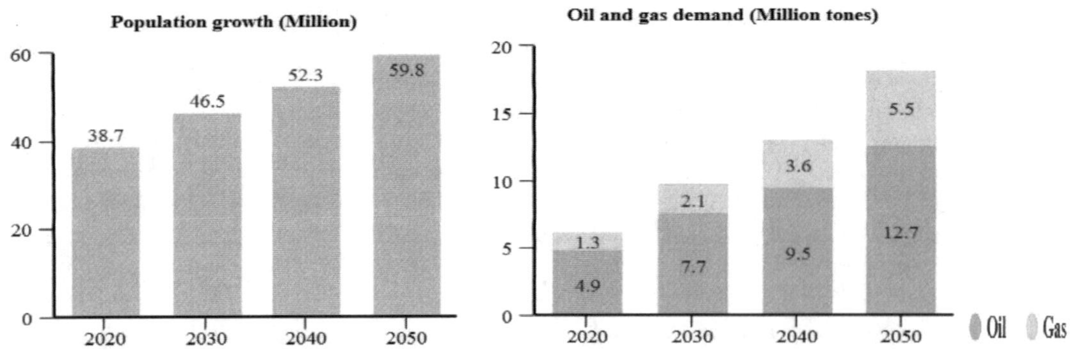

Figure 5. Afghanistan's estimated population growth and demand pattern for oil and gas from 2020-2050.

The development and utilization of natural resources, especially oil and gas resources of northern Afghanistan, is the key to improving the Afghan economy, decreasing dependency on international aid, and reducing energy imports. In the best scenario, through a proper 5-8 years of planning and investment, production of gas from the Sheberghan fields can meet the domestic necessities of the country, while in long term, it can be exported and ensure up to 50 percent of the public budget. It has been estimated that producing 12,328 tons of natural gas per day, can meet domestic demand, and Afghanistan will be able to save at least $ 800 million per year.

Moreover, considering the current oil price that is around 50 dollars a barrel, an output of 65,000 barrels per day that can be produced in Qashqari and Angot fields, would ensure Afghanistan's domestic oil demand and save about 1.15 billion dollars per year. Increasing the output to 125.000 barrels per day would earn Afghanistan 2.25 billion dollars, which is roughly 10 percent of the country's gross domestic product of 22 billion dollars in 2019.

4.4. Challenges.

The area observation and conducted interviews in northern basins have indicated that insecurity is the most challenging barrier against the development and utilization of oil and gas resources of Northern Afghanistan (table 3).

Table 3. The main barriers against the development and utilization of oil and gas resources of the Amu Darya and Afghan-Tajik basins.

Administrative barriers	Technical barriers
Insecurity	Poorly-maintained and limited infrastructures
Corruption	Lack of skilled workforce
Weak government performance in managing contracts	Lack of refineries and storage tanks
Lack of reliable data	Transportation cost and restrictions

5. Conclusion

The Amu Darya and Afghan-Tajik oil and gas resources can sufficiently meet Afghanistan's domestic demand. However, the basins have not developed, and the production level is limited due to different problems, such as insecurity, worn-out infrastructure, and corruption. Although the Afghan government has tried to exploit these resources by attracting international investors and internal private companies, it failed to achieve desirable results mainly due to the lack of experience in managing big projects and security instability. There are two potential scenarios for the future of the oil and gas resources of northern Afghanistan:

- If the Afghan government can guarantee the security of any potential big projects, have proper planning to deal with extensive corruption, and invest in infrastructures and capacity building, it will be able to attract high scale international investors and therefore, extensively exploit from its hydrocarbon reserves.
- By considering the current scenario, production in the Amu Darya and Afghan-Tajik basins is expected to be stagnant or very limited for a minimum of 3 to 5 years, as the majority of exploration operations have been postponed. The failure of national projects means that the country's infrastructure will not develop, and Afghanistan without infrastructure wills more likely experience economic problems and remains dependent on international aids for a long time.

References

[1] Taniwal Z 2014 *Economic Geography of Afghanistan* (Kabul: Maiwand Press) 118-9
[2] Shroder J 2014 *Natural Resources in Afghanistan: Geographic and Geologic Perspectives on Centuries of Conflict* (Afghanistan: Press) 436

[3] Safy L 2017 Afghanistan mines and petroleum; opportunities or conflict. *Research paper* **2** 21
[4] *Assessment of Undiscovered Petroleum Resources of Northern Afghanistan* Retrieved from: https://pubs.usgs.gov/of/2006/1253/of2006-1253.html
[5] *Afghanistan Ministry of Mines and Petroleum 2008 History of oil and gas in Afghanistan* Retrieved from: https://www.sigar.mil/pdf/audits/SIGAR-16-11-AR.pdf
[6] Mineral resources in Afghanistan 1992. *Nathan Berger Afghanistan studies project* 12-3
[7] Michael T K 2012 *The Race for What's Left: The Global Scramble for the World's Last Resources* (New York: Picador) 139-40
[8] *Afghanistan investment and business profile 2015. Basic information and contacts for successful investment and business activity* (Washington DC) **1** 52
[9] Afghanistan Ministry of Mines and Petroleum 2014. *Annual Report* 34-5
[10] *Afghanistan Overview* Retrieved from: https://www.worldbank.org/en/country/afghanistan/overview
[11] Dastyar A H, Bakhshi N A and Qadari M.K 1990 *Mesozoic and Cenozoic of the southern part of the Afghan-Tajik Basin* (New York) 24-32
[12] Malestani H 2016 The general geology of Afghanistan. *Research paper* 11-3
[13] Taniwal Z 2004 *The physical Geography of Afghanistan* (Kabul: Maiwand press) 17-9
[14] Arez G J 2007 *The Physical geography and structure of Afghanistan* (Kabul: University press) 46
[15] *Afghanistan Statistical Yearbook 2019* Retrieved from: https://nsia.gov.af/library

Raising funds for the development of distributed generation in the Far East and the Arctic

M M Gubanov[1], V V Potemkin[1], S V Shmanev[2] and D E Morkovkin[2]

[1]Far East Investment and Export Agency, 12, Presnenskaya naberezhnaya, Moscow, 123112, Russian Federation
[2]Financial University under the Government of the Russian Federation, 49, Leningradsky avenue, Moscow, 125993, Russian Federation

E-mail: m.m.gubanov@ya.ru

Abstract. The article discusses the issues of state support for the technological re-equipment of local energy facilities in the Far East and the Arctic. A study of the current system of state management of the industry in the macro-region was carried out, significant shortcomings were identified, possible ways to improve the management model were analyzed and proposals were formulated for further implementation.

1. Introduction

The development of remote territories of the Far East and the Arctic and their socio-economic development are impossible without ensuring reliable energy supply to consumers. At the same time, the construction of energy facilities is characterized by high capital intensity and low economic efficiency of investments, which is due to the historically established system of tariff regulation in the industry and its supporting nature in a market economy. The implementation of most projects in the field of construction of new or modernization of existing generating sources is carried out by energy companies on the basis of state support mechanisms established by the industry regulator as part of the implementation of a unified state energy policy defined by strategic planning documents within the framework of goal setting [1-2].

The current Decree of the President of the Russian Federation of 07.05.2018 No 204 "On national goals and strategic objectives of the development of the Russian Federation for the period up to 2024" as one of the tasks in the energy sector was determined by the guaranteed supply of affordable electricity, including through the development of distributed generation, including those based on renewable energy sources, primarily in remote and isolated energy regions. This task was included in the Energy Strategy of Russia for the period up to 2035, as well as the Comprehensive Plan for the modernization and expansion of the trunk infrastructure for the period until 2024.

Thus, at the highest level of government, the task was set to develop distributed generation in remote and isolated areas, which should be implemented through mechanisms of state support for energy companies specializing in this area and acting as the main investors in this industry. At the same time, now, the relevant mechanisms by the relevant department of the Ministry of Energy of Russia have not been developed, which calls into question the implementation of investment projects on time [3-4].

In this regard, the development of mechanisms of state support for the implementation of distributed generation projects is an urgent issue for scientific research in the field of state management of the fuel and energy complex.

2. Materials and methods

In order to determine rational mechanisms of state support for projects for the development of distributed generation in remote and isolated areas, statistical and analytical materials were used to review the current state of local energy facilities in the Far East and the Arctic, existing regulatory legal acts in the field of electricity, heat supply, energy conservation and improvement energy efficiency, macro-regional and regional socio-economic development, previously carried out scientific research on the development of energy in the Far North.

When performing the work, the following research methods were used [5-7]:

- Monitoring the process of development of the country's energy sector over a period of time;
- Comparative analysis of energy development directions in different countries and research of the level of energy infrastructure availability;
- Analysis of strategic planning documents and synthesis of strategies and programs for socio-economic and industry development;
- Abstraction from the solution of a specific industry problem of the development of distributed generation towards the solution of the broad problem of socio-economic development of the macro region;
- Deduction (power system - consumer), induction (consumer - power system).

3. Results

The results of the scientific work performed were the proposal of a list of measures of state support for private investors in the development of distributed generation in remote and isolated regions of the Far East and the Arctic, combining industry and macro-regional support measures.

The key proposal is to develop and approve the state subprogram "Development of distributed generation in remote and isolated areas" to state programs for the socio-economic development of the Far East and the Arctic zone in order to attract private investment in the implementation of projects for the modernization of inefficient diesel (fuel oil, coal) generation. As part of this subprogram, it is proposed to provide:

- Empowerment and responsibility of heads of local self-government bodies for the development of comprehensive schemes and programs for energy supply and increasing the energy efficiency of settlements in the area of decentralized power supply (figure 1) [5-7];
- Development and use of a mechanism for attracting private investments during the modernization of local energy facilities on the basis of a purchase and sale agreement, supply, transfer of energy resources, which includes the terms of an energy service agreement to ensure that a potential investor can choose the most rational solution for energy supply to a settlement for the long term [8];
- Empowering development institutions of the Far East and the Arctic to provide preferential long-term government funding for the implementation of projects to modernize local energy and determine the conditions for the allocation of this funding;
- Financing of the subprogram from the federal budget in the amount of up to 10 billion rubles [5-7];
- Expanding the powers of the Fund for Assistance to Housing and Utilities Reforms or the creation of a special Fund for the Development of Distributed Generation in Remote and Isolated Areas to conduct an examination of projects, determine the possibility of their co-financing from the federal budget, establish appropriate criteria and further support the implementation of projects by private investors [9-13].

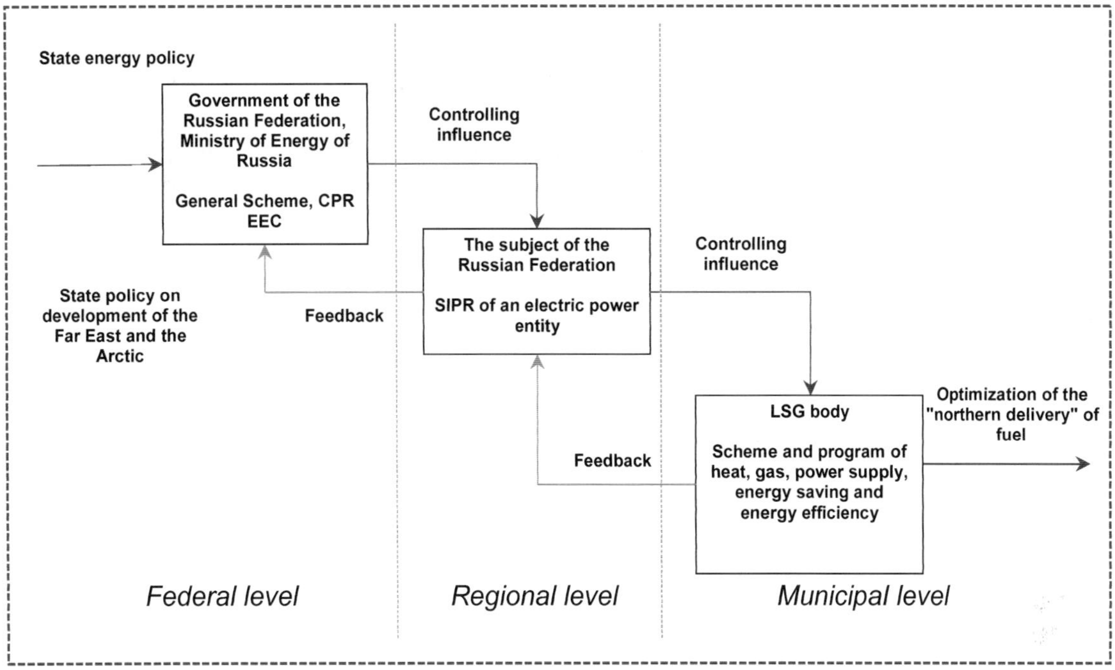

Figure 1. Diagram of a three-level distributed generation development management system.

4. Discussion

The research results were widely discussed at various expert platforms, including the Analytical Center under the Government of the Russian Federation and the Energy Committee of the State Duma of the Federal Assembly of the Russian Federation, and received the support of the professional and scientific community.

5. Conclusion

The result of the work was the preparation of the "Concept for attracting private investment in the development of distributed generation, including on the basis of renewable energy sources, in remote and isolated energy regions of the Far Eastern Federal District and the Arctic zone of the Russian Federation".

Acknowledgments

The scientific research was carried out because of the Far East Agency for Attracting Investments and Export Support in cooperation with the Financial University under the Government of the Russian Federation.

References

[1] Berdin V Kh, Kokorin A O, Yulkin G M and Yulkin M A 2017 *Renewable energy sources in isolated settlements in the Russian Arctic* (Moscow: World Wildlife Fund) 80
[2] *Economic problems of the energy complex* Retrieved from: https://ecfor.ru/wp-content/uploads/seminar/energo/z127.pdf
[3] Gibadullin A A, Sorokina V V, Sadriddinov M I, Petrusevich T V, Izzuka T B and Kokurina A D 2020 Assessment of factors ensuring industrial safety of enterprises of the electric power complex. *IOP Conference Series: Materials Science and Engineering* **862** 062029
[4] Saneev B G, Ivanova I Yu, Tuguzova T F and Izhbuldin A K 2018 Autonomous energy sources in the north of the Far East: characteristics and directions of diversification. *Spatial Economics* **1** 101-16

[5] *Generation facilities in isolated and hard-to-reach areas in Russia* Retrieved from: https://ac.gov.ru

[6] *Low carbon solutions for isolated regions of Russia with high energy costs* Retrieved from: http://www.cenef.ru/file/Low-Carbon_rus.pdf

[7] *Ministry of Energy of the Russian Federation* Retrieved from: https://minenergo.gov.ru/

[8] Sharipov F F, Krotenko T Y and Dyakonova M A 2021 Digital Potential of Economic Education: Information Technologies in a Management University. *Lecture Notes in Networks and Systems* **133** 561-72

[9] Romanova I N, Morkovkin D E, Nezamaikin V N, Gibadullin A A and Ivanova M A 2020 Formation of a policy to ensure environmental safety in modern economic conditions. *IOP Conference Series: Materials Science and Engineering* **734** 012166

[10] Gubanov M M et al *2020 IOP Conf. Ser.: Mater. Sci. Eng.* **837** 012011

[11] Gibadullin A A, Kerimova Ch V, Chorshanbiev S R, Shmanev S V, Romanova Ju A and Sozontov A A 2020 Management of mechanisms ensuring sustainable functioning and economic development of the electric power complex. *IOP Conference Series: Materials Science and Engineering* **734** 012170

[12] Salygin V I and Mustafinov R K 2019 Geopolitical aspects of electric power development: organizational and strategic approach. *Economic relations* **9** **1** 177-94

[13] Gubanov M M, Shmanev S V and Morkovkin D E 2019 The use of renewable energy sources in the implementation of the state strategy for the development of the macroregion (on the example of the Arctic and the Far East): problems and prospects. *OrelGIET Bulletin* **1(47)** 51-60

AUTHOR INDEX